Historical Guide to NASA and the Space Program

Ann Beardsley, C. Tony Garcia, and Joseph Sweeney

ROWMAN & LITTLEFIELD
Lanham • Boulder • New York • Toronto • Plymouth, UK

Published by Rowman & Littlefield
4501 Forbes Boulevard, Suite 200, Lanham, Maryland 20706
http://www.rowman.com

Unit A, Whitacre Mews, 26-34 Stannary Street, London SE11 4AB

British Library Cataloguing in Publication Information Available

Library of Congress Cataloging-in-Publication Data

Names: Beardsley, Ann, 1951–
Title: Historical guide to NASA and the space program / Ann Beardsley.
Other titles: NASA and the space program
Description: Lanham : Rowman & Littlefield, [2016] | Includes bibliographical references.
Identifiers: LCCN 2016019702 (print) | LCCN 2016022829 (ebook) | ISBN 9781442262867 (hardcover : alk. paper) | ISBN 9781442262874 (electronic)
Subjects: LCSH: Astronautics—United States—History. | United States. National Aeronautics and Space Administration—History. | Outer Space—Exploration—United States—History.
Classification: LCC TL789.8.U5 B427 2016 (print) | LCC TL789.8.U5 (ebook) | DDC 629.40973—dc23
LC record available at https://lccn.loc.gov/2016019702

™ The paper used in this publication meets the minimum requirements of American National Standard for Information Sciences Permanence of Paper for Printed Library Materials, ANSI/NISO Z39.48-1992.

Printed in the United States of America

Contents

Acknowledgments

First and foremost, thanks have to go to NASA for having so much of their history, missions, and data freely available, both online and in person. Writing a book about NASA without NASA's documents would have been an insurmountable task.

Thanks must also go to Mars One, because without the impetus to write the publisher and suggest a book about Mars, I would never have had the opportunity to write one about NASA. And a sincere thank you to Scott Hill, who helped with technical details; any resulting errors are, of course, my own.

I also have to guiltily thank all of my publishers for bearing with me while I said, "No, I can't edit that yet" month after month, especially Andrew Yoder, Melissa McNitt, Emily Natsios, Meredith Dias, Jessie McCleary, Laura Reiter, Karen Ackerman, Elaine McGarraugh, and anybody else I said no to at Rowman & Littlefield. A very grateful acknowledgment must also be made of our hardworking proofreader, Marianne Stone, whose invisible hand made this a much more consistent and accurate work; Andrew Yoder, our production editor, who graciously answered all my questions; and April Snider, our acquisitions editor.

And a sincere thank you to Elliott Walsh, who put up with me as I ignored hungry cats, messy counters, and dirty laundry while I worked to get just one more entry done. I couldn't have finished this without your support!

—Ann Beardsley

Thanks to the makers, designers, and engineers of the wondrous instruments, machines, and rockets that I read about, heard about, and viewed. The dusty passages and notes of creative designs and quests to perceive particles, waves, and light beyond the human condition made you the dreamers of dreams.

—C. Tony Garcia

To Stef, for supporting me in all my endeavors; to Devon, for inspiring me to always aim high; to Jake and Bob, for all the crazy adventures; and to NASA, for improving the lives of billions through their innovations and contributions to science.

—Joseph Sweeney

Foreword

As an aerospace engineer and scholar of spaceflight, it is my honor to provide this foreword to *Historical Guide to NASA and the Space Program* by Ann Beardsley, Joseph Sweeney, and Tony Garcia. Ann, Joseph, and I met through the Mars One program, which aims to send humans to Mars in the next decade. I think there is a growing sense of optimism about the future of humans in space: that we have been sitting on our laurels for the past few decades and that it is once more time to push our limits. It is an exciting time for spaceflight.

This book came about when Ann approached the publisher about writing a historical dictionary on Mars, a topic obviously near and dear to our hearts, and—given all the recent publicity—I suspect such a book is in our future. In the meantime, we thought it best to start with a broad and comprehensive resource on NASA and the space program. It contains pages of acronyms of the kind NASA loves to use, and details on NASA missions beyond anything I've seen in one place. This is a book that will resonate both with NASA-philes and citizen-scientists wanting to know more about what NASA has accomplished in its first nearly 60 years.

In the past, the best books about NASA have been written by NASA, with general reference books by outsiders rare, and those by students of science are even harder to find. We aim to change that. The conversational tone and lack of rocket-scientist jargon makes the book accessible to all. It stands apart from dry dictionaries of space programs and technical recounts of historical programs to provide an easily readable and fascinating history. This is not a book to be read from start to finish, but rather to be perused as the whim strikes: a bit like a coffee table book, but with a lot more substance.

It wasn't easy to edit down 60 years of spaceflight into a single reference. Choosing who and what to include was a difficult task, as NASA has launched hundreds of exciting missions over the years, and employed many thousands of talented people. Nevertheless, we think we have included the major highlights from across the decades.

Ann, Joseph, Tony, and I hope you enjoy this reference. If any of us make it to Mars, we'll be sure to update a few entries.

Ad astra a martem.

—Andrew Rader
Author, *Leaving Earth: Why One-Way to Mars Makes Sense*
December 29, 2014

Andrew Rader *is an author and aerospace engineer with a PhD in long-duration human spaceflight from MIT, and over five years' experience as a spacecraft systems engineer on over a dozen space missions. Andrew was a second-round candidate for the Mars One project, which aims to send human settlers to the Red Planet around 2024. He is an avid trivia player and public speaker, is active in social media, and designs educational card games.*

Acronyms and Abbreviations

Because of the abundance of acronyms in use by NASA as well as other space-science organizations, the list below is divided between the missions (and the organizations that run them) and the instruments in use during those missions. The first list of acronyms and abbreviations will be of primary use in the dictionary, while the instrument acronyms will be of best use when exploring appendix B, though they are by no means limited to those sections.

Many of the acronyms are also names of other objects, such as CERES (Cloud and Earth Radiant Energy Sensor) and Ceres (the dwarf planet), or MARS (Mid-Atlantic Regional Spaceport), MaRS (Mars Radio Science experiment), and Mars (the planet). Others are duplicated, such as NPP, which can be either NASA Postdoctoral Program or the National Polar-Orbiting Partnership. In most cases, either the context or the capitalization should make the distinction clear.

MISSIONS AND ORGANIZATIONS

ABMA	Army Ballistic Missile Agency
ACE	Advanced Composition Explorer
ACES	Advanced Crew Escape Suit
AEC	Atomic Energy Commission
AFISR	Air Force Intelligence, Surveillance, and Reconnaissance
AFRC	Armstrong Flight Research Center
AIM	Aeronomy of Ice in the Mesosphere
ALSEP	Apollo Lunar Surface Experiment Package
APL	Applied Physics Laboratory
ARC	Ames Research Center
ARCTAS	Arctic Research of the Composition of the Troposphere from Aircraft and Satellites
ARGOS	Active Response Gravity Offload System
ARIES	Airborne Research Integrated Experiments System
ARM	Asteroid Redirect Mission

ARMD	Aeronautics Research Mission Directorate
ARPA	Advanced Research Projects Agency
AS	Apollo-Saturn
ASI	Agenzia Spaziale Italiana (Italian Space Agency)
ASP	Airspace Systems Program
AST	Astrium Space Transportation
ASTP	Apollo-Soyuz Test Project
ATP	Aeronautics Test Program
ATTREX	Airborne Tropical Tropopause Experiment
ATV	Agena Target Vehicle
AU	Astronomical units
AvSP	Aviation Safety Program
BARREL	Balloon Array for RBSP [Radiation Belt Storm Probes] Relativistic Electron Losses
C/NOFS	Communications/Navigation Outage Forecasting System
C3PO	Commercial Crew and Cargo Program Office
CAIB	Columbia Accident Investigation Board
CALIPSO	Cloud-Aerosol LIDAR and Infrared Pathfinder Satellite Observation
CAS	Cross Agency Support
CCP	Commercial Crew Program
CCtCap	Commercial Crew Transportation Capability
CECR	Construction and Environmental Compliance and Restoration
CEV	Crew Exploration Vehicle
CGRO	Compton Gamma-Ray Observatory
CHAMP	Challenging Mini-Satellite Payload
CINDI	Coupled Ion-Neutral Dynamics Investigations
CME	Coronal Mass Ejection
CNES	Centre National D'études Spatiales
CNSA	China National Space Administration
COBE	Cosmic Background Explorer
COPUOS	Committee on the Peaceful Uses of Outer Space

COSPAR	Committee on Space Research
COTS	Commercial Orbital Transportation Services
CRS	Commercial Resupply Services
CSLI	CubeSat Launch Initiative
CSM	Command and Service Module
CXC	Chandra X-ray Center
DARPA	Defense Advanced Research Projects Agency
DFRC	Dryden Flight Research Center
DLR	German Aerospace Center (Deutsches Zentrum für Luft- und Raumfahrt eV)
DoD	Department of Defense
DS1/2	Deep Space 1/2
DSCS	Defense Satellite Communications System
DSN	Deep Space Network
DSP	Defense Support Program
DSX	Demonstration and Science Experiments
EASEP	Early Apollo Scientific Experiment Package
ED	Engineering Directorate
EDL	Entry Descent and Landing
EO-1	Earth Observing-1
EPO	Education and Public Outreach
ERBS	Earth Radiation Budget Satellite
ESA	European Space Agency
ESPO	Earth Science Project Office
EUVE	Extreme Ultraviolet Explorer
EVA	Extravehicular Activity
FAA	Federal Aviation Administration
FAST	Fast Auroral Snapshot Explorer
FCR	Flight Control Room
FKA	Roscosmos State Corporation for Space Activities/ Russian Space Agency
FUSE	Far Ultraviolet Spectroscopic Explorer
GALEX	Galaxy Evolution Explorer

GE	General Electric
GGS	Global Geospace Science
GISS	Goddard Institute for Space Studies
GLAST	Gamma-Ray Large Area Space Telescope
GOES	Geostationary Operational Environmental Satellites
GOSAT	Greenhouse Gases Observing Satellite
GP-B	Gravity Probe-B
GPM	Global Precipitation Measurement
GPS	Global Positioning System
GRACE	Gravity Recovery and Climate Experiment
GRAIL	Gravity Recovery and Interior Laboratory
GRC	Glenn Research Center
GSFC	Goddard Space Flight Center
HEO	Human Exploration and Operations Directorate
HETE-2	High-Energy Transient Explorer-2
HST	Hubble Space Telescope
HUNCH	High School Students United with NASA to Create Hardware
HYSICS	Hyperspectral Imager for Climate Science
IAU	International Astronomical Union
IBEX	Interstellar Boundary Explorer
ICESAT	Ice, Cloud, and Land Elevation Satellite
ICON	Ionospheric Connection Explorer
IDSCS	Initial Defense Satellite Communications System
IGY	International Geophysical Year
IMAGE	Imager for Magnetopause-to-Aurora Global Exploration
InSIGHT	Interior Exploration Using Seismic Investigations, Geodesy, and Heat Transport
INTEGRAL	International Gamma-Ray Astrophysics Laboratory
ISA	Israel Space Agency
ISEE	International Sun-Earth Explorer
ISRO	Indian Space Research Organization
ISRP	Integrated Systems Research Program

ISS	International Space Station
ISTP	International Solar-Terrestrial Physics
IV&V	Independent Verification and Validation Facility
JAXA	Japan Aerospace Exploration Agency
JPL	Jet Propulsion Laboratory
JSC	Johnson Space Center
JWST	James Webb Space Telescope
KSC	Kennedy Space Center
L1, L2, L3, . . .	Lagrange Points 1, 2, 3, . . .
LADEE	Lunar Atmosphere and Dust Environment Explorer
LAGEOS	Laster Geodynamics Satellite
LaRC	Langley Research Center
LAS	Langley Research Center
LDEF	Long Duration Exposure Facility
LIDAR	Light Detection and Ranging
LCC	Launch Control Center
LCROSS	Lunar Crater Observation and Sensing Satellite
LDCM	Landsat Data Continuity Mission
LEO	Low Earth Orbit
LLRV	Lunar Landing Research Vehicle
LM	Lunar Module
LOR	Lunar Orbit Rendezvous
LRO	Lunar Reconnaissance Orbiter
LSP	Launch Services Program
LWS	Living with a Star
M3	Moon Mineralogy Mapper
MAF	Michoud Assembly Facility
MARS	Mid-Atlantic Regional Spaceport
MAVEN	Mars Atmosphere and Volatile Evolution
MCC	Mission Control Center
MCC-21	Mission Control Center for the 21st Century
MEP	Mars Exploration Program

MESSENGER	Mercury, Surface, Space Environment, Geochemistry, and Ranging
MGS	Mars Global Surveyor
Mini-RF	Miniature Radio Frequency
MISS	Man in Space Soonest
MMM	Moon Mineralogy Mapper
MMS	Magnetospheric Multiscale
MOL	Manned Orbiting Laboratory
MOV	Manned Orbital Vehicles
MPCV	Multi-Purpose Crew Vehicle
MRO	Mars Reconnaissance Orbiter
MSFC	Marshall Space Flight Center
MSFN	Manned Space Flight Network
MSL	Mars Science Laboratory
MTPE	Mission to Planet Earth
NAA	North American Aviation
NACA	National Advisory Committee for Aeronautics
NASC	National Aeronautics and Space Council
NASP	National Aerospace Plane
NEAR	Near Earth Asteroid Rendezvous
NEN	Near Earth Network
NEO	Near Earth Objects
NEOWISE	Neo-Wide-Field Infrared Survey Explorer
NERVA	Nuclear Engine for Rocket Vehicle Application
N-G	Northrop-Grumman
NIAC	NASA Innovative Advanced Concepts
NISAR	NASA-ISRO Synthetic Aperture Radar
NMP	New Millennium Program
NOAA	National Oceanic and Atmospheric Administration
NOAA-NEEMO	NOAA Extreme Environment Mission Operations
NPP	NASA Postdoctoral Program
NPP	National Polar-Orbiting Partnership

NRL	Naval Research Laboratory
NRO	National Reconnaissance Office
NSC	National Space Council
NSF	National Science Foundation
NSSC	NASA Shared Services Center
NuSTAR	Nuclear Spectroscopic Telescope Array
OCO-2	Orbiting Carbon Observatory-2
ORI	Office of Research and Intelligence
OSIRIS-REx	Origins Spectral Interpretation Resource Identification Security Regolith Explorer
PARASOL	Polarization and Anisotropy of Reflectances for Atmospheric Sciences Coupled with Observations from LIDAR
PCM	Pressurized Cargo Module
PEAP	Personal Egress Air Pack
POES	Polar Operational Environmental Satellite
R&D	Research and Development
RKA	Roscosmos State Corporation for Space Activities/ Russian Federal Space Agency
ROSAT	Roentgen Satellite
SABoR	Ship-Aircraft Bio-optical Research
SAC	Office of Defense Mobilization Science Advisory Committee
SAGE	Stratospheric Aerosol and Gas Experiment
SAMPEX	Solar Anomalous and Magnetospheric Particle Explorer
SCaN	Space Communication and Navigation
SDO	Solar Dynamics Observatory
SDOGS	Solar Dynamics Observatory Ground Station
$SEAC^4RS$	Studies of Emissions and Atmospheric Composition, Clouds, and Climate Coupling through Regional Surveys
SEI	Space Exploration Initiative
SET	Space Environment Testbeds
SETI	Search for Extraterrestrial Intelligence
SLI	Space Launch Initiative

SLS	Space Launch System
SMD	Science Mission Directorate
SMPD	Office of Space Science Mission and Payload Development Division
SNC	Sierra Nevada Corporation
SNPO	Space Nuclear Propulsion Office
SOFIA	Stratospheric Observatory for Infrared Astronomy
SOHO	Solar and Heliospheric Observatory
SORCE	Solar Radiation and Climate Experiment
SpaceX	Space Exploration Technologies
SRTM	Shuttle Radar Topography Mission
SSC	Stennis Space Center
SSEP	Student Spaceflight Experiments Program
SSS	Sea Surface Salinity
SSTP	Small Spacecraft Technology Program
STADAN	Spacecraft Tracking and Data (Acquisition) Network
STEREO	Solar Terrestrial Relations Observations
STGT	Second Tracking and Data Relay Ground Terminal
STMD	Space Technology Mission Directorate
STP	Solar Terrestrial Probes
SWAS	Submillimeter Wave Astronomy Satellite
TAL	Transoceanic Abort Landing
TDRS	Tracking and Data Relay Satellites
THEMIS	Time History of Events and Macroscale Interactions during Substorms
TIMED	Thermosphere, Ionosphere, Mesosphere Energetics, and Dynamics
TISP	Teacher in Space Project
TOMS-EP	Total Ozone Mapping Spectrometer—Earth Probe
TRACE	Transition Region and Coronal Explorer
TRMM	Tropical Rainfall Measuring Mission
UARS	Upper Atmosphere Radiation Satellite

UK-SERC	United Kingdom—Science and Engineering Research Council
ULA	United Launch Alliance
UNCOPUOS	UN Committee on the Peaceful Uses of Outer Space
VAB	Vehicle Assembly Building
VAC	Vertical Assembly Center
VIIRS	Visible Infrared Imaging Radiometer Suite
VIM	Voyager Interstellar Mission
WISE	Wide-Field Infrared Survey Explorer
WMAP	Wilkinson Microwave Anisotropy Probe
WSC	White Sands Complex
WSGT	White Sands Ground Terminal
WSSH	White Sands Space Harbor
WSTF	White Sands Test Facility
XMM-Newton	X-ray Multi-Mirror Mission—Newton

INSTRUMENTS

3DP	Three-Dimensional Plasma and Energetic Particle Investigation
ABI	Advanced Baseline Imager
ACC	Accelerometers
ACIS	Advanced CCD Imaging Spectrometer
ACRIM2	Active Cavity Radiometer Irradiance Monitor II
ACS	Advanced Camera for Surveys
ADP	Axial Double Probe
AERONET	Aerosol Robotic Network
AFG	Analog Fluxgate
AIA	Atmospheric Imaging Assembly
AIRS	Atmospheric Infrared Sounder
ALI	Advanced Land Imager
ALICE	Ultraviolet Mapping Spectrometer

AMR	Advanced Microwave Radiometer
AMSR-E	Advanced Microwave Scanning Radiometer for EOS
AMSU	Advanced Microwave Sounding Unit
APS	Aerosol Polarimetry Sensor
APXS	Alpha Particle X-ray Spectrometer (MSL Curiosity/ Mars Exploration Rovers)
APXS	Alpha Proton X-ray Spectrometer
ASI	All Sky Imager
ASI	Atmospheric Structure Instrument (Galileo Atmospheric Probe)
ASI/MET	Atmospheric Structure Instrument/Meteorology Package
ASPERA	Analyzer of Space Plasmas and Energetic Atoms
ASPOC	Active Spacecraft Potential Control Experiment
ASTER	Advanced Spaceborne Thermal Emission and Reflection Radiometer
ATLAS	Advanced Topographic Laser Altimeter System
ATMS	Advanced Technology Microwave Sounder
AVHRR	Advanced Very High Resolution Radiometer
BATSE	Burst and Transient Source Experiment
BBXRT	Broadband X-ray Telescope
CAMMICE	Charge and Mass Magnetospheric Ion Composition Experiment
CAPS	Cassini Plasma Spectrometer
CARMEN-2	Environment Characterization and Modelization-2
CCD	Charge-Coupled Device
CDA	Cosmic Dust Analyzer
CDE	Cosmic Dust Experiment
CDS	Coronal Diagnostic Spectrometer
CELIAS	Charge, Element, Isotope Analysis
CEPPAD	Comprehensive Energetic Particle Pitch Angle Distribution
CERES	Cloud and Earth Radiant Energy Sensor
CHEMCAM	Chemistry and Camera

CHEMIN	Chemistry and Mineralogy
CHIPS	Cosmic Hot Interstellar Plasma Spectrometer
CIPS	Cloud Imaging and Particle Size
CIRS	Composite Infrared Spectrometer
CIS	Cluster Ion Spectrometry Experiment
CLAES	Cryogenic Limb Array Etalon Spectrometer
COMPTEL	Compton Telescope
COS	Cosmic Origins Spectrograph
COSTAR	Corrective Optics Space Telescope Axial Replacement
COSTEP	Comprehensive Suprathermal and Energetic Particle Analyzer
CPI	Charged Particle Instrument
CPI	Comprehensive Plasma Instrument (Geotail)
CPR	Cloud Profiling Radar
CRaTER	Cosmic Ray Telescope for the Effects of Radiation
CRIS	Cosmic Ray Isotope Spectrometer
CRISM	Compact Reconnaissance Imaging Spectrometer for Mars
CRT	Cosmic Ray Telescope
CTX	Context Camera
DAN	Dynamic Albedo of Neutrons
DDS	Dust Detection System
DFG	Digital Fluxgate
DIRBE	Diffuse Infrared Background Experiment
DLRS	Diviner Lunar Radiometer Experiment
DMR	Differential Microwave Radiometers
DOD SPHERES-CSAS	Department of Defense Synchronized Position, Hold, Engage, Reorient, Experimental Satellites—Chip Scale Atomic Clock
DORIS	Doppler Orbitography and Radio-Positioning Integrated by Satellite
DWP	Digital Wave Processing Experiment
ECT	Energetic Particle, Composition, and Thermal Plasma Suite

EDI	Electron Drift Instrument
EFD	Electric Field Detector
EFI	Electric Fields Instrument
EFW	Electric Field and Wave Experiment
EFW	Electric Field and Waves Suite (Van Allen Probes)
EGRET	Energetic Gamma Ray Experiment Telescope
EIS	Energetic Ion Spectrometer
EIS	Extreme Ultraviolet Imaging Spectrometer (Hinode)
EIT	Extreme Ultraviolet Imaging Telescope
ELT	Emergency Locator Transmitter
EMFISIS	Electric and Magnetic Field Instrument Suite and Integrated Science
EPAC	Ulysses Energetic Particle Composition Experiment
EPACT	Energetic Particles: Acceleration, Composition, and Transport
EPAM	Electron, Proton, and Alpha Monitor
EPD	Energetic Particles Detector
EPHIN	Electron Proton Helium Instrument
EPI	Energetic Particle Investigation
EPIC	Energetic Particles and Ion Composition (Geotail)
EPIC	European Photon Imaging Camera
EPIRB	Emergency Position Indicating Radio Beacons
EPPS	Energetic Particle and Plasma Spectrometer
EPS	Energetic Particles Sensor
ERBE	Earth Radiation Budget Experiment
ERNE	Energetic Analyzer
ESA	Electrostatic Analyzer
EUV	Extreme Ultraviolet
EVE	Extreme Ultraviolet Variability Experiment
EXIS	Extreme Ultraviolet and X-ray Irradiance Sensors
FC	Framing Camera
FEEPS	Fly's Eye Energetic Particle Sensor
FGM	Flux Gate Magnetometer

FIRAS	Far Infrared Absolute Spectrophotometer
FOC	Faint Object Camera
FOS	Faint Object Spectrograph
FPI	Fast Plasma Instrument
FREGATE	French Gamma Telescope
FUSE	Far-Ultraviolet Spectroscopic Explorer
FUV	Far Ultraviolet
GBM	Gamma-ray Burst Monitor
GC/MS	Gas Chromatography/Mass Spectrometry
GHRS	Goddard High-Resolution Spectrograph
GLAS	Geoscience Laser Altimetry System
GLG/SWICS	Solar Wind Ion Composition Spectrometer
GLM	Geostationary Lightning Mapper
GMAGS	Ground-Based Magnetometers
GOLF	Global Oscillations at Low Frequencies
GPM	Global Precipitation Monitor
GPSDR	Global Positioning System Demonstration Receiver
GPSP	Global Positioning System Payload
GRAND	Gamma Ray and Neutron Detector
GRS	Gamma Ray Spectrometer
GTT	Geiger Tube Telescope
GUVI	Global Ultraviolet Imager
HAD	Helium Abundance Detector
HALOE	Halogen Occultation Experiment
HENA	High-Energy Neutral Atom Imager
HEP	High-Energy Particles
HETGS	High-Energy Transmission Grating Spectrometer
HFI	High-Frequency Instrument
HIC	Heavy Ion Counter
HIFI	Heterodyne Instrument for the Far Infrared
HIRDLS	High-Resolution Dynamics Limb Sounder
HIRES	High-Resolution Camera

HiRISE	High-Resolution Imaging Science Experiment camera
HIRS	High-Resolution Infrared Sounder
HMI	Helioseismic and Magnetic Imager
HP3	Heat Flow and Physical Properties Probe
HPCA	Hot Plasma Composition Analyzer
HRC	High-Resolution Camera
HRDI	High-Resolution Doppler Imager
HRI	High-Resolution Imager
HRI	High-Resolution Instrument
HRSC	High-Resolution Stereo Camera
HSB	Humidity Sounder for Brazil
HSP	High-Speed Photometer
HUS/GRB	Solar X-rays and Cosmic Gamma-ray Bursts
HUT	Hopkins Ultraviolet Telescope
HVM	Helium Vector Magnetometer
HXD	Hard X-ray Detector
HYDRA	Hot Plasma Analyzer
IBIS	Imager on Board INTEGRAL Satellite
IBEX-Hi	Interstellar Boundary Explorer High
IBEX-Lo	Interstellar Boundary Explorer Low
IES	Ion and Electron Sensor
IMP	Imager for Mars Pathfinder
INMS	Ion and Neutral Mass Spectrometer
IPP	Imaging Photopolarimeter
IRAC	Infrared Array Camera
IREM	INTEGRAL Radiation Environment Monitor
IRIS	Infrared Interferometer Spectrometer (Voyager)
IRIS	Interface Region Imaging Spectrograph
IRS	Infrared Spectrograph
IRTM	Infrared Thermal Mapper
ISAMS	Improved Stratospheric and Mesopheric Sounder
ISS	Imaging Science Subsystem

ITS	Impact Targeting Instrument
IUVS	Imaging Ultraviolet Spectrograph
IVM	Ion Velocity Meter
JADE	Jovian Auroral Distribution Experiment
JEDI	Jovian Energetic Particle Detection Instrument
JEM-X	Joint European X-Ray Monitor
JIRAM	Jovian Infrared Auroral Mapper
JMR	Jason-1 Microwave Radiometer
KBR	H-Band Ranging System
LAMP	Lyman Alpha Mapping Project
LAN	Low-Energy Ions and Electrons
LAPC	Large-Area Proportional Counter
LAT	Large Area Telescope
LASCO	Large-Angle Spectrometer Coronagraph
LDEX	Lunar Dust Experiment
LECP	Low-Energy Charged Particles
LEISA/LAC	Linear Etalon Imaging Spectrometer Array/ Atmospheric Corrector
LENA	Low-Energy Neutral Atom Imager
LEND	Lunar Exploration Neutron Detector
LEP	Low-Energy Particles
LETGS	Low-Energy Transmission Grating Spectrometer
LFI	Low-Frequency Instrument
LGRS	Lunar Gravity Ranging System
LICA	Low-Energy Ion Composition Analyzer
LION	Low-Energy Ion and Electron Instrument
LIS	Lightning Imaging Sensor
LLCD	Lunar Laser Communications Demonstration
LOLA	Lunar Orbiter Laser Altimeter
LORRI	Long-Range Reconnaissance Imager
LPT	Light Particle Telescope
LPW	Langmuir Probe and Waves

LRA	Laser Retroreflector Array
LRD	Lightning and Radio Emission Detector
LROC	Lunar Reconnaissance Orbiter Camera
LWIR	Long-Wave Infrared Camera
MAE	Material Adherence Experiment
MAG	Magnetometer
MAG/ER	Magnetometer and Electron Reflectometer
MAHLI	Mars Hand Lens Imager
MARCI	Mars Color Imager
MARDI	Mars Descent Imager
MARIE	Mars Radiation Environment Experiment
MaRS	Mars Radio Science Experiment
MARSIS	Mars Advanced Radar for Subsurface and Ionospheric Sounding
MASCS	Mercury Atmospheric and Surface Composition Spectrometer
MAST	Mass Spectrometer Telescope
MASTCAM	Mast Camera
MAWD	Mars Atmospheric Water Detector
MBR	Mars Balloon Relay
MCS	Mars Climate Sounder
MDI	Michelson Doppler Imager
MDIS	Mercury Dual Imaging System
MEDLI	Mars Science Laboratory EDL Instrument
MELFI	Minus Eighty-Degree Laboratory Freezer for the ISS
MENA	Medium Energy Neutral Atom Imager
MET	Meteorological Package
MFE	Magnetic Fields Experiment
MFI	Magnetic Field Investigation
MGF	Magnetic Fields Measurement
MHS	Microwave Humidity Sounder
MI	Microscopic Imager

MIGHTI	Michelson Interferometer for Global High-Resolution Thermospheric Imaging
MIPS	MultiBand Imaging Photometer
MIR	Mid-Infrared Camera
MIRI	Mid-Infrared Instrument
MIMI	Magnetospheric Imaging Instrument
MINI-RF	Miniature Radio-Frequency Instrument
MINI-TES	Miniature Thermal Emission Spectrometer
MIRO	Microwave Instrument for Rosetta Orbiter
MISR	Multi-Angle Imaging Spectro-Radiometer
MLA	Mercury Laser Altimeter
MLS	Microwave Limb Sounder
MOC	Mars Observer Camera
MOC	Mars Orbiter Camera
MODIS	Moderate Resolution Imaging Spectroradiometer
MOLA	Mars Orbiter Laser Altimeter
MOPITT	Measurement of Pollution in the Troposphere
MRI	Medium Resolution Instrument
MSI	Multispectral Imager
NEP	Nephelometer
NFR	Net-Flux Radiometer
NGIMS	Neutral Gas and Ion Mass Spectrometer Package
NICMOS	Near-Infrared Camera/Multi-Object Spectrometer
NIMS	Near-Infrared Mapping Spectrometer
NIR	Near-Infrared Camera
NIRCam	Near-Infrared Camera
NIRISS	Near-Infrared Imager and Slitless Spectrograph
NIRSpec	Near-Infrared Spectrograph
NIS	Near-Infrared Spectrograph
NISP	Near-Infrared Spectrometer and Photometer
NLR	Near Laser Rangefinder
NMS	Neutral Mass Spectrometer

NS	Neutron Spectrometer
NWM	Neutral Wind Meter
OCAMS	OSIRIS-REx Camera Suite
OCPP	Orbiter Cloud Photopolarimeter
OGBD	Omnidirectional Gamma-ray Burst Detector
OGPE	Gas-Plasma Environment-Dual Frequency Experiment
OEFD	Orbiter Electric Field Detector
OETP	Orbiter Electron Temperature Probe
OIMS	Orbiter Ion Mass Spectrometer
OIR	Orbiter Infrared Radiometer
OLA	OSIRIS-REx Laser Altimeter
OM	Optical Monitor
OMC	Optical Monitoring Camera
OMAG	Orbiter Magnetometer
OMEGA	Infrared Mineralogical Mapping Spectrometer
OMPS	Ozone Mapping Profiler Suite
OMI	Ozone Monitoring Instrument
ONMS	Orbiter Neutral Mass Spectrometer
OPA	Orbital Plasma Analyzer
ORAD	Orbiter Surface Radar Mapper
ORPA	Orbiter (Charged Particle) Retarding Potential Analyzer
OSSE	Oriented Scintillation Spectrometer Experiment
OSTM	Ocean Surface Topography Mission
OTES	OSIRIS-REx Thermal Emission Spectrometer
OUVS	Orbiter Ultraviolet Spectrometer
OVIRS	OSIRIS-REx Visible and Infrared Spectrometer
PACS	Photodetector Array Camera and Spectrometer
PANCAM	Panoramic Camera
PEACE	Plasma Electron and Current Experiment
PEM	Particle Environment Monitor
PEPSSI	Pluto Energetic Particle Spectrometer Science Investigation

PET	Proton Electron Telescope
PFS	Planetary Fourier Spectrometer
PIXIE	Polar Ionospheric X-ray Imaging Experiment
PLS	Plasma Detector
PMIRR	Pressure Modulated Infrared Radiometer
PPR	Photopolarimeter Radiometer
PR	Precipitation Radar
PSPC	Position Sensitive Proportional Counter
PWI	Plasma Waves Investigation
PWS	Plasma Wave Spectrometer
QuikSCAT	Quick Scatterometer
RAD	Radiation Assessment Detector
RALPH	Visible Imager and Imaging Spectrometer
RAPID	Research with Adaptive Particle Imaging Detectors
RAC	Robotic Arm Camera
RAT	Rock Abrasion Tool
RBSPICE	Radiation Belt Storm Probes Ion Composition Experiment
REMS	Rover Environmental Monitoring Station
REX	Radio Science Experiment
REXIS	Regolith X-ray Imaging Spectrometer
RGS	Reflection Grating Spectrometer
RHESSI	Reuven Ramaty High-Energy Solar Spectroscopic Imager
RISE	Rotation and Interior Structure Experiment
RPA	Retarding Potential Analyzer
RPI	Radio Plasma Imager
RPI	Radio Plasma Instrument (IMAGE)
RPS	Relativistic Proton Spectrometer
RPWS	Radio and Plasma Wave Science
RS	Radio Science Experiment
RS	Remote Sensing
RSS	Radio Science Subsystem

SABER	Sounding of the Atmosphere Using Broadband Emission Radiometry
SAGE II	Stratospheric Aerosol Gas Experiment
SAM	Sample Analysis at Mars
SAM	Stratospheric Aerosol Measurement (I & II)
SAR	Synthetic Aperture Radar
SBUV	Solar Backscatter Ultraviolet Spectral Radiometer
SCA	Star Camera Assembly
SCM	Search Coil Magnetometer
SDP	Spin-Plane Double Probe
SEE	Solar EUV Experiment
SEIS	Seismic Experiment for Interior Structure
SEISS	Space Environment in-situ Suite
SEM	Space Environment Monitor
SEP	Solar Energetic Particles
SEPICA	Solar Energetic Particle Ionic Charge Analyzer
SHARAD	Shallow Subsurface Radar
SIS	Solar Isotope Spectrometer
SIM/COSPIN	Cosmic Ray and Solar Particles
SOFIE	Solar Occultation For Ice Experiment
SOLSTICE	Solar Stellar Irradiance Comparison Experiment
SOT	Solar Optical Telescope
SPI	Spectrometer on INTEGRAL
SPICAM	Spectroscopic Investigation of the Characteristics of the Atmosphere of Mars
SPIRE	Spectral and Photometric Imaging Receiver
SSALT	Single-Frequency Poseidon Altimeter
SSI	Solid State Imaging (Galileo)
SSI	Stereo Surface Imager
SST	Solid State Telescope
STAFF	Spatio-Temporal Analysis of Field Fluctuation experiment
STATIC	Suprathermal and Thermal Ion Composition

STIS	Space Telescope Imaging Spectrograph
SUMER	Solar Ultraviolet Measurement of Emitted Radiation
SUSIM	Solar Ultraviolet Spectral Irradiance Monitor
SUVI	Solar Ultraviolet Imager
SWAN	Solar Wind Anisotropies
SWAP	Solar Wind Analyzer around Pluto
SWE	Solar Wind Experiment
SWEA	Solar Wind Electron Analyzer
SWEPAM	Solar Wind Electron, Proton, and Alpha Monitor
SWIA	Solar Wind Ion Analyzer
SWICS	Solar Wind Ion Composition Spectrometer
SWIMS	Solar Wind Ion Mass Spectrometer
SXC	Soft X-ray Camera
T2L2	Time Transfer by Laser Link
TAGSAM	Touch and Go Sample Acquisition Mechanism
TEAMS	Time of Flight Energy Angle Mass Spectrograph
TEGA	Thermal and Evolved Gas Analyzer
TES	Thermal Emission Spectrometer
TES	Tropospheric Emission Spectrometer
TGRS	Transient Gamma-Ray Spectrometer
THEMIS	Thermal Emission Imaging System
TIDE	Thermal Ion Dynamics Experiment
TIDI	TIMED Doppler Interferometer
TIMAS	Toroidal Ion Mass Spectrographs
TIM	Total Irradiance Monitor
TLP	Total Luminance Photometer
TMI	TRMM Microwave Imager
TMR	TOPEX Microwave Radiometer
TOMS	Total Ozone Mapping Spectrometer
TRD	Trapped Radiation Detector
TSI	Total Solar Irradiance
UIT	Ultraviolet Imaging Telescope

ULEIS	Ultra-Low-Energy Isotope Spectrometer
UPS	Unique Payload Services
URAP	Unified Radio and Plasma Waves
USO	Ultra Stable Oscillator
UV/Vis	Ultraviolet/Visible Camera
UVCS	Ultraviolet Coronagraph and Spectrometer
UVI	Ultraviolet Imager
UVIS	Ultraviolet Imaging Spectrograph
UVS	Ultraviolet Imaging Spectrometer
UVS/EUV	Ultraviolet Spectrometer/Extreme Ultraviolet Spectrometer
VENETIA	Venetia Burney Student Dust Counter
VIIRS	Visible Infrared Imaging Radiometer Suite
VIMS	Visible and Infrared Mapping Spectrometer
VIRGO	Variability of Solar Irradiance and Gravity Oscillations
VIR	Visible and Infrared Spectrometer
VIRS	Visible and Infrared Scanner
VIS	Visible Imaging Spectrometer
VIS	Visible Imaging System
VLF	Very Low Frequency
VSP	Visible Spectrometer
WAE	Wheel Abrasion Experiment
WBD	Wide-Band Data Instrument
WFC	Wide-Field Camera
WFC3	Wide-Field Camera 3
WFPC1	Wide-Field and Planetary Camera 1 (also WF/PC)
WFPC2	Wide-Field and Planetary Camera 2
WHISPER	Waves of High Frequency and Sounder for Probing of Electron Density by Relaxation Experiment
WINDII	Wind Imaging Interferometer
WUPPE	Wisconsin Ultraviolet Photopolarimetry Equipment
WXM	Wide Field X-ray Monitor
XIS	X-ray Imaging Spectrometer

XPS	X-ray Ultraviolet Photometer System
XRFS	X-ray Fluorescence Spectrometer
XRS	X-ray Spectrometer
XRS-GRS	X-ray-Gamma-Spectrometer
XRT	X-ray Telescope
XUV	X-ray Ultraviolet
ZOD	Zero-Order Detector Camera

Chronology

ca. 250 Chinese use hot air balloons (little more than lanterns) as military signaling devices.

1490 Leonardo da Vinci pens his plans for human-powered flight in a drawing inspired by the legend of Icarus.

1590 Giordano Bruno suggests that the stars are far-away suns; he is later burned at the stake for the belief.

1784 First powered balloon flight, by Gen. Jean Baptiste Marie Meusnier, followed by the first balloon flight by a woman, Madame Thible, both in France.

1852 24 September: Henri Giffard flies his steam-engine-powered dirigible from Paris to Trappe, a distance of 17 miles.

1890s Otto Lilienthal promotes heavier-than-air flight in hang gliders, reaching up to 49 feet above the ground during flights lasting up to five hours, before falling from his glider and breaking his neck.

1903 Konstantin Tsiolkovsky publishes *The Exploration of Cosmic Space by Means of Reaction Devices*, the first scientific work on space travel using liquid hydrogen and oxygen. **December 17:** Orville and Wilbur Wright make their first flight in a controlled and fixed-wing powered flying machine for a grand total of 120 feet at speeds up to 6.8 miles per hour and up to 10 feet above the ground.

1912 Independently, French aircraft designer Robert Esnault-Pelteri uses the same equations as Tsiolkovsky and calculates the energy required to travel to the moon, the eight known planets, and the return trip back to Earth.

1914 Robert Goddard applies for and receives patents on multistage and liquid-fueled rockets.

1923 Herman Orlieth publishes *By Rocket into Planetary Space*.

1924 The Society for Studies of Interplanetary Travel is founded in the USSR.

1926 16 March: Goddard launches the first liquid-fueled rocket.

1927 Inspired by advances in space science, the Society for Space Travel (Verein für Raumschiffahrt) is formed by many of the world's top rocket scientists.

1928 Herman Potočnik (Hermann Noordung) publishes his *The Problem of Space Travel—The Rocket Motor* in Austria; the book discusses space and scientific experiments.

1929 Oberth sends off his first liquid-fueled rocket; one of his students is Wernher von Braun.

1931 Germany develops its liquid-fueled rocket engines for military use.

1933 Wernher von Braun begins work on the Aggregate rockets (precursor to the V-2 rockets). **17 August:** The Soviet Union's Group for the Study of Reactive Motion launches the first liquid-fueled rocket in that country.

1942 3 October: Nazi Germany launches its V-2 rocket, which is the first spacecraft to cross the Kármán line (62 miles above the Earth).

1946 10 May: The United States launches a modified V-2 and begins cosmic radiation experiments. **22 May:** The first rocket designed in the United States reaches the edge of space. **24 October:** U.S. scientists receive the first pictures of Earth from 105 km (65 mi) altitude.

1947 20 February: The United States sends fruit flies into space.

1948 11 June: Albert, a rhesus monkey, rides a V-2 rocket to 63 km (39 mi) but suffocates during the flight.

1949 14 June: Albert II ascends into space and reaches 134 km (83 mi), but dies on impact after a parachute failure. **16 September:** Albert III dies in an explosion at 35,000 ft. **8 December:** Albert IV dies on impact after another parachute failure.

1957 21 August: The Soviet Union launches the R-7 Semyorka/SS-6 Sapwood, the first intercontinental ballistic missile. **4 October:** *Sputnik 1* reaches orbit to become the first artificial satellite and sends the first signals from space. **3 November:** The Soviet Union launches *Sputnik 2* and its passenger, Laika (a dog), who died during the flight.

1958 31 January: *Explorer 1* (US) is the first satellite launched into orbit by the United States and is the first to detect the near-Earth region later known as the Van Allen belts. **1 October:** The National Aeronautics and Space Administration (NASA) begins official operation, with T. Keith Glennan as administrator, appointed by President Dwight Eisenhower. **11 October:** *Pioneer 1* (US) becomes NASA's first launch from Cape Canaveral, Florida. **7 November:** NASA research pilot John McKay makes the last flight in the X-

1E. **6 December:** *Pioneer 3* (US) is launched and attains altitude of 63,580 miles. **18 December:** An Air Force Atlas booster is placed into orbit as a communications relay satellite. **19 December:** President Dwight Eisenhower's Christmas message is beamed from this satellite, making it the first voice sent from space.

1959 2 January: *Luna 1* (USSR) is the first firing of a rocket in Earth's orbit; it reaches Earth's escape velocity and provides first detection of solar wind. **4 January:** *Luna 1* becomes the first artificial satellite to reach the moon. **17 February:** *Vanguard 2* (US), an International Geophysical Year scientific satellite, launches from Cape Canaveral, Florida. **28 February:** *Discoverer 1* (US), a U.S. Air Force reconnaissance satellite, is launched. **3 March:** *Pioneer 4* (US) successfully makes the first U.S. lunar flyby. **9 April:** NASA identifies the Mercury astronaut corps. **28 May:** The United States launches Abel and Baker (monkeys) aboard a Jupiter missile and recovers them alive. **8 June:** Scott Crossfield makes the first unpowered glide flight in the joint X-15 hypersonic research program. **7 August:** *Explorer 6* (US) provides the first photograph of Earth from orbit. **13 September:** *Luna 2* (USSR) crashes into the moon, delivering Soviet flags. **4 October:** *Luna 3* (USSR) provides the first photos of the far side of the moon.

1960 1 April: *TIROS 1*, a meteorological satellite, begins monitoring Earth's weather. **13 April:** The United States launches *Transit 1B*, an experimental orbital navigation system. **1 July:** The Army Ballistic Missile Agency becomes part of NASA, and its name is changed to the George C. Marshall Space Flight Center. **19 August:** *Sputnik 5* brings the first live plants and dogs back from orbit.

1961 21 January: Hugh L. Dryden becomes acting administrator of NASA. **31 January:** The United States sends Ham (a chimpanzee) into space; Ham performs trained tasks. **12 February:** *Venera 1* (USSR) launches from Earth's orbit as an upper-stage rocket into a heliocentric orbit, and responds to midcourse corrections and spin stabilization. **14 February:** James E. Webb takes over from Dryden as administrator, appointed by President John Kennedy. **12 April:** Yuri Gagarin becomes the first human in space on board the *Vostok 1*. **5 May:** Alan Shepard becomes the first American to fly in suborbital space. **19 May:** *Venera 1* performs the first planetary flyby, less than 100,000 km from Venus. **25 May:** President Kennedy commits the United States to landing a man on the moon. **19 September:** NASA announces that Houston, Texas, will be their manned space flight center. **25 October:** NASA announces the establishment of a test facility in Mississippi for the Saturn boosters that will be used in the Apollo projects. **27 October:** First successful test of the Saturn I rocket.

1962 20 February: John Glenn becomes the first American to be launched into Earth orbit. **7 March:** The United States launches OSO-1, the first orbital solar observatory. **14 December:** *Mariner 2*, launched on the 27 August, gets within 35,000 km of Venus.

1963 15–16 May: The last NASA astronaut to travel solo, Gordon Cooper, did so on the last Mercury flight. **16 June:** Valentina Tereshkova becomes the first woman in space on board the *Vostok 6*. **19 July:** The first reusable crewed spacecraft, the X-15 Flight 90, is launched by the United States.

1964 29 January: NASA's largest launch vehicle, the Saturn SA-5, lifts 19 tons into orbit, a record weight. **8 April:** Gemini makes four unpiloted orbits around the Earth and is successfully recovered.

1965 18 March: The Soviet Union's crew on board the *Voskhod 2* performs the first extravehicular activity (EVA). **23 March:** *Gemini 3* (US) launches its first piloted mission. **3–7 June:** *Gemini 4* (US) launches and stays aloft for four days; first EVA by an American, Ed White. **14 July:** *Mariner 4* (US), the first successful mission to Mars, approaches within 10,000 km of the Red Planet and transmits its first photographs of Mars. **21–29 August:** *Gemini 5* (US) sets record with an eight-day orbital flight. **15 December:** *Gemini 6A* and *Gemini 7* (both US) perform the first orbital rendezvous by flying parallel.

1966 3 February: *Luna 9* (USSR) soft lands on the moon and sends photos to Earth. **1 March:** *Venera 3* (USSR) crash lands on Venus. **16 March:** *Gemini 8* and an Agena target vehicle (both US) successfully dock while in orbit. **3 April:** *Luna 10* (USSR) becomes the first artificial satellite around the moon. **2 June:** *Surveyor 1* (US) soft lands on the moon and returns more than 10,000 photos to Earth. **11–15 November:** *Gemini 12* (US), the last Gemini flight, launches; astronauts complete three EVAs and dock with Agena target vehicle.

1967 27 January: During a simulation exercise aboard *Apollo-Saturn 204*, a flash fire kills the three astronauts on board, shocking the nation. **25 April:** NASA conducts first XB-70A flight, studying delta-wing flying at supersonic speeds. **3 October:** The X-15 experimental rocket plane sets a speed record for piloted vehicles when it reaches 4,534 mph (Mach 6.72). **22 October:** *Apollo 7* launches to become the first manned Apollo launch. **30 October:** *Cosmos 186* and *Cosmos 188* (both USSR) perform the first automated docking. **21 December:** *Apollo 8* launches atop the Saturn V booster from Kennedy Space Center and becomes the first crewed vehicle to orbit the moon.

1968 8 October: Dr. Thomas O. Paine becomes acting administrator of NASA. **7 December:** NASA launches OAO-2, the first orbital ultraviolet observatory.

1969 21 March: Paine is confirmed as administrator by President Richard Nixon. **21 July:** The *Apollo 11* (US) crew lands on the moon and successfully launches back to space. **15 September:** The Space Task Group issues its report on the post-Apollo space program and recommend, among other things, a human expedition to Mars. **19 November:** *Apollo 12* and *Surveyor 3* (both US) rendezvous on the surface of the moon, for a second moon landing.

1970 5 March: In a joint program with the U.S. Air Force, NASA conducts extensive tests on thermal stress, aerodynamics, high-altitude environments, propulsion, and flight control systems in high-speed flights. **11–17 April:** *Apollo 13*'s mission is compromised; all crew return safely to Earth. **16 September:** Dr. George M. Low takes over as acting administrator of NASA. **24 September:** *Luna 16* (USSR) receives the first automatic sample return from the moon. **23 November:** The *Lunokhod 1* (USSR) deposits the first lunar rover on the moon. **21 December:** NASA launches the *Uhuru*, the first X-ray orbital observatory. **15 December:** *Venera 7* (USSR) soft lands on Venus and returns signals from another planet for the first time.

1971 31 January–9 February: NASA conducts the third lunar landing mission, returning with almost 100 lbs. of lunar material. **9 March:** NASA completes initial tests on Langley researcher Richard Whitcomb's supercritical wing. **23 April:** *Salyut 1* (USSR) becomes the first space station. **27 April:** President Nixon appoints Dr. James C. Fletcher administrator of NASA. **June:** *Orion 1* is the first manned observatory. **7 August:** *Apollo 15* returns from the moon with the "Genesis Rock." **14 November:** *Mariner 9* (US) orbits Mars. **27 November:** *Mars 2* (USSR) crash lands on Mars. **2 December:** *Mars 3* (USSR) soft lands on Mars and returns the first signals from Mars's surface.

1972 5 January: NASA administrator James Fletcher meets with President Richard Nixon, who later decides to proceed with development of the space shuttle. **3 March:** *Pioneer 10* (US) launches toward Jupiter. **16–27 April:** *Apollo 16* is the fifth American landing on the moon. **25 May:** The first Digital Fly-by-Wire program is tested, laying the groundwork for later use by the space shuttle. **15 July:** *Pioneer 10* enters the asteroid belt (leaves the inner solar system). **23 July:** *Landsat 1* is launched, providing extensive image-mapping photos of Earth. **15 November:** SAS 2 (US) becomes the first orbital gamma ray observatory. **7–19 December:** *Apollo 17* is the last of the six Apollo missions to the moon. **11 December:** The *Apollo 17* crewmembers become the last humans to walk on the moon (so far).

1973 14 May: *Skylab*, an orbital space platform, is launched. **3 December:** *Pioneer 10* flies by Jupiter.

1974 5 February: *Mariner 10* (US) performs the first gravitational assist maneuver and flies by Venus within 6000 km. **29 March:** *Mariner 10* performs the first Mercury flyby within 1000 km. **1 September:** *Pioneer II* (US) sends back first polar photos of Jupiter.

1975 15 July: The United States (Apollo) and the Soviet Union (Soyuz) complete the first multinational human space flight, despite the Cold War. **5 August:** NASA performs initial landing tests on a space shuttle–like vehicle, proving it's possible to land safely. **20 August:** *Viking 1* is launched on a trip to Mars. **20 October:** *Venera 9* (USSR) performs the first orbit around Venus. **22 October:** *Venera 9* sends photos from the surface of Venus.

1976 18 February: The space shuttle *Enterprise* (named after Star Trek spacecraft) is flown on test flights atop a Boeing 747 ferrying aircraft. **17 April:** The United States and West Germany coordinate on the *Helios 2*; the vehicle gets within 27 million miles (43 million km) of the sun and sets a maximum speed record of 252,792 km/h. **20 July:** The *Viking* lander (US) sends photos and soil sample analysis from Mars. **September:** *Viking 2* discovers water frost on Mars's surface.

1977 2 May: Dr. Alan M. Lovelace is appointed acting administrator of NASA and serves 49 days. **21 June:** President Jimmy Carter appoints Dr. Robert Frosch NASA's next administrator.

1978 26 January: The International Ultraviolet Explorer (a cooperative effort of NASA, the European Space Agency, and UK-SERC) is the first real-time, remotely operated ultraviolet orbital observatory. **20 May:** NASA launches the *Pioneer Venus* orbiter. **26 June:** *SeaSat-A* is launched to observe the oceans; its batteries die after 106 days. **24 October:** *Nimbus 7* is launched, and later confirms evidence of Antarctic ozone depletion in the 1980s. **4 December:** The *Pioneer Venus* orbiter begins its multiyear orbital exploration of Venus (1978–1992).

1979 5 March: *Voyager 1* (US) flies within 350,000 km of Jupiter. **11 July:** A deteriorated *Skylab* impacts Earth; debris stretches from the southeastern Indian Ocean to western Australia. **1 September:** *Pioneer 11* (US) flies by Saturn at 21,000 km.

1980 14 February: Solar Maximum Mission, the first U.S. launch to study the sun during heavy sunspot activity, is initiated. **7 March:** The *Gossamer Albatross*, a human powered aircraft, makes several test flights. **12 November:** *Voyager 1* flies by Saturn at 124,000 km.

1981 21 January: With the election of President Ronald Reagan, Dr. Lovelace is reappointed acting administrator. **12 April:** The space shuttle *Columbia* flies as a part of the Space Transportation System (STS-1). **11 July:** President Reagan confirms James M. Beggs as administrator of NASA. **26 August:** *Voyager 2* begins transmitting images of Saturn.

1982 1 March: *Venera 13* (USSR) provides analysis of Venus soil samples and the first sound recording of another world. **11–16 November:** STS-5 (*Columbia*) launches two commercial communications satellites during its mission. **11 November:** *Viking 1*'s last transmission reaches Earth.

1983 25 January: IRAS (a joint NASA/UK-SERC/Netherlands project) is the first infrared orbital observatory. **23 March:** The USSR and France launch Astron, the first ultraviolet orbital observatory. **4–9 April:** STS-6 (*Challenger*) deploys the first Tracking and Data Relay Satellites (TDRSs) into geostationary orbit. **13 June:** *Pioneer 10* (US) becomes the first vehicle to travel beyond Neptune. **18 June:** Sally Ride becomes the first American woman in space, almost 20 years after the Soviet Union's Valentina Tereshkova. **30 August:** Guion S. Bluford becomes the first African American astronaut. **28 November:** The *Columbia* carries the first non-US astronaut, Ulf Merbold, from West Germany; *Columbia* also carries *Spacelab 1* (joint NASA/ESA), the first laboratory in space.

1984 25 January: President Reagan announces the construction of a space station within a decade. **7 February:** U.S. astronaut Bruce McCandless takes the first untethered spacewalk on board the STS-41-B. **30 August:** First flight of space shuttle *Discovery* (STS-41-D).

1985 3 October: First flight of the space shuttle *Atlantis* (STS-51), which deploys a classified satellite during its Department of Defense mission. **4 December:** Dr. William Graham is appointed acting administrator of NASA.

1986 24 January: *Voyager 2* (US) gets within 82,000 km of Uranus and performs the first flyby of that planet. **28 January:** The space shuttle *Challenger* explodes 73 seconds into the flight, killing all aboard. **19 February:** *Mir* (USSR) becomes the first long-term research space station. **May:** The National Commission on Space advocates aggressive exploration and colonization of the moon and other planets. **12 May:** Dr. James Fletcher is reappointed NASA administrator by President Reagan. **25 August:** *Voyager 2* flies within 30,000 km of Neptune.

1988 August: *Voyager 2* begins transmitting images from Neptune. **29 September:** *Discovery* marks the successful return to flight for the space shuttle.

1989 8 April: Dale Myers is appointed acting administrator; he serves 35 days. **4 May:** *Magellan* (US) launches on its way to Venus. **14 May:** Richard Truly is appointed acting administrator; he serves 47 days before his confirmation on 1 July. **20 July:** President George H. W. Bush announces plans for the Space Exploration Initiative. **18 October:** *Galileo* is launched from STS-34 toward Jupiter, but its antenna does not fully deploy.

1990 14 February: *Voyager 1* takes the first photograph of the entire solar system. **24 April:** The Hubble Space Telescope launches from the space shuttle *Columbia* (STS-31). **September:** Using radar, *Magellan* maps 99 percent of the surface of Venus. **17 December:** The Advisory Committee on the Future of the U.S. Space Program issues its report recommending a balanced space program, to include human spaceflight, robotic probes, space science, applications, and exploration, all within a limited budget.

1991 15 July: Initial flight of the F/A-18 High Angle of Attack Research Vehicle (HARV). **21 October:** *Galileo* is the first craft to do an asteroid flyby, of 951 Gaspa, within 1600 km.

1992 8 February: Ulysses (joint US/ESA) completes the first polar orbit around the sun. **1 April:** Daniel S. Goldin is appointed administrator by President George H. W. Bush. **2 May:** First flight of the space shuttle *Endeavor*, which captured and repaired a private satellite. **25 September:** The *Mars Observer* is launched but fails to respond after 21 August 1993, three days before it was scheduled to enter Mars's orbit.

1993 2 December: The space shuttle *Endeavour* (STS-61) successfully repairs the optics of the Hubble Space Telescope.

1994 25 January–3 May: *Clementine* (US) maps the lunar surface at various resolutions, suggesting ice might be found at the moon's poles. **3 February:** Sergei Krikalev becomes the first Russian cosmonaut to fly on a space shuttle.

1995 2 February: Eileen Collins becomes the first female shuttle pilot. **22 March:** Valeri Polyakov (Russia) completes the longest time in space (437.7 days). **27 June:** The space shuttle *Atlantis* (US) docks to the *Mir* space station (Russia). **11–20 November:** The space shuttle *Atlantis* carries up and attaches a Russian-built docking port to the *Mir* space station. **7 December:** *Galileo* completes the first orbit of Jupiter and sends an atmospheric entry probe into the gas giant's atmosphere.

1996 22–31 March: Space shuttle *Atlantis* docks at *Mir* and leaves Shannon Lucid on board, the first U.S. woman to fly on the space station, for five months. **7 August:** NASA announces evidence that microscopic life might once have existed on Mars. **13 August:** *Galileo* reveals that Europa (one of Jupiter's moons) may harbor "warm ice" or even liquid water.

1997 13 January: NASA announces discovery of three black holes, indicating that nearly all galaxies once had these. **11–21 February:** *Discovery* (US) performs a second servicing mission on the Hubble telescope. **20 February:** *Galileo* discovers icebergs on Europa, suggesting subsurface oceans. **25 June:** Docking accident at *Mir* endangers the crew, who refuse to abandon the vessel, which is later resupplied and repaired. **4 July:** NASA lands the Mars *Pathfinder* onto the surface of Mars. **11 September:** The Global Surveyor space probe reaches orbit around Mars and begins to provide 3-D maps of the Red Planet. **25 September–6 October:** *Atlantis* delivers repair equipment subsequent to the 25 June docking accident on Mir as well as other supplies. **15 October:** *Cassini-Huygens* (US/ESA/ASI) launches on its mission to Saturn from Cape Canaveral, Florida.

1998 29 January: Representatives from 15 countries meet in Washington regarding the framework for the International Space Station (ISS). **28 May:** Images from the Hubble Space Telescope indicate the first direct image of a planet outside our solar system. **29 October–7 November:** John Glenn, age 77, returns to space on STS-95. **20 November:** Assembly of the International Space Station begins.

1999 27 May–6 June: The space shuttle *Discovery* is the first spacecraft to dock with the ISS in orbit. **22 July:** Eileen Collins becomes the first woman to command a shuttle mission (STS-93). **23 September:** The *Mars Climate Orbiter* (US) fails to emerge from behind the planet and contact is lost; subsequent examination showed a failure to convert critical data from American units to metric. **3 December:** The Mars *Polar Lander* (US) is lost during the landing sequence, causes unknown. **18 December:** *Terra* (US) is launched from Vandenberg Air Force Base to monitor Earth's environment over 15 years.

2000 14 February: NASA's *NEAR Shoemaker* (US) orbits the 433 Eros asteroid. **19–29 May:** *Atlantis* (US) delivers supplies to the ISS and boosts its orbit from 370 km to 400 km (200–230 miles). **8–20 September:** STS-106 delivers 6,600 lbs. of supplies and prepares the ISS for its first "permanent" crew. **31 October:** Expedition 1, the first permanent crew for the ISS, launches from Baikonur Cosmodrome in Kazakhstan.

2001 12 February: *NEAR Shoemaker* lands on 433 Eros. **8–21 March:** *Discovery*, STS-102, retrieves Expedition 1 and delivers Expedition 2 from the ISS; they also deliver the contents of the Leonardo Multipurpose Logistics Module, a reusable cargo carrier built for the Italian Space Agency by what is now Thales-Alenia. **23 March:** No longer needed, *Mir* is returned to Earth, its debris splashing into the South Pacific Ocean. **29 April–7 May:** Denis Tito spends $20 million to become the first space tourist. **30 June:** The *Microwave Anisotropy Probe* is launched from Cape Canaveral to investigate origins of the universe. **8 August:** *Genesis* is launched to collect solar wind samples. **10-21 August:** Expedition 3 from the ISS is launched from Kennedy Space Center. **19 November:** Dr. Daniel Mulville is appointed acting administrator. **5–17 December:** Expedition 4 begins. **21 December:** Sean O'Keefe is appointed NASA administrator by President George W. Bush.

2002 5 February: NASA launches the *High-Energy Solar Spectroscopic Imager* to study solar flares. **February:** The *Mars Odyssey 2001* begins mapping the surface of Mars in detail. **8–19 April:** *Atlantis* carries with it significant sections to extend the ISS, which it successfully deploys. **5–19 June:** Space shuttle *Endeavour*, STS-111, delivers Expedition 5 as well as logistics and equipment. **3 July:** *Comet Nucleus Tour* is launched to meet the path of two comets, but breaks up shortly after launch. **7–18 October:** STS-112 continues in-orbit construction of the ISS. **23 November–7 December:** Expedition 6 is launched on the space shuttle *Endeavor*, STS-113.

2003 13 January: A Cosmic Hot Interstellar Spectrometer is launched by NASA to study nebulae; NASA also launches a Geoscience Laser Altimeter System, enabling accurate measurements of ice sheets. **1 February:** The space shuttle *Columbia*, STS-107, breaks up in the atmosphere 15 minutes before landing back on Earth, killing all seven astronauts aboard. **28 April:** NASA launches the *Galaxy Evolution Explorer* to study evolution of galaxies. **25 August:** The largest-diameter infrared telescope ever in space, the Spitzer Space Telescope, is launched. **21 September:** *Galileo* crashes into the surface of Jupiter. **5 October:** The People's Republic of China sends its first astronaut in space, becoming the third nation to do so.

2004 2 January: *Stardust* (US) returns a sample from a comet (81P/Wild). **3 and 24 January:** *Spirit* and *Opportunity* land on the surface of Mars. **14 January:** President George W. Bush announces the Vision for Space Exploration, which includes the Constellation program. **20 April:** *Galaxy Probe B* is launched to test Einstein's famous hypothesis. **1 July:** *Cassini-Huygens* (US/ESA/ASI) orbits Saturn to become the first vehicle to go into orbit around Saturn. **3 August:** *MESSENGER* is launched to conduct in-depth study of Mercury. **8 September:** *Genesis* reports the first sample return beyond lunar orbit (solar wind). **12 November:** The X-43A vehicle reaches Mach 10 in a test flight.

2005 14 January: *Huygens*, after being deployed from *Cassini*, soft lands on Titan, one of Saturn's moons. **11 February:** Frederick Gregory is appointed acting administrator of NASA. **14 April**: President Bush appoints Michael Griffin NASA's newest administrator. **26 July:** *Discovery* marks NASA's first return to human spaceflight after the *Columbia* tragedy. **12 August:** *Mars Reconnaissance Orbiter* is launched to study the atmosphere and subterranean layers of Mars. **19 November:** Japan enters space with the first asteroid ascent by the *Hayabusa* (Falcon) after it rendezvoused with the asteroid 25143 Itokawa, the first to land on an asteroid and then return to Earth powered by xenon ion engines.

2006 15 January: *Stardust* returns, carrying samples of interstellar dust. **19 January:** *New Horizons* (US) is launched toward Pluto. **7 May:** The *Venus Express* (ESA) begins an extended orbital exploration of Venus. **24 August:** Pluto is demoted to dwarf planet status, leaving us with eight planets in our solar system. **3 October:** Dr. John Mather of NASA receives Nobel Prize for physics along with the University of California's George Smoot.

2007 4 August: *Phoenix*, a small Mars scout lander, is launched by NASA. **14 September:** Japan launches *Kaguya* (Selene), a lunar orbiter. **27 September:** NASA launches *Dawn* to study asteroids. **24 October:** The PRC launches *Chang'e 1*, a lunar orbiter.

2008 14 January: *MESSENGER* takes photos of unknown areas on Mercury. **25 May:** The *Phoenix* Mars lander descends safely to the planet. **25 September:** The PRC sends three astronauts into space in their *Shenzhou-7* spacecraft. **22 October:** ISRO (India) launches *Chandrayaan-1*, a lunar orbiter.

2009 21 January: With the election of President Barack Obama, Christopher Scolese is appointed acting administrator. **6 March:** NASA launches the Kepler mission, the first space telescope designed to search for Earth-like planets. **28 June:** The *Lunar Reconnaissance Orbiter* (US) launches, as well as *LCROSS*, the lunar impactor. **17 July:** President Obama confirms Charles Bolden, a former astronaut, as NASA administrator.

2010 11 February: The *Solar Dynamics Observatory* is launched and will observe the sun for five years. **15 April:** President Obama cancels the Vision for Space Exploration (President George Bush's plan, then called the Constellation program) and sets his own policy. **May:** The GPS 2F-1, the United States' first new GPS satellite, launches. **21 May:** JAXA launches Venus orbiter *Akatsuki*. **13 June:** The *Hayabusa* returns the first sample analysis from an asteroid. **25 September:** The U.S. Air Force launches a space-based surveillance satellite to record other spacecraft and space debris. **1 October:** *Chang'e 2* is launched from China to orbit the moon. **10 October:** Virgin Galactic announces the successful human glide suborbital spaceflight of the VSS *Enterprise.* **3 December:** The U.S. Air Force's X-37B unmanned returns from seven months in orbit. **7 December:** *Akatsuki* misses Venus's orbit but will retry on its return in 2020. **8 December:** SpaceX successfully launches the Falcon 9 rocket with Dragon capsule.

2011 March: *Discovery* delivers NASA's final contribution to the International Space Station, a permanent multipurpose module. **18 March:** *MESSENGER* completes the first orbit of Mercury. **24 June:** Kerbal Space Program, a video game and spaceflight simulator, is released as version Alpha 0.7.3. **18 July:** The *RadioAstron* (Russia) launches and becomes the largest space observatory. **21 July:** Space shuttle *Atlantis* lands at Kennedy Space Center, the last of the space shuttle flights. **2 November:** China docks to robotic spacecraft in low Earth orbit. **26 November:** NASA launches the *Mars Science Laboratory* to deliver the rover *Curiosity.*

2012 22 May: SpaceX launches its Dragon C2+ mission to resupply the International Space Station. **August:** *Voyager 1* (US) transitions into interstellar space. **6 August:** NASA's *Curiosity* rover successfully lands on Mars.

2013 15 February: A large meteor explodes over Russia's Chelyabinsk region. **6 September:** *LADEE* (US), the third lunar probe in five years, is launched. **9 September:** Mars One announces 200,000 interested parties in proposed one-way trip to Mars. **5 November:** *Mangalyaan* (India) Mars orbiter is launched from India. **18 November:** *MAVEN* (US), a mission to study the atmosphere of Mars, is launched. **1 December:** *Chang'e 3*, China's lunar lander, is launched. **30 December:** Mars One announces Round Two candidate selection of 1,058 individuals.

2014 9 January: Orbital Sciences launches its first robotic resupply mission to the ISS. **6 August:** The ESA's *Rosetta* reaches its comet and becomes the first spacecraft to orbit a comet. **12 September:** NASA unveils its new Vertical Assembly Center at its Michoud facility. **15 September:** NASA's own Office of the Inspector-General issues critical report on NASA's program to identify near-Earth objects. **16 September:** NASA awards contracts to SpaceX and Boeing for crew transportation; Sierra Nevada Corp. protests nine days later. **22 September:** *MAVEN* reaches Mars and begins orbiting the Red Planet. **16 October:** India launches the third of its seven communications satellites. **17 October:** The U.S. Air Force's secret, unmanned X-37B lands quietly in California after 22 months in orbit. **19 October:** Siding Spring, a large comet, flies past Mars for the first time in 1 million years. **28 October:** Orbital Sciences's unmanned resupply mission to the International

Space Station is lost when its Antares launch vehicle explodes seconds after launch at Wallops Island. **12 November:** *Rosetta*'s *Philae* lander (ESA) is deployed and lands in shadows on the comet 67P/Churyumov-Gerasimenko and becomes unable to recharge its solar batteries. **11 December:** NASA successfully launches and retrieves its Orion capsule, testing avionics, attitude control, parachutes, and heat shield. **17 December:** The *Curiosity* rover on Mars makes the first definitive detection of organic molecules on Mars.

2015 31 January: The *Soil Moisture Active Passive* satellite is launched into orbit by United Launch Alliance's Delta II rocket. **5 February:** *New Horizons* returns its first new images of Pluto. **8 February:** NASA releases a video showing the "dark" side of the moon. **13 February:** Mars One narrows its pool to 100 candidates. **9 March:** *Dawn* arrives at the dwarf planet Ceres. **12 March:** NASA's Magnetospheric Multiscale Mission, comprised of four identically instrumented spacecraft, is launched. **14 June:** ESA's *Philae* reactivates on Comet 67P/Churyumov-Gerasimenko as it emerges from shadow. **16 June:** Traces of methane in Martian meteorites reignites speculation about life on Mars. **14 August:** Astronomers detect the smallest alien planet (51 Eridani b) yet to be photographed by an Earth telescope.

Introduction

The National Aeronautics and Space Act of 1958 declared "it is the policy of the United States that activities in space should be devoted to peaceful purposes for the benefit of all mankind . . . [and that] the general welfare and security of the United States require that adequate provision be made for aeronautical and space activities."

NASA—the National Aeronautics and Space Administration created in the wake of the Space Act—has and continues to accomplish those precepts every day. With many hundreds of satellites launched into space and close to 200 human spaceflights, NASA is a proven leader in space exploration.

However, to fully understand NASA's emergence and partnership with space programs throughout the world, we must first explore the environment into which NASA was created.

PRE-NASA SPACE EXPLORATION

World War II provided the impetus for not just new inventions but also for innovative modes of production that virtually guaranteed mass deliveries. By the end of World War II, the aircraft business had become the single largest industry in the world, much of it in the United States. In 1939, fewer than 6,000 planes had been produced; by the time of the attack on Pearl Harbor, on December 7, 1941, more than 25,000 had been manufactured, and in the few short years from then until the end of the war in 1945, another 275,000 were built. This was also the beginning of California's rise to industrial power, as 44 percent of the aviation industry was centered there. Aviation industry meant not just planes but also development of newer instruments, more responsive controls, and modern techniques.

In 1947, the United States modified V-2 rockets that had been captured from Nazi Germany and sent the first animals—fruit flies—into suborbital space. Suborbital flights are those where the vehicle crosses the Kármán line (62 miles, 100 km above sea level). Because the vehicles' speeds didn't reach orbital velocity, however, those initial spacecraft fell to Earth before completing a full orbit. To stay in orbit, a vehicle needs a speed of around 18,000 miles per hour when it reaches the Kármán line, and initial rockets didn't carry enough fuel to reach such levels.

The National Advisory Committee for Aeronautics (NACA), which had been founded in 1915 before U.S. participation in World War I, began experimenting with "rocket planes" such as the supersonic Bell X-1 in the mid-1940s. Other countries were experimenting with their own versions of rockets. When the International Geophysical Year (1957–1958) was declared, international competition stretched to 67 countries in fields of study ranging from aurora at the poles to meteorology, and gravity to solar activity. The capstone, however, and the match that lit the US/USSR space race, was the USSR's launch of the world's first artificial satellite, *Sputnik 1*, on October 4, 1957.

Concerned over possible threats to national security and with national pride at stake, the U.S. Congress proposed immediate action. After consultation with President Eisenhower, a new federal agency—NASA—was proposed to conduct nonmilitary operations in space. Military operations were to be retained by the appropriate Department of Defense organizations, in close cooperation with the newly created Advanced Research Projects Agency (ARPA), which began operations in February 1958 to develop space technology for military applications.

Five months later, on July 29, 1958, President Eisenhower signed the National Aeronautics and Space Act into being, creating NASA, which began operations on October 1, 1958. To consolidate departments and work within the limited budget, NACA was absorbed into NASA. In many ways, it was simply a change of name; in others, it was a deeper purpose, a change of mind-set. All NACA employees, including its three research laboratories (Langley Aeronautical, Ames Aeronautical, and Lewis Flight Propulsion), the High-Speed Flight Research Facility, and Wallops Flight Facility, became key parts of NASA, as did other smaller institutions.

EARLY DAYS OF NASA

Coming out of World War II and the Korean War, the United States was in far better shape than most of the world. The only substantial attack on its shores had been in 1941, more than 15 years earlier by the time NASA was founded. There was no limit to the belief in American ingenuity, no bounds to hopes for the future. Project Vanguard, an International Geophysical Year project led by the Naval Research Laboratory intended to launch a satellite in 1955 and 1956, was popular. The U.S. Army's Redstone ballistic missile launch system provided backup to national goals of space exploration.

Sputnik 1, a 184-lb., battery-operated satellite that fell to Earth some three months after it achieved orbit in 1957, shattered the illusion of American superiority. The United States sent *Explorer 1* into space three months after

Sputnik 1, but second place did not sit well with Americans. Disappointment happened again on April 12, 1961, when the Soviet Union sent Yuri Gagarin into space, the first man to orbit the Earth. Alan Shepard followed a month later. The national mood with regard to space became grim and fearful, afraid of a second-place finish.

President John F. Kennedy changed that with his speech before a joint session of Congress on May 25, 1961, less than a month after Shepard's flight, and again on September 12, 1962, in a speech at Rice University: "This generation does not intend to founder in the backwash of the coming age of space. We mean to be a part of it . . . we mean to lead it." "Americans are leaders" was the sentiment of the time, "and leaders lead." Sending a man to the moon would surely regain our rightful spot at the top.

NASA's first administrator was T. Keith Glennan, who had immediate supervision of the National Advisory Committee for Aeronautics—its 8,000 employees, its research laboratories and two other facilities, and its annual budget of almost $100 million.

The Space Act's section 102 set the following eight goals for NASA:

1. The expansion of human knowledge of phenomena in the atmosphere and space;
2. The improvement of the usefulness, performance, speed, safety, and efficiency of aeronautical and space vehicles;
3. The development and operation of vehicles capable of carrying instruments, equipment, supplies, and living organisms through space;
4. The establishment of long-range studies of the potential benefits to be gained from the opportunities for, and the problems involved in the utilization of aeronautical and space activities for peaceful and scientific purposes;
5. The preservation of the role of the United States as a leader in aeronautical and space science and technology and in the application thereof to the conduct of peaceful activities within and outside the atmosphere;
6. The making available to agencies directly concerned with national defense of discoveries that have military value or significance, and the furnishing by such agencies, to the civilian agency established to direct and control nonmilitary aeronautical and space activities, of information as to discoveries which have value or significance to that agency;
7. Cooperation by the United States with other nations and groups of nations in work done pursuant to this act and in the peaceful application of the results thereof; and
8. The most effective utilization of the scientific and engineering resources of the United States, with close cooperation among all interested agencies of the United States in order to avoid unnecessary duplication of effort, facilities, and equipment.

Though the act has been amended, these goals have remained virtually unchanged over the years. In 2012, a ninth goal was added: "The preservation of the United States preeminent position in aeronautics and space through research and technology development related to associated manufacturing processes."

Within two years of its founding, NASA had incorporated the other groups that would make up its core: Jet Propulsion Laboratory, a part of the California Institute of Technology; the Army Ballistic Missile Agency in Huntsville, Alabama; and the space science group (soon to be the Goddard Space Flight Center) of the Naval Research Laboratory in Maryland. It also acquired several satellite programs, two lunar probes, and the charge to develop a million-pound-thrust rocket engine. Glennan, who was president-on-leave from Case Institute of Technology, oversaw the rapid expansion that included all space activities with the exception of reconnaissance satellites, ballistic missiles, and some Department of Defense projects.

GOLDEN AGE OF NASA

Glennan left in early 1961, but the programs he initiated were ably administered by James Webb, a consummate politician and manager appointed by President Kennedy: Project Mercury came first, highlighted by Alan Shepard's suborbital flight in May 1961 and John Glenn's orbit around Earth in February 1962. Next was Project Gemini, with 10 flights that extended our knowledge of space and of humans' physical response to it. As an interim goal before landing men on the moon, NASA researchers gathered more data on weightlessness and its effect on humans, and its scientists and engineers perfected reentry and splashdown methods, docking maneuvers, and rendezvous procedures.

And then the goal toward which NASA had been aiming for more than a decade—landing a man on the moon—was imminent. James Webb provided the direction and administration for NASA until 1968, dealing with the U.S. Congress after the *Apollo 1* (*AS-204*) fire in early 1967 that left three astronauts dead and stunned the nation. Webb convinced President Johnson to let NASA lead the investigation; the subsequent enquiry—by all accounts fair and complete—left NASA's image largely intact. Webb left just one year shy of *Apollo 11*'s historic accomplishments when Neil Armstrong, Michael Collins, and Buzz Aldrin fulfilled President Kennedy's goal.

Webb was succeeded by Dr. Thomas O. Paine, a former manager at General Electric's Center for Advanced Studies, in January 1968. Given the task of continuing the Apollo program through the moon landing, Paine saw 14 astronauts travel to the moon on six separate missions. Flushed with success, he also presented plans for a lunar base and a giant space station before the end of the 1970s, leading up to a human spaceflight mission to Mars in 1981. These grandiose yet viable plans were not approved by President Nixon, and Paine resigned three years after taking office.

George M. Low became acting administrator on September 16, 1970, just over a year after *Apollo 11* but before the end of the Apollo flights. He had direct experience in all of the human spaceflight programs and was deputy administrator to both Paine and his successor administrator, James C. Fletcher. Fletcher, a political conservative like the president who appointed him, had more political savvy than his two predecessors, and oversaw *Skylab*, a small orbiting platform that hosted three missions ranging from 28 to 84 days. *Skylab*, a counterpart to the Soviet Union's *Salyut 1* that had already hosted cosmonauts for 23 days, proved to the world that long-duration spaceflight and a safe return to Earth was possible. Fletcher strongly believed in life beyond Earth and was therefore a huge proponent of space science and of NASA. He decentralized operations, assigning projects to field centers as needed, but maintained NASA headquarters as the hub around which the field centers revolved.

The 1970s also saw a new generation of scientific research and exploratory spacecraft: the two Pioneer spacecraft (10 and 11) which traveled as far as Jupiter and Saturn, at the time an extraordinary distance, as well as *Voyagers 1* and *2*, dispatched on their interstellar voyage that continues today. Probes explored the moon, others ventured to Venus. Still more satellites began their dance around Earth, improving communications, weather information, intelligence gathering, and solar research.

NASA's first 20 years were incredible, transitioning from the unsuccessful launch of a single satellite to the landing of men on another celestial body:

- Human spaceflight programs: Mercury (1961–1963), Gemini (1965–1966), and Apollo (1968–1972)
- Robotic missions to the moon (*Ranger*, *Surveyor*, and the Lunar Orbiter), to Venus (*Pioneer Venus*), to Mars (*Mariner 4*, *Viking 1* and *2*), and to the outer planets (*Pioneer 10* and *11*, *Voyager 1* and *2*)
- *Skylab*, an orbital laboratory
- Satellites for communications (*Echo 1*, *Telstar*) and weather
- Satellites for information gathering (*LandSat* and others)
- Aeronautics research for air safety, reliability, and efficiency. The decades saw the exponential growth of our air transport system, and NASA participated with aircraft research, propulsion technology study, airport safety measures, and much more
- Development of the space shuttle program since 1972

SPACE SHUTTLE ERA

Now, with the knowledge that living in space was possible (and confirmed by further trips sponsored by the Soviet Union and Russia), two things occurred to push the United States to the next steps in space: (1) the space shuttle, which began flying in 1981; and (2) in 1984 Congress gave permission to NASA to construct a new space station, an International Space Station (ISS), that would keep America and its partners in space for 20 years or more.

By the time President Ronald Reagan proposed the ISS on January 25, 1984, NASA had been through five more administrators: Acting Administrator Alan Lovelace (May–June 1977), Administrator Robert Frosch (1977–1981), Acting Administrator Alan Lovelace (January–July 1981), Administrator James Beggs (1981–1985), and Acting Administrator William Graham (December 1985–May 1986). Frosch inherited cost overruns on the space shuttle, almost causing President Carter to cancel the shuttle program. Beggs later backed NASA's growing involvement in global environmental issues, leading NASA to a scientific leadership position in determining the causes of ozone depletion in the Antarctic. He also got the go-ahead for what was to eventually become the International Space Station. When he resigned, former administrator James Fletcher was called back into duty to oversee NASA in 1986 and to deal with the aftermath of the *Challenger* tragedy earlier that year. By 1986, the designs for the ISS were declared complete.

It took a further five years for NASA and Congress to reconcile the station's purpose and budget, but Space Station *Freedom*, the U.S. portion of the ISS, was approved. Russia had launched *Mir*, which became the longest-occupied space station until that point, and space shuttle missions began to ferry American astronauts to *Mir* in an international effort allowed by Russia. With revisions, Space Station *Freedom* became Space Station *Alpha*, and building the ISS commenced in 1998, going through another three administrators: Acting Administrator Dale Myers (April–May 1989), Administrator Richard Truly (1989–1992), and Administrator Daniel Goldin (1992–2001), NASA's longest-serving administrator. Truly, a former astronaut, had considerable experience in space. When the National Space Council declared a priority of a return to the moon followed by a piloted mission to Mars, he gamely helped promote the plan. However, after notifying Congress of the half-a-trillion-dollar budget, the program was never funded. After Truly's resignation in 1992, Goldin followed the "faster, better, cheaper" theory by instigating multiple missions and by promoting the International Space Station, his major achievement. He also remade the Mars program, which had suffered since 1976 when Viking died on its way to the red planet. Not all his programs succeeded. The X-33, a space shuttle successor, was terminated by NASA itself after almost $1 billion in development costs.

By 2000, NASA had weathered four decades of significant accomplishments, but many of them were continuations of programs started before the mid-1980s—the space shuttle and the ISS in particular. NASA's portion of the federal budget in those last two decades of the 20th century became a smaller and smaller percentage of the whole, except for a small blip during the building of the ISS. And unlike the *Apollo 1* fire where NASA emerged unscathed, management took a beating in the wakes of the *Challenger* and *Columbia* disasters. Though the two events were 17 years apart, in the public's mind it showed that NASA didn't learn from its mistakes, that nothing had changed—and there were no significant new projects in the works to keep the public's fickle attention from focusing on NASA's mistakes.

Turnover at the highest level of NASA continued. Daniel R. Mulville was acting administrator for just over a month, followed by Sean O'Keefe, a former deputy director of the Office of Management and Budget. O'Keefe's tenure had its controversial moments. During his first year in office, he eliminated $5 billion in cost overruns on the ISS—a natural result, given his background in OMB. His second year was consumed by the *Columbia* tragedy, forcing the nation to realize that the shuttle itself was old, flawed, and potentially dangerous. His third year, O'Keefe focused on President George Bush's Vision for Space Exploration plan to return humans to the moon and then on to Mars. After the criticism of NASA's risk policies in the wake of the *Columbia* accident, O'Keefe made the unfortunate decision to cancel a repair mission for the Hubble Space Telescope as too dangerous. Then—part culture clash, part political maneuverings, probably part misunderstandings—NASA climatologist James Hansen reported in the *New York Times* that O'Keefe had cautioned him to soft-pedal humanity's part in global warming. The matter of O'Keefe's lack of a scientific or engineering background became a political hot potato.

O'Keefe was replaced by former astronaut Frederick Gregory as acting administrator before the appointment of Michael Griffin some two months later. Griffin, a scientist and engineer, wasted no time in coordinating a repair plan for the Hubble Space Telescope and in cancelling the moon and Mars plans set in motion by the former administrator. Griffin's priority, he stated, was to send humans out into the solar system, and his aim was to have humans on Mars by 2037. Unfortunately, his tenure ran into trouble as well when his statements about global warming were taken as "arrogance and ignorance," and when budgets for science and research field centers (Ames, Glenn, Langley, and so forth) were cut far more than those concentrating on human spaceflight.

By this time, it was obvious that political turmoil in Washington and an out-of-control federal budget increased by Hurricane Katrina and the war in Iraq as well as other unexpected events spilled over into NASA to a high degree. Budgets, personnel, and NASA priorities had become political tools to wield when needed. Christopher Scolese was appointed acting administrator for six months, followed by the presidential appointment of Major General Charles F. Bolden, a former astronaut, who took office on July 17, 2009. He, too, sees a human settlement on Mars as within the realm of possibility, though acknowledges NASA's limited budget does not encourage a speedy settlement there. It has been under his watch that the space shuttles were terminated after 30 years (an ending that had been planned since well before he stepped up as administrator) and the Space Launch System (SLS) and Orion spacecraft promoted. His leadership has included the creation of the Space Technology Mission Directorate and the privatizing of ferry service to and from the ISS and beyond by 2017. He has continued earlier programs of the Curiosity rover, the James Webb Space Telescope, the enlargement of Earth-observing satellite network, and the launch of a spacecraft to Jupiter. In aeronautics, he has promoted a greener, faster, quieter fleet of airplanes, with aeronautical safety at our airports a high priority. But all is not rosy for Bolden either.

With President Obama's cancellation of the Constellation program, Bolden was accused of bad management when the reasons for the cancellation seemed vague to NASA supporters. When Bolden stated in 2010 that he had been charged with reaching out to the Muslim world and to help them feel good about their science contributions, he played into the hands of those looking for a reason to dismantle or at least discredit NASA, perhaps to defund it. He has redeemed NASA's reputation with the Boeing/SpaceX/private contracts to return launches to American soil instead of the expensive launch contracts with Russia. Bolden at this point has served the second-longest term of any NASA administrator.

NASA does not operate in a vacuum. Issues in Washington either trickle or flood down to NASA missions regardless of the mission's worth. It is a tribute to NASA personnel—including the administrators and acting administrators who were all too aware of their limited time in office—that missions were as continuous and efficient as they were allowed to be, and this was due in large part to the able deputy administrators who were not quite so susceptible to political whims and who managed NASA on a day-to-day basis. Each administrator was a presidential appointee and although many were familiar with NASA, some were from outside the aeronautical fields. All were intelligent, capable men, but the average administrator served under three years, leaving the continuation of programs they favored subject to the approval of their successors. The sheer number of administrators made the likelihood of a cohesive program a harder task for all.

OVERLOOKED NASA PROGRAMS

Beginning with the incorporation of the National Advisory Committee for Aeronautics at its founding, NASA has conducted various cutting-edge aeronautics research programs, including wind shear, wind tunnel, and flight testing research and computer simulations. Today, NASA's Aeronautics Research Mission Directorate continues to solve the problems of air traffic congestion, safety, and environmental impacts. Green aviation goals enable fuel-efficient flight planning for private, commercial, and governmental operations, reduce fuel consumption, reduce emissions, and lessen noise. Five programs—Aeronautics Test, Airspace Systems, Aviation Safety, Fundamental Aeronautics, and Integrated Systems Research—operate primarily out of four of NASA's ten field centers: Ames Research Center, Armstrong Flight Research Center, Glenn Research Center, and Langley Research Center. Research centers—as opposed to spaceflight centers—are the "behind-the-scenes" engineers and researchers who support programs that are just as important as the more publicly visible space programs.

Less obvious are also the launch and support services for the weather and communications satellites that support the scientists, run the GPS systems on our phones and watches, and let us watch television and run our computers from cloud-based systems.

Spin-offs from all sorts of programs cover everything from invisible dental braces to memory foam (but not including ballpoint pens, Velcro, and Tang, as popularly believed). NASA freely makes its inventions and discoveries available to those who would use them, resulting in a number of products that today we consider "necessary"—but which also do not provide financial remuneration for NASA, to any significant degree.

BUDGETS AND BUREAUCRACY

NASA's budget, less as a percentage of federal budget by more than 1,000 percent since its heyday in the 1960s, is still a fairly substantial $17.7 billion (see figure I.1). Continuing science and aeronautics programs take up roughly 20–25 percent of that, leaving 75–80 percent (or roughly $13 billion) for the various human spaceflight programs NASA manages or participates in.

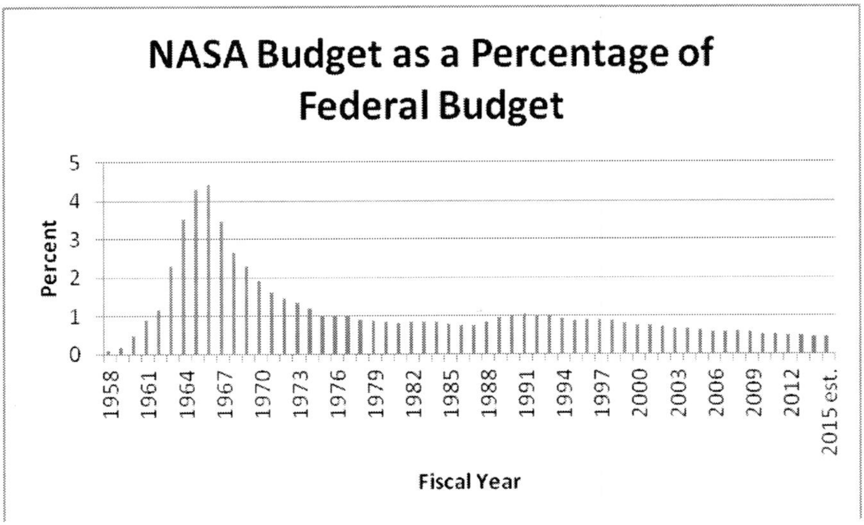

Figure I.1 Note the peak during the Apollo years (1962–1970) and the bulk of the International Space Station building projects (1988–1994). Source: Office of Management and Budget.

Although the FY 2015 budget has not yet been approved at the time of this writing, the president's proposed budget includes 22 percent ($3.905 billion) for space operations and 21 percent ($3.976 billion) for exploratory programs, the two largest expenditures other than science missions (Earth science, planetary science, astrophysics, heliophysics, and space telescopes) and the ones the public seems most interested in (although that, of course, is subjective). If an overall vision is to come from NASA that will engage the public, it will need to be in the human spaceflight realm. The balance of the budget is composed of education ($89.8 million), cross-agency support ($2.8 billion, which includes management and operations), new facilities and maintenance ($446 million), and the inspector-general's office ($37 million). (See figure I.2; also see appendix A and the breakdown of the budget into individual missions at www.nasa.gov/sites/default/files/files/508_2015_Budget_Estimates.pdf.)

It is a mistake to believe that NASA has always enjoyed strong popular support. Historical data show that at the time the Apollo program was being conducted, the majority of Americans did not feel Apollo was worth the cost—except during July 1969 at the time of the moon landing. A fairly consistent 10–20 percent of the public feels that the nation spends too little on NASA, according to Pew Research Center polls. At the same time, 30–40 percent of the public feels we spend too much. That leaves 40–60 percent who have no opinion or who feel spending is "just right." The numbers vary depending on public reaction to current events: A poll taken just after it was revealed that the Hubble Space Telescope mirror had been faulty, for exam-

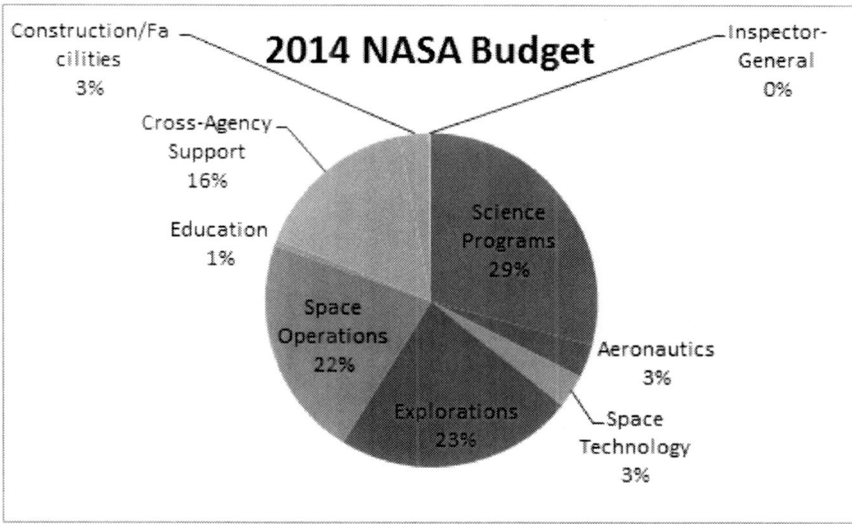

Figure I.2 Breakdown of NASA's 2014 budget divided into categories. Source: NASA.

ple, showed lowered support for NASA. And yet a 2011 poll showed that 60 percent of Americans feel that it is essential for the United States to continue space exploration efforts, with almost the same amount feeling that NASA does either a "fair" or a "poor" job. Does this mean that the public would prefer private corporations (Boeing, SpaceX, Orbital Sciences, Sierra Nevada, and so forth) to conduct our space exploration efforts? Not necessarily. And given the support NASA has provided to private corporations for research and development, it could well be that private corporations may be overpromising what they can deliver.

Another way of looking at trends is shown in the following chart (see figure I.3):

Over 15 years, education, aeronautics, construction (data begins in 2010), and space technology (2010 as well) have remained low and fairly stable. Cross-agency support (begun in 2007) is fairly stable as well. But space operations have declined considerably, exploration is on a general upward trend, and science is slowly rising.

Naturally, there are those who feel the budget is way too high, for a program not considered necessary to our nation. At the opposite end are those who feel NASA works with too few financial resources, is underappreciated by the general public, and is vital to our nation's security.

Space exploration will always be risky, and to judge NASA on the tragedies that have occurred would be a mistake as well. NASA and its astronauts know that the people it sends to space have an approximate 1-in-60 (or

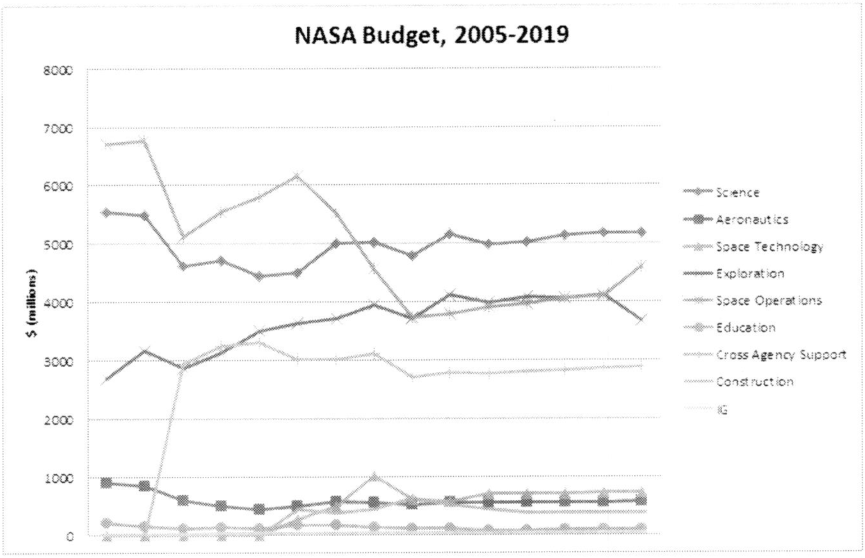

Figure I.3 NASA organizational budget by general categories, 2004–2015. Source: NASA.

roughly 2 percent) chance of death during spaceflight. Of the roughly 600 people who have traveled in space (many of them have made more than one voyage), only 19 have perished as a direct result of their spaceflight since the moon landing. To be associated with the spaceflight—the repair workers, crane operators, construction workers, and so forth—is slightly *more* dangerous: 24 people have died on the job in such activities in the same amount of time. (To put this in perspective, your odds of perishing in an auto accident are about 1 in 90.) To be as safe as possible, to put the crew members and the people who work on the rockets in as safe a position as they can, is expensive. Redundancy becomes paramount, which adds to complications and expenses.

SETBACKS FOR NASA

Setbacks for NASA tend to be in two types: the tragic deaths of astronauts and crewmembers, which is by far the hardest for NASA personnel to deal with, and the budgetary/political ups and downs that cancel missions already started and with which personnel have dedicated hours and years of their lives.

The *Apollo 1* (*AS-204*) fire of January 27, 1967; the *Challenger* tragedy of January 28, 1986; and the *Columbia* tragedy of February 1, 2003: all were heart-rending events that caught NASA by surprise and slowed the forward momentum of their respective space programs. The 17 or more years between accidents seemed to give NASA just enough time to recover, regroup, and then relax. The first, *Apollo 1*, caused postponed flights for the following four months, but because it was an on-the-ground accident, space flight soon resumed. The second, *Challenger*, caused space shuttle flights for 32 months to be cancelled as investigators debated and changes were made; and the last, *Columbia*, resulted in a 29-month delay. Each time, NASA and investigators' resources concentrated on determining what had happened and how to eliminate such problems in the future.

Public and political reactions to the latter two events were different in many ways. President Ronald Reagan in 1986 was much more forward looking in his speech after the *Challenger* tragedy, but at the time the United States was involved in the last of the Cold War, and the space program (or at least the Strategic Defense Initiative) was deemed essential to our national security. In addition, the public had identified strongly with Christa McAuliffe, the teacher in space, and her death (along with the rest of the crew) left a mark. Even now, almost 30 years later, hers is the name most people remember. By the time of the *Columbia* tragedy, the space program was less essential to national pride and far less important to President George W. Bush, who had rarely mentioned NASA or its programs in his speeches. Space shuttle flights had become fairly routine. Bush's remarks after the event, although heartfelt and sorrowful, were less focused on our future in space and more on the risks of such activities. This is not to say that one reaction was more important or better than the other but to acknowledge the political exigencies of the times.

Budgetary setbacks are taken much more matter-of-factly. Although no one likes to have his or her $900+-million project simply discarded, it is a known risk going into any new mission. The president submits his budget to Congress, the House and Senate work to agree on what should take priority, and NASA must work with whatever it has been given. How much or how little that is, is in some ways dependent on how well the administrator has formulated and pushed the programs that NASA would like to see developed over the coming years. Any change in administration—whether presidential or in Congress—can affect NASA's priorities, despite what an administrator might want. Still, the cancellation of a program has direct results not only on the engineers and scientists working on it but on the programs that were never funded so that NASA could focus on the chosen ones that were later cancelled. Many people still believe that if we had continued to fund and focus on the post-Apollo plans to go to Mars, we would have been there by 1990 and would be well on our way to having a human outpost there.

NASA ADMINISTRATION

NASA's administrator (currently, Maj. Gen. Charles F. Bolden Jr.) operates out of NASA Headquarters in Washington, D.C. The overall direction of NASA's policies and priorities are set by this office, with funding provided by the U.S. Congress. Its current vision statement reads: "To reach for new heights and reveal the unknown so that what we do and learn will benefit all humankind." Approximately 1000 employees join the administrator, deputy administrator, and associate administrator in fulfilling this vision (see figure I.4).

In addition to the general administration of NASA, the ten field centers concentrate on four main mission directorates: Aeronautics Research, Science, Human Exploration and Operations, and Space Technology, each of which is headed by its own associate administrator.

Aeronautics Research Mission Directorate (ARMD) tends to be the forgotten stepchild, at least as far as the public in concerned. By concentrating on the challenges existing in our passenger-based air transportation system—air traffic congestion, safety, and environmental impacts—ARMD works behind the scenes and rarely comes to the notice of the flying public. For example, in July 2014, ARMD turned over to air traffic controllers a new, NASA-developed computer software tool called Terminal Sequencing and Spacing that will allow air traffic controllers to more effectively manage the spacing between aircraft and to fly efficient approaches into airports by around 2018. Pilots will have fewer course and altitude changes as they approach destination airports, allowing a more gradual descent arrival procedure.

The Science Mission Directorate's (SMD's) goal with regard to our own planet is to "meet the challenges of climate and environmental change," which involves six main areas:

- Atmospheric composition
- Weather
- Carbon cycles and ecosystems
- Water and energy cycles
- Climate variability and change
- Earth's surface and the interior of the planet

Not to be forgotten are the satellite tracking systems around Earth, available to the public to track the weather, communication, and other satellites at any given time, as well as the Joint Agency Satellite Division. However, it is SMD's other departments that garner the vast majority of public attention: heliophysics (the study of the sun), astrophysics (how our universe works),

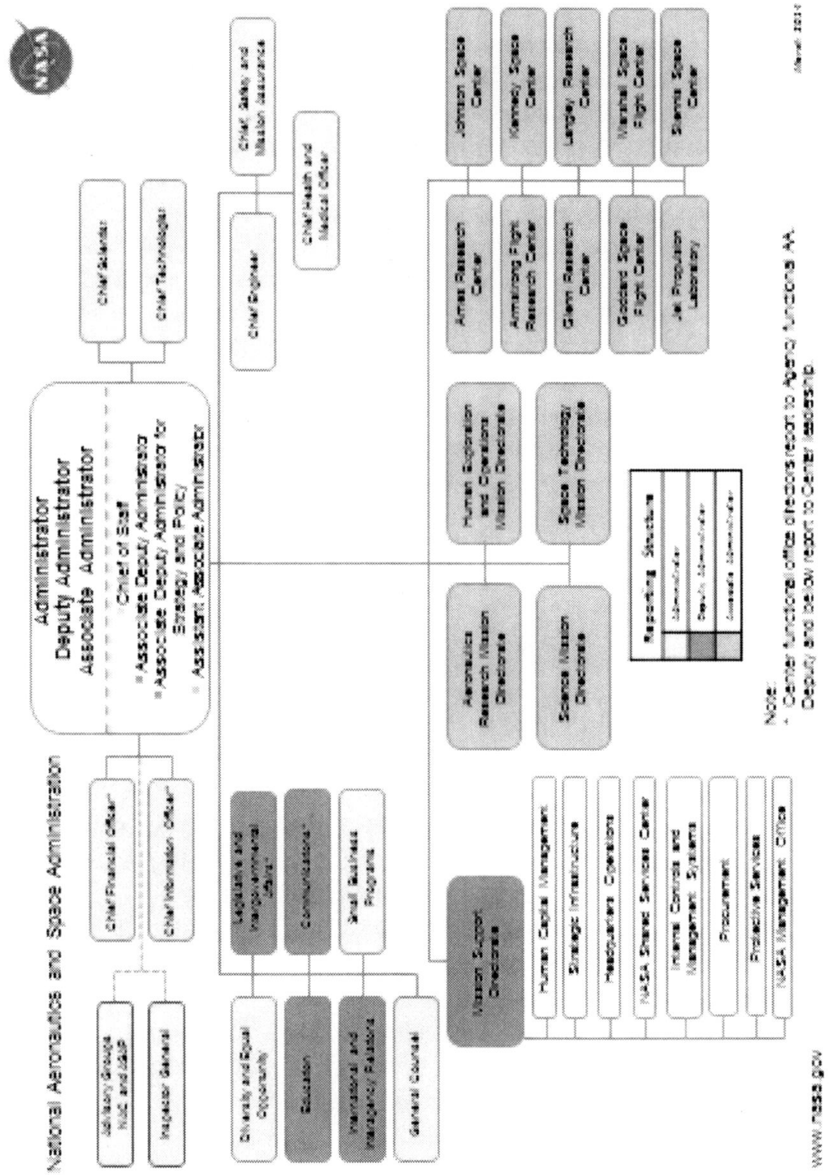

Figure I.4 NASA organizational chart. Source: NASA, www.nasa.gov/sites/default/files/atoms/files/org_chart_agencywide_april_2015.pdf.

and the other planets, particularly Mars. Still, the James Webb Space Telescope comes in for its share of attention, especially as it nears its launch date in 2018.

Human Exploration and Operations (HEO) is the poster child of NASA's work, represented currently by the International Space Station, and all that such spacefaring activities entail: launch services, space transportation, and space communication for both human and robotic exploration programs. In addition to current programs, HEO is also responsible for developing the next generation of technology that will assist humans into an ever-wider presence in space—a costly endeavor in experimentation and research, not only to discover what doesn't work but to invent what does work.

Finally, Space Technology Mission Directorate (STMD) is responsible for identifying the technology NASA will need in future missions, through competitions, research and development, investments in start-up businesses as well as established corporations, partnerships with colleges and universities, and collaborations with private industry. The recent (August 2014) Sample Return Robot Challenge, suborbital and orbital launch partnership proposals (September 2014), and advanced oxygen recover proposals (October 2014) are samples of the broad-based, technological goals of the STMD.

NEXT STEPS FOR NASA

NASA's budget is a good place to start when searching for NASA's future plans. (An itemized listing can be found online at www.nasa.gov/sites/default/files/files/508_2015_Budget_Estimates.pdf). The 2015 budget proposal and its extensions through 2019 highlight the following:

- *OSIRIS-REx*. An asteroid retrieval and exploration plan with partner Deep Space Industries, a private mining company, with plans to launch a spacecraft for research and analysis in 2015, OSIRIS-REx and other asteroid redirect missions are preliminaries to Mars exploration.
- *Orion Multipurpose Crew Vehicle*, its related *Space Launch System* (the launch vehicle), and the *Exploration Ground System* (which will handle the launches). NASA is budgeting more than $2.5 billion a year for the next five years on these. This is where the space exploration dreams get funded and the plans that will get humans to Mars by 2030 are formulated.
- *Commercial Crew Program*. By partnering with private companies such as SpaceX, Boeing, and Sierra Nevada, NASA hopes to reduce the costs of each space launch.

At the same time, identifying those programs whose budgets are holding steady or decreasing indicates a lesser emphasis. These include:

- *Science.* Overall, these programs are reduced by about 3.5 percent and then expected to hold steady for the next few years, with the exception of heliophysics studies, which is expected to go up by 1 percent.
- *Aeronautics Research.* Down 2.5 percent.
- *Education.* Down by almost 25 percent.

Taking just a two-year analysis is somewhat misleading, however, since a high-dollar satellite one year might only require oversight and communication in following years. This is somewhat hampered because missions come and go (the space shuttle, for example) and budgetary classifications can change over time (10 years ago, there was no "Cross Agency Support" line in the budget). However, very roughly, some trends can be tracked over the years (see table I.1).

As a general rule, then, we can expect space exploration to expand in the coming years, with aeronautics and education of lesser significance for the future. This seems eminently logical and probably follows the public perception that NASA is synonymous with human spaceflight and space exploration. Space Technology, a line item in the 2015 budget, did not exist in 2005. Construction and Environmental Compliance and Restoration, initiated in 2010, was probably split between all sorts of departments in 2005.

Table I.1. Ten-Year Budget Analysis.

Budget Category	2005	2015	Percent Change
Earth Science	1485	1770	+16%
Planetary Science	2148	1280	–34%
Astrophysics	1405	1252	–10%
Heliophysics	746	669	–10%
Aeronautics	949	551	–42%
Exploration	1782	3976	+223%
Education	169	88	–48%
ISS	1863	3050	+61%

So where does all this leave NASA in the coming years?

- More and more, NASA's role will likely be that of a sponsor or supervisor, with private industries developing a greater share of space efforts and with NASA guiding the direction.
- As private space industries grow, personnel challenges for NASA could include keeping talented men and women who might receive greater financial benefits or less structured careers working in private industry.

- NASA's bureaucracy and risk aversion have slowed down its rate of innovation and it will need to figure out a way to continue balancing the fine line between too much and too little of oversight and hazard acceptance.
- NASA will need to find a way to reach those citizens and legislators who feel its missions (or selected parts of them) are unimportant or unnecessary—while at the same time determining a vision that garners public support.
- NASA has a strong record in investing in and developing technology, but that technology does not always lead NASA in the right direction or the project gets cancelled before it can actually be used. NASA will need to figure out how to aim in a direction most likely to be fulfilled.

In short, NASA needs a viable and visible goal that the legislators can get behind and must put most of its efforts toward achieving that goal; it must lead the space effort, not wander around in it. NASA is a phenomenal organization that has accomplished much; it is our sincere hope that it will continue to do so in a logical and coherent manner.

THE DICTIONARY

A complete guide of all things NASA is beyond the scope of this historical dictionary. This single book has by necessity condensed many subjects and failed to include some entries that others might deem important. There are many unsung heroes and forgotten programs that couldn't fit between the covers of this book. Just as NASA treads a fine line between innovation and risk, we too balance between readers familiar with NASA and those for whom NASA's history is vague and uncertain. Many worthwhile individuals and entries were written, deleted, added again, and finally removed. Others popped in at the last minute as new discoveries and amendments required the updating of previously complete entries. Even in the short time it takes to publish this book, some entries will be outdated.

We chose our entries based on three criteria: (a) the program or individual's importance to NASA and to the public perception of NASA; (b) the program or individual's place in U.S. history or to future U.S. space efforts; and (c) items we found not generally known but of quirky interest.

As with other historical dictionaries, there are numerous cross-references between entries. Within each entry, terms that are found elsewhere in the dictionary are in boldface, while related terms that are not found in the entry are indicated by *See also*. *See* refers to separate entries that deal with the topic.

We hope you find the material in this book as fascinating and informative as we do.

A

A-TRAIN CONSTELLATION. The A-Train (or Afternoon Train, so named because it crosses the equator northward about 1:30 p.m. local time) satellite **constellation** is a set of four to seven **satellites** that fly together 438 miles above the Earth. Initial components were NASA's **Aqua** and **Aura** satellites and **CNES**'s **PARASOL** (since retired); inserted later were the **CALIPSO** and **CloudSat** satellites, Shizuku (a Japanese spacecraft), and in July 2014, **Orbiting Carbon Observatory-2**. **Glory**, scheduled for 2011, failed to reach orbit. All of them travel in formation within 30 minutes from first to last at speeds over 15,000 miles per hour (keeping up with Earth's rotation as it orbits the sun). As a group, the spacecraft study atmospheric particles and cloud properties as they change depending on the environment below and around them. The timing of the satellites from first to last is critical, as clouds and other air composition qualities can form, grow, and dissolve in less than 15 minutes.

See also MORNING CONSTELLATION.

ACTIVE RESPONSE GRAVITY OFFLOAD SYSTEM (ARGOS). An extensive, coordinated, computer-controlled system of cables, weights, sensors, and supports, ARGOS is designed to simulate lower-gravity environments such as those found in space and on the **moon** and **Mars**. The current, second-generation ARGOS system at **Johnson Space Center** measures 41′ × 24′ × 25′ and permits variable training scenarios within and outside spacecraft as astronauts and equipment "float" on cables suspended from the supports.

ADVANCED COMPOSITION EXPLORER (ACE). ACE, launched on August 25, 1997, from **Kennedy Space Center**, measures and compares samples of **solar wind**, solar corona and flares, and other interplanetary particles. ACE provides one-hour advance warning of geomagnetic solar surges that present problems for **astronauts**, power grids, and communications systems on Earth. It is expected to operate through 2024 and is the longest running mission of the **Explorer** series currently in operation. Refer to appendix B for a list of its instruments.

See also CORONAL MASS EJECTION (CME); SOLAR FLARE; SOLAR STUDIES.

ADVANCED CREW ESCAPE SUIT (ACES). The current ACES (also called a "pumpkin suit" for its orange Nomex color, which is easy to spot on recovery) is worn during ascent and reentry times and during spacewalks. It is fully pressurized and includes a parachute and a cooling system; pockets include flares, a radio, and an assortment of small personal items. Prior to crewed launches, the limited time for **astronaut** preparation includes only 45 minutes for the astronaut (with the aid of one or two helpers) to get into the suit, connect helmet, gloves, and boots by way of interlocking metal rings, and test the system. Generally, the astronaut then removes the gloves and helmet on the way to the launch pad, and reconnects them once in place. Modified ACES allow complete integration with the **Orion** system.

ADVANCED RESEARCH PROJECTS AGENCY (ARPA). Established as a new **Department of Defense** research organization in 1958 under President **Dwight D. Eisenhower**, ARPA was initially intended to cut through governmental bureaucracy and free up funds for the **Army Ballistic Missile Agency**'s (ABMA) proposal of a clustered engine model to achieve large payload capabilities, and had authority over all space activities. Its first director, Roy W. Johnson, led initial efforts in outer space and the antimissile/missile projects until Congress could establish a national space agency (eventually NASA) that would take over nonmilitary aspects. As part of the **International Geophysical Year**, ARPA launched **low Earth orbit satellites** and three lunar **probes**, and coordinated ABMA's deep space probes for scientific investigations. To compensate for extended duties and partly in response to fears of **Soviet** domination in space, ARPA's 1959 **budget** was raised from $340 million to $520 million. As a member of the Senate, **Lyndon B. Johnson** was a key player in initial ARPA–NASA division of responsibilities with regard to **manned orbital vehicles**, and in 1960 all civilian space programs were transferred to NASA and all military space programs to their respective military branches.

In 1962, ARPA was tasked with research and development in the new field of computers and computer technology, and formed the Information Processing Techniques Office (IPTO). Working closely with advanced education institutions (particularly the Massachusetts Institute of Technology, Stanford, and University of California–Berkeley), IPTO instigated the computer science field of study and created ARPANET (precursor to the current version of the Internet), the largest computer network in the world at the time, with 29 routers in 1972.

By 1976, ARPA's focus had become more defense and security related (tactical armor and anti-armor programs, laser technology for space-based defense, stealth applications, and others), although scientific investigations—especially in charged particle beam research—continued.

In 1972, the **Defense Advanced Research Projects Agency** (DARPA) was created as a separate defense agency with a priority of military applications, and by 1976 it had incorporated virtually all aspects of ARPA. Then, in 1993, DARPA was redesignated ARPA, but only until 1996 when Public Law 104-106 changed the name back to DARPA. It is currently known as DARPA.

AERONAUTICS RESEARCH MISSION DIRECTORATE (ARMD). ARMD is a NASA directorate geared toward finding solutions to the problems in U.S. air transportation systems and advancing the aeronautics field of study. ARMD's goals are to improve gate-to-gate mobility while reducing noise, emissions, and fuel use in the surrounding communities as safely as possible in an increasing complex system. Four NASA research centers—**Ames**, **Dryden**, **Glenn**, and **Langley**—concentrate on flight simulation, aerodynamics, atmospheric operations, new propulsion technology, studies of ice formation, wind tunnels, and aviation weather, among others. Each mission directorate is headed by an associate administrator (AA).

AERONAUTICS TEST PROGRAM (ATP). Headquartered at **Ames**, **Dryden**, **Glenn**, and **Langley Research Centers** and serving government, corporate, and institutional clients, ATP is a NASA program concentrating on aerospace testing and engineering. Started in 2005, initial focus on maintaining testing flight capabilities continues with subsonic (up to 0.3 Mach), supersonic (up to 4.6 Mach), transonic (up to 1.2 Mach), and spin (up to 85 ft/s) tunnels as well as an icing research tunnel with capabilities up to 395 mph and an aerothermodynamic lab up to 10 Mach. Hurt by **budget** cuts for several years, funds from the recent American Reinvestment and Recovery Act (2009) have updated facilities and equipment, improved environmental controls, and reinvigorated research in aerodynamics.

AERONOMY OF ICE IN THE MESOSPHERE (AIM). AIM is an **Explorer** mission launched by **Orbital Sciences** on April 25, 2007, from **Vandenberg Air Force Base** to study noctilucent clouds (thin, colored polar clouds that appear near twilight about 50 miles above the Earth). The clouds are of interest because they may be an indication of climate change, and at the very least are clues to our changing mesospheric meteorology. Originally expected to last 26 months, the study has been extended and is still operating as of mid-2015. Refer to appendix B for a list of its instruments.

AGENA TARGET VEHICLE (ATV). A series of seven uncrewed spacecraft used for rendezvous and docking practice by the **Gemini** astronauts in 1965 and 1966 as NASA prepared for the **Apollo moon** landing missions. The ATVs, launched shortly before the companion Gemini launch, would fly in a **low Earth orbit**. After docking, the Gemini **astronauts** would fly the combined vehicles, perform aerial maneuvers, and conduct **extravehicular activities** (EVAs) between the two spacecraft. In general, Agenas were the upper-stage launch vehicles used with Thor and Atlas first stages; the Thors launched scientific and applications **satellites**, whereas the Atlas launched the larger Earth satellites and lunar and interplanetary space **probes**. The Agena D was primarily used with Gemini projects. Agena was named after the star Agena in the Centaurus **constellation**.

AGENZIA SPAZIALE ITALIANA (ASI). ASI, the Italian Space Agency, was founded in 1988. It cooperates with the **European Space Agency** on many projects. Employing about 200 people, it has headquarters in Rome as well as a Center for Space Geodesy in Matera and a balloon launch site in Trapani (both in Italy). It has its own spaceport in coastal Kenya, the Broglio Space Center, named after the father of Italian astronautics, Luigi Broglio, who died in 2001. The ASI is a frequent partner with NASA, beginning in the 1960s, including the **Mars Reconnaissance Orbiter**, **Venus Express**, **Dawn**, and the **Cassini-Huygens** spacecraft, among many others. Seven ASI **astronauts** have participated in **International Space Station** missions.

AIR FORCE INTELLIGENCE, SURVEILLANCE, AND RECONNAISSANCE (AFISR). Begun in 1948 as the **U.S. Air Force** Security Service, AFISR was responsible for cyber and geospatial intelligence resources and applications in both national and international civilian and military operations. Its National Air and Space Intelligence Center worked in cooperation with NASA, especially with regard to meteorological modeling and tracer capabilities. In October 2014, it was reorganized into the 25th Air Force assigned to Air Combat Command.

AIRBORNE RESEARCH INTEGRATED EXPERIMENTS SYSTEM (ARIES). Aries was NASA's **Boeing** 757-200 aircraft (called the "Flying Lab") that specialized in the research and development of aeronautical subsystems and airframe safety. ARIES was the second Boeing 757 made and was modified to serve NASA's needs after its purchase from Eastern Airlines in 1994. It replaced an aging Boeing 737. Stationed at **Langley Research Center** for many years, it was decommissioned in 2005 and sent to **Dryden Flight Research Center** (now **Armstrong Flight Research Center**) in 2006/2007.

ARIES is also a sounding rocket flown by NASA between 1974 and the early 1990s, after which it was used for military purposes as a target launch vehicle until late 2004.

AIRBORNE TROPICAL TROPOPAUSE EXPERIMENT (ATTREX). Coordinated by NASA's **Earth Science Project Office**, ATTREX is an ongoing **educational and public outreach** program to inspire young adults to follow a career in the science field by giving them actual experiences with field data. Using NASA's long-range Global Hawk uncrewed aircraft system, ATTREX brings data to the classroom and allows students and researchers to track changes in the upper atmosphere by studying water vapor, ozone concentrations, and chemical composition. Begun in December 2011 and with flights initiating in 2013, ATTREX's team of researchers includes scientists from **Ames Research Center**, **Langley Research Center**, and **Goddard Space Flight Centers** as well as **Jet Propulsion Laboratory**. The program also completed two months in the upper atmosphere over the Pacific Ocean in March 2014 with outreach to Guam schools. Refer to appendix B for a list of ATTREX's instruments.

AIRSPACE SYSTEMS PROGRAM (ASP). NASA's ASP is involved with air traffic management research needs to maintain and expand air traffic considerations to and on airports. Because airport infrastructure development and modification can take decades to complete, ASP coordinates with other government and corporate users to predict, design, and incorporate research ideas to keep up with the challenges of increasing air traffic flow.

See also AVIATION SAFETY PROGRAM (AvSP).

ALDRIN, EDWIN "BUZZ" (January 20, 1930–). Graduating from West Point in 1951 with a degree in mechanical engineering, Aldrin joined the **U.S. Air Force** and flew combat missions during the Korean War. Selected as an **astronaut** in 1963, the same year he earned a doctorate from the Massachusetts Institute of Technology in astronautics, Aldrin flew in space on November 11, 1966, when ***Gemini** 12* launched from **Cape Kennedy**, and he became the second astronaut to walk on the **moon** on July 20, 1969, following **Neil Armstrong**, during the ***Apollo** 11* mission. In March 1972, he retired from the space program, though he continued in air force management as the commander of the Air Force Test Pilots School. He received patents for his design of a permanent space station, the Starbooster reusable rocket, and multicrew modules for space flight. His own eloquent account of his experiences can be found in *Return to Earth* (1973) and *Magnificent Desolation* (2009). Among the numerous honors and awards he has received are the Presidential Medal of Freedom, the Langley Gold Medal, and the Congressional Gold Medal. NASA awards include the NASA **Distinguished Service Medal**, the NASA **Exceptional Service Medal**, and the NASA **Space Flight Medal**. He continues to promote space exploration through computer games, interviews, speeches, music, films, and videos, and through research and development of space vehicles and equipment.

ALICE. Alice (not an acronym) is NASA's instrument on board the **European Space Agency**'s **Rosetta** orbiter. In September 2014, Alice surveyed Comet 67P/Churyumov-Gerasimenko and recorded the first far-ultraviolet light spectra of the comet's surface. Based on Alice's data, it appears the comet is unusually dark and nonreflective—scientists had been expecting more water-ice and a lighter color. The reasons for this anomaly are yet unknown. Rosetta dropped a lander, Philae, in November 2014, to further investigate the comet. Unfortunately, Philae landed in the shadows near a cliff and its solar batteries were soon extinguished; as the comet traveled, however, it would again be facing the sun and it was hoped Philae's batteries would recharge. Philae awoke on 13 June after 211 days of hibernation and communicated intermittently with Rosetta until 9 July 2015, when communication was again lost, probably due to dust accumulation. Initial reports examine the molecules that make up the comet and provide new information to scientists regarding the comet's origins; research is still underway.

ALPHA. *See* SPACE STATION ALPHA.

AMES AERONAUTICAL LABORATORY. *See* AMES RESEARCH CENTER (ARC).

AMES RESEARCH CENTER (ARC). ARC, one of 10 NASA **field centers**, is situated at Moffett Field in California, and was incorporated into NASA in 1958 when the **National Advisory Committee for Aeronautics**, of which it was part, was designated part of NASA. Ames had been founded in 1939 as the Moffett Field Laboratory, but was renamed Ames Aeronautical Laboratory in honor of Dr. Joseph S. Ames, an aerodynamicist. When NASA took over, Ames managed the **Pioneer** and **Biosatellite** space projects and participated in propeller-aircraft wind tunnel research. Currently, much of its work centers on astrobiology, nanotechnology, the search for habitable planets (exoplanets, such as Kepler 62e), airborne science, NextGen air traffic, and lunar exploration. Ames is mission control for **Kepler**, the **Lunar**

Crater Observation and Sensing Satellite mission, the **Stratospheric Observatory for Infrared Astronomy**, and a member of the **Orion** vehicle team. Ames also participates in **Cassini** (Titan), **Iris** (solar coronas), **Mars Science Laboratory**'s ***Curiosity***, and **SpaceX**'s **Dragon** capsule. The center's current director is Dr. Eugene Tu.

ANGRY BIRDS. Rovio Entertainment, creator of the popular Angry Birds game, partnered with NASA to provide Angry Birds Space to teach users about space exploration, physics in space, and planetary science. The game was released in 2012 and became the fastest-growing mobile game in history at that time, reaching 50 million downloads in 35 days. NASA continues to use the Angry Birds characters in children's promotions, displays, and interactive elements at **Kennedy Space Center**.

ANIMALS IN SPACE. The first animals to be sent into space and recovered alive were fruit flies, launched in test tubes on board a V-2 rocket on February 20, 1947. Two years later, the United States sent Albert 2, a monkey, who died after his parachute failed to open on reentry. An earlier Albert was also part of a failed mission. Several other Alberts were sent into space, and almost all were fatal. In order to determine the effects of space on mammals, some of the monkeys had been implanted with medical sensors to allow scientists to monitor their vital signs. Yorick (also called Albert 6) managed to live through his flight and its recovery in September 1951, having reached an altitude of 43 miles, though he survived it by just two hours. During the 1950s, mice were also evaluated by U.S. scientists, and in 1958 a squirrel monkey, Gordo, was launched. Data showed that Gordo survived the 10G of launch, weightlessness 600 miles from Earth in a **Jupiter** rocket, and (temporarily at least) 40G at 10,000 mph when the recovery parachute failed to deploy on reentry. Finally, on May 28, 1959, from **Cape Canaveral**, two monkeys, Abel and Baker, reached an altitude of 300 miles and survived their spaceflight to much acclaim; unfortunately, Abel died four days later from an adverse reaction to anesthesia while doctors removed an infected sensor. Baker, somewhat more fortunate, spent the next 25 years at the U.S. Space and Rocket Center in Huntsville, Alabama. Finally, in January 1961, Ham, a chimpanzee, flew on board a **Mercury** capsule for 16 minutes 39 seconds, setting the path for **Alan Shepard**'s successful flight five months later. The United States also sent frogs, parasitic wasps, and beetles into space in the 1960s, and fish and spiders in the 1970s.

The **Soviet Union** meanwhile had sent two dogs into suborbital space in July 1951, and both survived. In 1957, they sent Laika, another dog, aboard ***Sputnik** 2* (nicknamed "Muttnick"), and she became the first animal to enter orbit; technology did not exist then to return her from orbit and she died inflight. Two years later, the Soviets launched a pair of dogs and a rabbit on *Sputnik 5* (the first to send animals into orbit and return them alive), followed by mice, rats, chimpanzees, frogs, a tortoise, and a guinea pig. In late August 2014, the Russians retrieved geckos sent into space on a six-week **satellite** voyage to test weightlessness in reproduction, but unfortunately the geckos had died during a brief temperature loss; fruit flies, however, survived and reproduced.

Other countries joined in the animal-in-space effort. France sent rats, a cat, and monkeys in the 1960s. **China** launched mice, rats, and dogs, and Argentina a rat and a monkey. Japan joined the others in 1995 when it sent newts into space on board the Space Flyer Unit, which launched from Tanegashima Space Center.

After **Yuri Gagarin** became the first human in space in 1961, animal expeditions focused more on physical responses to spaceflight as opposed to mere survival. For example, bullfrogs were launched to study motion sickness in space in 1970, and newts were studied for limb regeneration in space in 1985. In 2007, a cockroach became the first Earth creature to conceive and give birth in space; all the baby cockroaches were successfully returned to Earth for further study. In September 2014, **SpaceX** launched mice and flies into space in preparation for commercial launch of **astronauts** to the **International Space Station**. Numerous other projects have studied the effects of microgravity and zero gravity on development, long-term weight loss, nutrition, immunology, vision, and so on.

Astronauts themselves are often used in studies, taking part in before and after measurements of bone density, vision, muscle loss, and so forth. In March 2015, NASA sent astronaut Scott Kelly into space for one year on the International Space Station, leaving his identical twin brother Mark on Earth; both men will be examined for everything from telomere loss to muscle mass before, during, and after the mission.

ANTARES. Orbital Science's uncrewed Antares is an expendable launch two-stage rocket used to launch Orbital's **Cygnus** spacecraft into **low Earth orbit**. As part of NASA's **Commercial Orbital Transportation Services**, Orbital was awarded a **contract** to rendezvous and deliver cargo to the **International Space Station** (ISS). Its first test flight was on April 21, 2013, from the Mid-Atlantic Regional Spaceport near **Wallops Flight Facility**; its next flight carried four **CubeSats** among its other commercial cargo to the ISS five months later. Antares is the largest liquid-fueled rocket to fly from Wallops Island and the largest rocket in Orbital's collection. Although the first four launches (two of them as resupply missions) were completely successful, the fifth launch failed on liftoff. The vehicle and cargo (including another CubeSat project, RACE) were intentionally destroyed by the **range safety officer** to protect people on the ground on October 28, 2014. Investigation is still underway, but initial reports suggest that the failure of a turbopump on one of its engines was a leading cause of the accident. Orb CRS-4, with an Atlas V engine, is scheduled for late 2015.

APOLLO. Apollo refers to both a 19-mission program (Project Apollo) that led to and incorporated **Neil Armstrong**'s **moon** landing and the subsequent exploration of the moon, as well as (colloquially) the combined **command and service module** (CSM) atop the rocket assembly of **booster** (Little Joe II and **Saturn** launch vehicles) and instrument units as part of the three-stage spacecraft. Apollo was NASA's third human spaceflight program, following Project **Mercury** (1950–1963) and Project **Gemini** (1962–1966). During Apollo's 11 years (1961–1972), 12 men in six separate excursions walked on the moon, at a cost of $24.5 billion. (Only the Panama Canal had come equal to the size and scope of a non-wartime project.) Construction of the **Johnson Space Center** and the **Kennedy Space Center** were funded primarily under Apollo's budget in the wake of ***Sputnik 1***'s historic achievement by the **Soviet Union**.

NASA managed both uncrewed (launch-escape system tests and **Apollo-Saturn**) and crewed Apollo missions. After the tragic *Apollo-Saturn 204* (*AS-204*, renamed ***Apollo 1***) fire that killed astronauts **Gus Grissom**, **Ed White**, and **Roger Chaffee**, NASA launched *Apollo 4*, *5*, and *6* as uncrewed missions, testing and retesting all aspects of space flight. With the conclusion of *Apollo 6*, NASA declared the **Saturn** V rocket as approved for crewed flights. In 1968, *Apollo 7* saw astronauts **Wally Schirra**, Donn Eisele, and Walt Cunningham enter **low Earth orbit** on a Saturn 1B rocket for 11 days, which included the first live television broadcast from a spacecraft. Two months later, *Apollo 8* entered orbit around the moon after launching with the more powerful Saturn V booster. *Apollo 9* tested the lunar module as well as portable life support systems in March 1969, and in May of that year, *Apollo 10* held the "dress rehearsal" for the lunar landing, descending to 8.4 miles above the moon. Finally, on July 16, 1969, *Apollo 11* made its historic flight and landed on the moon four days later. Astronauts Neil Armstrong and **Buzz Aldrin** crewed the lunar module *Eagle* while Michael Collins remained on board the CSM *Columbia*. Subsequent Apollo missions (*Apollo 12* through *17*) carried out other lunar excursions, with the exception of *Apollo 13*, which had to abort after an oxygen tank exploded; the crew safely returned to Earth, reinforcing the country's belief that astronauts were somewhat equivalent to superheroes. *Apollo 17* also included the first scientist-astronaut: Jack Schmitt, a geologist. That was the last Apollo flight: *Apollo 18*, *19*, and *20* were cancelled, the first two to **budget** cuts and the last saw the launch vehicle used instead for **Skylab**. Refer to appendix B for a typical list of instruments carried on board Apollo flights.

APOLLO 1. The first crewed mission of the **Apollo-Saturn** series—*AS-204* and its crew of **Gus Grissom**, **Ed White**, and **Roger Chaffee**—was targeted to launch on February 21 from **Cape Kennedy** in an Apollo **command and service module** (CSM) atop a **Saturn** 1B rocket to **low Earth orbit**. Delays had already pushed the launch date from late 1966 to early 1967, partly because the new CSM was larger and more complex than earlier versions. In a rush to beat the **Soviets**, and to some extent a victim of its own success, NASA was impatient to keep its schedule. When the CSM arrived from North American Aviation, the *Report of the Apollo 204 Review Board* later stated that "The number of the open items at the time of shipment of the Command Module 012 was not known. There were 113 significant Engineering Orders not accomplished at the time Command Module 012 was delivered to NASA; 623 Engineering Orders were released subsequent to delivery. Of these, 22 were recent releases which were not recorded in configuration records at the time of the accident." Responsibility for these issues was a matter that would later take years to conclude and was by no means one-sided. At the time, however, matters were so bad that Grissom had removed a lemon from the tree at his house and hung it in the module as a message. There were additional problems with flammable materials, a high-pressure oxygen environment, and hatch door design, among others, but most of these did not come to light as problems until the accident investigation occurred.

In any event, on the evening of January 27, 1967, the three **astronauts** were conducting a launch simulation. Because no fuel was on board either the launch vehicle or the spacecraft, the pyrotechnics systems were not thought to be necessary and were therefore disabled. Later investigations concluded that a probable short in an exposed wire near Grissom's seat sparked, igniting the fire. From the first notice of the fire to the hull rupture was 17 seconds. The astronauts had initiated their planned emergency escape, but time ran out. Death for all three came from cardiac arrest due to the high levels of carbon monoxide. By the time the hatches were opened from the outside, some five to eight minutes later, there was no hope of recovery.

Congressional investigations as well as NASA's thorough internal investigation revealed many problems, and set back the **Apollo** mission by almost two years. However, by the end of that time, methods were in place to eliminate most, if not all, of the issues with regard to system design, operations, management, and procedures.

North American Aviation merged with Rockwell-Standard a few months after the accident, its name was changed to **Rockwell International**, and the company was later sold to Boeing in 1996.

Originally referred to as *Apollo-Saturn 204* as part of a series of uncrewed missions, the name was redesignated *Apollo 1* in honor of the astronauts.

APOLLO-SATURN (AS). The Apollo program included a series of uncrewed launches between 1961 and 1968 that included Apollo spacecraft launched by **Saturn** rockets. Saturn rocket testing had begun in 1961 (designated SA, for Saturn-Apollo, by **Marshall Space Flight Center**). The uncrewed AS mission began in 1966 with the launch from **Cape Canaveral** of the Apollo-Saturn 1B three-stage rocket, although years of significant work had already been accomplished by that point. The initial flight tested the structural integrity of the command and service module (CSM), launch loads, and compatibility with the launch vehicle; separation of the various units at the correct moment; guidance; electrical subsystems; and heat shield reentry. The rocket reached an altitude of 303 miles and lasted almost 37 minutes. Not all portions of the flight were successful, and serious malfunctions occurred. Missions *AS-202* (August 25, 1966) and *AS-203* (July 5, 1966) were more successful, testing in addition the fuel cell power system and fluid dynamics, among others, though the Saturn IB rocket in *AS-203* technically carried no Apollo spacecraft.

Further uncrewed launches were part of the Apollo mission (designated *Apollo 4*, *Apollo 5*, and *Apollo 6* after the fire that killed the ***Apollo-Saturn 204*** crew).

***APOLLO SATURN 204* (*AS-204*).** *See APOLLO 1.*

APOLLO-SOYUZ TEST PROJECT (ASTP). International cooperation had been discussed between nations for decades, starting with the **International Geophysical Year** 1957–1958 where scientists from 57 countries participated in cooperative projects. Still, it wasn't until 1970 that actual planning began for a US–USSR joint venture that culminated in the Apollo-Soyuz mission in 1975. At a time when Cold War tensions had eased somewhat, both parties signed an Agreement Concerning Cooperation in the Exploration and Use of Outer Space for Peaceful Purposes, committing themselves to the joint venture. The United States—with **Soviet** help—designed and built a docking module that would allow **astronauts** to move between the two capsules. On July 15, 1975, both capsules were launched from their respective countries. Two days later, they docked. Over the next 44 hours, the two crews conducted scientific experiments, shared meals and gifts, and practiced undocking and docking. The Soyuz returned to Earth after six days aloft, the Apollo after nine days. The APAS-89 docking hardware continued to be used throughout the **space shuttle** program to dock with **Mir** and then to the **International Space Station**.

This was the last **Apollo** flight, as the United States' crewed program turned toward the space shuttle. The National Air and Space Museum in Washington, D.C., has a model of the two linked capsules.

APPLIED PHYSICS LABORATORY (APL). During World War II, the Office of Scientific Research and Development oversaw the creation of APL in order to energize science and engineering programs in the nation's universities. It is currently a nonprofit research center affiliated with Johns Hopkins University and has many governmental clients, primarily the **U.S. Navy**, but also NASA. APL designed and built the **MESSENGER** spacecraft, launched on August 3, 2004, to complete a yearlong orbit of **Mercury** as part of NASA's **Discovery** program. In addition, APL manages and operates the **Van Allen probes**, studying how the sun interacts with radiation belts on Earth; and has been involved with **Near Earth Asteroid Rendezvous**, **New Horizons**, **Stereo**, and **Solar Probe Plus**. NASA maintains a management office at APL.

AQUA. NASA's Aqua Project, part of the **Earth Observing System** as EOS PM-1, launched on May 4, 2002, to study the Earth's water cycle, as well as fluxes of radiation from the sun, aerosols, and trace gases in the atmosphere. On board the **satellite** are six instruments to measure ocean circulation and evaporation, atmospheric water vapor, clouds, precipitation, soil moisture, ice on land and over water, and snow cover to help scientists understand global ecosystems and determine weather patterns. Aqua (water, in Latin) was intended to send data back to Earth for six years, but is still operating and expected to be so until the early 2020s. It is part of an international (U.S., Japan, France, and Brazil), NASA-centered group of satellites and is a "sister satellite" to NASA's **Terra** spacecraft (launched in December 1999, part of the **Morning Constellation**). Aqua flies in formation about 15 minutes ahead of **Aura** in a sun-synchronous, near-polar orbit, and as of July 2014, about 15 minutes behind the **Orbiting Carbon Observatory-2**. Refer to appendix B for a list of its instruments.

AQUARIUS. (a) Launched on June 10, 2011, Aquarius is a NASA-built instrument aboard the Argentinian SAC-D **satellite** as part of an international satellite program designed to measure the Earth's sea surface salinity (SSS) and allows scientists to track the circulation of the oceans' huge water masses. The extremely accurate radiometers (capable of detecting one-sixth of a teaspoon of salt in a gallon of water from 408 miles above sea level) were built by **Goddard Space Flight Center**, and **Jet Propulsion Laboratory** designed the scatterometer that minimizes errors due to ocean surface waves. Within two months of its launch, Aquarius had collected more SSS measurements than in the previous 125 years, many of them from places that had never been measured before. Designed as a three-year mission, Aquarius was to run through mid-November 2014 and possibly much longer. Unfortunately, the SAC-D satellite on which Aquarius flew was lost after a power supply failure on June 7, 2015. Refer to appendix B for a list of Aquarius's instruments.

(b) Aquarius was also the call sign for the ***Apollo 13*** **lunar module**, although that mission was aborted after oxygen tanks ruptured. The astronauts used the *Aquarius* as a lifeboat to return to Earth.

(c) The underwater laboratory that is used during **NASA Extreme Environment Mission Operations** for space simulations is also called Aquarius and is owned by the **National Oceanic and Atmospheric Administration**.

ARCTIC RESEARCH OF THE COMPOSITION OF THE TROPOSPHERE FROM AIRCRAFT AND SATELLITES (ARCTAS). As the Earth's lands, waters, and atmosphere change, the polar regions have become vulnerable harbingers of those effects. To study those environmental changes, the **Earth Science Project Office** at **Ames Research Center** coordinates the projects moving through ARCTAS, whether they are in Finland, Canada, Greenland, the United States, or elsewhere. Recent projects have included studies of arctic haze (caused by the accumulation of pollutants like mercury), arctic pollution (in part from massive forest fires), and loss of snow albedo (reflection). ARCTAS is part of the International Polar Year (IPY) umbrella.

ARES. (a) Ares was part of the **Constellation** program, which was cancelled in October 2010, primarily due to **budget** limitations. At the time of its proposal in June 2006, it was intended to be a crew launch vehicle (Ares 1) and a cargo launch vehicle (Ares 5) that would support **commercial space transportation** to and from the **International Space Station**. The Ares 1 rocket was developed to carry the **Orion** spacecraft into **low Earth orbit**; it could lift more than 55,000 lbs. and was a stepping stone to the systems needed to carry cargo to the **moon** and **Mars**. Although the program was cancelled in October 2010, government inaction forced NASA to continue paying almost $500 million for the program, mainly to its contractors as contract termination fees, until Congress took action to pass a budget in March 2011. Ares has been replaced by the **Space Launch System**, however, which may end up using much of the Ares equipment and technology.

(b) ARES is also an acronym for Advanced **Rover** Engineering and Science tool, which is the software that allows scientists to manipulate the **Mars Exploration Rovers**.

(c) ARES is Astromaterials Research and Exploration Science, the directorate at **Johnson Space Center** that performs the physical science research to support the overall mission.

(d) ARES is also an acronym for Aerial Regional-Scale Environmental Survey of Mars, proposed by **Langley Research Center** as a Mars scouting mission intended to study magnetism in Mars's upper crust; the composition, chemistry, and dynamics of the atmospheric boundary layer; and the potential near-surface water. It was proposed to NASA as part of a competition but was never acted upon; the **Mars Atmosphere and Volatile Evolution Mission** was instead chosen.

ARIANE. Ariane is a family of expendable launch vehicles manufactured in Europe, generally for the **European Space Agency**, and managed by Arianespace, a European consortium of 24 countries and industries. The rockets launch from French Guyana and generally deploy **satellites** for its European members as well as other countries, although they have also been used as resupply vehicles for the **International Space Station**. The Ariane 5 is its current delivery vehicle; Ariane 6 is in development.

ARIEL. Ariel is an ionospheric research **satellite** and represents a joint project (1961–1979) between the United Kingdom and NASA. Two of the satellites were built in the UK, but all were launched by NASA. Ariel 1 was the world's first international satellite. Refer to appendix B for a list of its instruments.

ARMSTRONG FLIGHT RESEARCH CENTER. As of March 1, 2014, **Dryden Flight Research Center** was renamed the **Neil A. Armstrong** Flight Research Center in honor of the **astronaut** who took the first steps on the **moon**.

ARMSTRONG, NEIL ALDEN (August 5, 1930–August 25, 2012). Remembered primarily as the first human to walk on the **moon**, Armstrong was also an accomplished aerospace engineer, test pilot, and university professor. Born in Wapakoneta, Ohio, he spent his early years moving around the state with his family. His first airplane flight was when he was five years old, and he learned to fly while still in high school. He entered the **U.S. Navy** as part of a college degree scholarship program and qualified for flight training at age 18, flying combat missions in Korea two years later in 1951 and 1952. Armstrong received a BS degree in aeronautical engineering from Purdue University in 1955, and an MS in aerospace engineering from the University of Southern California 15 years after that.

After a short stint working for the **National Advisory Committee for Aeronautics**, the precursor to NASA, Armstrong started work as a pilot at Edwards Air Force Base for **Dryden Flight Research Center** (later renamed Neil A. **Armstrong Flight Research Center** in his honor). His first flight in a rocket was in 1957, and then in 1958 he was selected for the **Man in Space Soonest program** under the auspices of the **U.S. Air Force**. NASA, meanwhile, was selecting its second round of astronauts, and Armstrong was selected in 1962 as the first American civilian to go into space. On March 16, 1966, he served as command pilot for the ***Gemini*** *8* mission scheduled to perform the first-ever docking between two spacecraft when he caught up with the unmanned **Agena** target. Despite problems with the orbital altitude and maneuvering system on board the Gemini capsule, causing the mission to be cut short, Armstrong and fellow astronaut David Scott landed safely back on Earth. He served on the ground and in the air for two more missions before he was named commander of ***Apollo*** *11*. On July 21, 1969, Armstrong descended from the **lunar module** *Eagle* and said, "That's one small step for a man, one giant leap for mankind"—unrehearsed and unplanned—to approximately 450 million listeners. After an uneventful splashdown landing, the crew then spent 18 days in quarantine to make sure they hadn't inadvertently returned with unknown pathogens from the moon.

After his historic mission, Armstrong spent a few years working for NASA, and for the **Advanced Research Projects Agency**, served on spaceflight accident investigations (including the shortened *Apollo 13* mission), was a spokesman for assorted U.S. companies, a professor of aerospace engineering at the University of Cincinnati, and in 2010, a voice actor for a **Jet Propulsion Laboratory**/NASA animated film, *Quantum Quest: A Cassini Space Odyssey*. He died in 2012 after complications from heart bypass surgery at the age of 82.

ARMY BALLISTIC MISSILE AGENCY (ABMA). Formed in February 1956 with **Wernher von Braun** as technical director, the ABMA developed the **U.S. Army**'s first large ballistic missiles at the **Redstone** Arsenal in Huntsville, Alabama. Based on work von Braun had done in Germany during World War II as well as earlier experimentation and research by Dr. **Robert H. Goddard**, the facility focused on military missiles that could also be used to launch **satellites**. In March 1958, ABMA (as well as other agencies such as **Jet Propulsion Laboratory** and **White Sands Proving Ground**) came under the purview of the new Army Ordnance Missile Command, which was then transferred to the newly formed NASA. The southern half of the Redstone facility was renamed the **Marshall Space Flight Center** in 1960.

See also ADVANCED RESEARCH PROJECTS AGENCY (ARPA).

ARTEMIS. ARTEMIS—Acceleration, Reconnection, Turbulence and Electrodynamics of the Moon's Interaction with the Sun—is a repurposing of two earlier **THEMIS** microsatellites (**probes**) that had been launched into **high Earth orbit** in 2007 and completed their mission in 2009/2010. Now operating from lunar orbit, the two ARTEMIS spacecraft (Artemis P1 and P2) are studying how turbulence develops and changes while in the path of **solar wind** and in Earth's magnetosphere, as well as particle acceleration in the **moon**'s wake. The two probes are the first to ever reach orbit around the Earth–moon's **Lagrangian points**. The mission was approved by NASA's Heliophysics Senior Review Panel in 2008, began operating on July 20, 2009, and is expected to operate for several more years. Refer to appendix B for a list of their instruments.

ASTEROID MINING. Asteroids—from the size of small cars to those of small planets—are made up primarily of rock, metals, carbon, other elements such as oxygen, magnesium, calcium and silicon, and possibly even water. From examining meteorites, scientists have an idea of what that particular asteroid was made of. Iron meteorites, for example, can be more than 90 percent iron and will give clues as to where they originated. NASA not only monitors asteroids (especially if their orbit will intersect near that of Earth), it is actively planning on exploring them—which leads to international issues regarding ownership of the asteroid and its resources. Private companies are also viewing asteroids as potential resources. Current UN treaties prohibit individual countries from claiming ownership of any celestial bodies, but does removing resources from them violate that claim? What exactly is a celestial body? Deep Space Industries and Planetary Resources, both U.S. companies, are currently investigating the possibilities of asteroid mining, with the former gearing up to send out exploratory spacecraft to locate potential asteroid candidates before 2017. This is a very new field, and legal precedents are virtually nonexistent.

ASTEROID REDIRECT MISSION (ARM). ARM (also known as the Asteroid Retrieval and Utilization mission) is a three-part mission from NASA paving the way toward future solar system exploration (primarily **Mars**), still in its planning stages. The first part is to identify qualified asteroids that pass near enough to Earth for its purposes. The second is to redirect the asteroid by solar electric propulsion to a stable orbit around the moon. Sending humans to land on, explore, and return from the asteroid is the final segment. Coordinated by **Jet Propulsion Laboratory**, NASA is still in the first stages of identifying near Earth asteroid candidates.

ASTRIUM SPACE TRANSPORTATION (AST). AST, a division of the European Aeronautic Defence and Space Company's Astrium (developer of the **Ariane** rocket launcher, among many other contracts), is working with **Lockheed Martin** on a project for NASA's **Crew Exploration Vehicle**. In addition, AST has partnered with NASA and others to develop equipment for use on the **International Space Station**. With Astrium–North America offices near all of NASA's land-based launch and recovery sites, AST is poised to be a part of future NASA/international efforts.

ASTRO OBSERVATORY. Astro 1 (launched December 2, 1990, STS 35, returned December 11, 1990) and Astro 2 (launched March 2, 1995, STS 67, returned March 18, 1995) were sets of telescopes (ultraviolet and broadband X-ray) used for recurring flights of the **space shuttle** for astronomical observation. Mission and payload specialists (**astronauts**) controlled the operations after mounting, and Astro 1 was also the first shuttle mission partly controlled from **Spacelab** Mission Operations Control at **Marshall Space Flight Center**. Despite technical difficulties with Astro 1, both missions were successful in exploring origins of the universe, dark energy, black holes, stars, and galaxies. Refer to appendix B for a list of the instruments carried on each of the spacecraft.

ASTRO-E2. Known as Suzaku, formerly called **NeXT**, Astro E2 was a joint U.S./**Japan** astronomical observation mission launched in July 2005. Astro-E, an earlier attempt in 2000, failed on launch. The U.S. facility was located at **Goddard Space Flight Center** and allowed U.S. astronomers access to the Suzaku mission via Japan's Institute of Space and Astronautical Science Guest Observer program. Suzaku was an orbiting observation **satellite** made up of two major instruments, an X-ray imaging spectrometer and a hard X-ray detector, as part of its equipment, and four foil X-ray telescopes. This mission ended on 2 September 2015, eight years after its designed lifespan, when **JAXA** turned off the radio transmitters because communications since June 2015 had deteriorated. Refer to appendix B for a list of its instruments.

ASTRONAUT. NASA's astronaut corps is currently composed of about 50 active individuals and 35 management astronauts. Management astronauts are those astronauts who, for whatever reason, are no longer eligible to fly missions but are still employed by NASA. Active astronauts tend to be either payload specialists or pilot astronauts. The mission specialist astronauts conduct the experiments, maintain the spacecraft, and perform all sorts of duties to keep the mission going. Pilot astronauts command and pilot the spacecraft, so they must be familiar with all aspects of the craft and mission and be able to act decisively in an emergency. Recently, NASA has also added educator astronauts, who go through all aspects of **astronaut training** with the hopes of encouraging interest in and development of young students.

NASA selected its first crew of seven men in 1959 (referred to as the **Mercury Seven**), all of whom had extensive engineering and jet aircraft flight experience. Because of space restrictions in the **Mercury** program's

Freedom *7* capsules, that first crew could be no taller than 5'11" and weigh no more than 180 lbs. **Sally Ride** was the first female astronaut in June 1983, though she had been preceded by a **Soviet** woman, **Valentina Tereshkova**, almost exactly 20 years earlier. More recent groups have been international mixes of men and women scientists, pilots, and mission specialists. Twenty-eight have received the **Congressional Space Medal of Honor**.

NASA also works with international astronauts and with payload specialists. Payload specialists are employees (usually scientists from private companies or universities) of the owner of the payload—the **satellite** or other equipment being launched. Although they do not work for NASA, NASA holds final approval of allowing their presence on the mission.

As of 2013, there had been 282 men and 48 women astronauts, and 201 military and 129 civilian astronauts. Almost 200 astronauts have retired or resigned, and 49 have died, 23 of them in the line of duty. Astronauts have ranged in age from 26 to 46, though once accepted as an astronaut there is no upper age limit. **John Glenn** flew his final mission in space at the age of 77.

NASA has a highly competitive astronaut candidate selection process that takes place approximately every two years, though applications are continuously accepted. In 2013, more than 6,100 people applied for eight candidate positions. Candidates must have a bachelor's degree, suitable work experience, be U.S. citizens (dual citizenship is allowed), meet specific vision and blood pressure requirements, and must complete astronaut training before they are allowed to fly.

Russian astronauts are referred to as cosmonauts.

See also APOLLO 1; *CHALLENGER* TRAGEDY; *COLUMBIA* DISASTER.

ASTRONAUT TRAINING. Astronauts begin their training with approximately two years of classroom learning at **Johnson Space Center** in Houston, Texas. Any field that NASA feels might be useful is included, but in general these are specialized courses in engineering, meteorology, space science, and Earth sciences. Potential astronauts also have to complete land and sea survival training, including the ability to swim and tread water while wearing a flight suit. Simulators play a significant part in training, particularly with pilot astronaut candidates. A second phase of training begins when the astronaut is assigned a mentor, an experienced astronaut who will ensure the student is competent in all aspects of on-the-job training. Astronauts then received mission and crew assignments, beginning another phase of training—10 months to 5 years of training specific to the vehicle, mission, **extravehicular activities**, equipment, and experiments to which they are assigned. Astronauts assigned to the **International Space Station**, for example, need to be trained in several fields. For longer term missions, the ability to work well with others is of higher concern, as many months or years will be spent in the company of only a few other people.

Since the end of the **space shuttle** program, more astronauts have been training at Star City, Russia; fears that recent events in Ukraine may complicate that training have not yet materialized.

ASTRONOMICAL UNIT (AU). An astronomical unit is the average distance from the Earth to the sun, or approximately 93,000,000 miles (149,597,871 km). Distances within the solar system are often measured in AU, whereas interstellar units of distance are measured in parsecs. For example, ***Voyager 1*** is currently traveling at 3.5 AU per year (or 325,345,325.43 miles per year, approximately 37,000 miles per hour).

ATLANTIS. *Atlantis* was the fourth orbiter (OV-104) in the **space shuttle** series to become operational (October 3, 1985) and the third to be retired. During its years of service, it sent **probes** to **Jupiter** and **Venus**, and delivered the **Destiny laboratory** to the **International Space Station**. *Atlantis* completed its final flight in July 2011, having completed 33 missions, before it was moved to the **Kennedy Space Center**, where it is on display.

ATOMIC ENERGY COMMISSION (AEC). The **National Advisory Committee for Aeronautics** (NACA) worked with the Atomic Energy Commission to develop a test nuclear reactor in 1956. Subsequent to World War II, scientists were working to develop nuclear energy in the hope that it could be used for all manner of machines, including spacecraft propulsion. NACA began construction of a test reactor in **Plum Brook**, Ohio; by the time it was completed in 1961, NACA had, for the most part, become NASA, who inherited the reactor and, to promote confidence in the community, applied for and received their AEC license. NASA's hopes were to develop a nuclear rocket but, by the end of the **Apollo** program, **budget** constraints led to setting the reactor facility in standby mode. A final decommissioning program for the reactor facility was determined in 2008, but the other test facilities remain in use.

NASA and AEC also jointly operated the Space Nuclear Propulsion Office in 1961 to develop the nuclear thermal rocket created by Project **Rover**. The office was closed in 1972.

ATOMS FOR PEACE. President **Dwight D. Eisenhower** presented his "Atoms for Peace" speech on December 8, 1953, to the UN General Assembly in order to promote the peaceful use of nuclear energy. Prior to his speech, much of the world's nuclear research was shrouded in secrecy, as most of it was concerned with weapons and the military. After the speech, a more public acceptance of nuclear energy spurred research into other areas, including that of nuclear thermal rockets.

See also ATOMIC ENERGY COMMISSION (AEC); PLUM BROOK STATION (PBS); ROVER.

AURA. Aura is a **satellite** launched on July 25, 2004, as part of the Earth Observing System to measure trace aspects of the chemicals in our troposphere (ozone, aerosols, and gases) with state-of-the-art instrumentation. Equipped with a high-speed data link, Aura can relay data almost instantaneously or can store data for later broadcast. Aura flies in formation with **Aqua** and the **A-Train** in a sun-synchronous, near-polar orbit 438 miles (705 km) above sea level on Earth. Like its sister satellites, it was designed for a six-year mission that has since been renewed and will be reviewed again in 2015. Aura's Flight Operations Team operates out of **Goddard Space Flight Center** with downlink stations in Norway and Alaska. Refer to appendix B for a list of its instruments.

AVIATION SAFETY PROGRAM (AvSP). In partnership with the Federal Aviation Administration, the **Department of Defense**, and the aviation industry, NASA's AvSP is working to prevent accidents of commercial and private, smaller aircraft as part of its Next Generation Air Transportation System (NextGen). Part of their early accident mitigation research involved dropping airplanes from heights so that they could study how to improve the safety of seats, seat belts, and structures; these studies are now primarily modeled on computers. The $500 million program operated between 2000 and 2004 with the goal of reducing aviation accidents by 80 percent over 10 years and 90 percent over 25 years. Research continues through the Commercial Aviation Safety Team (CAST). Although the commercial aviation accident rate was very low, it had not significantly improved in over 20 years. Projected growth in air travel, however, represents potential higher risks, and so CAST is working to eliminate even more aviation accidents.

B

BALLOON ARRAY FOR RBSP RELATIVISTIC ELECTRON LOSSES (BARREL). A balloon-based mission, BARREL measures the precipitation of relativistic electrons over the course of summers 2013, 2014, and 2015. Lightweight but durable balloons carried instruments and floated over Antarctica (2013, 2014) and Sweden (2015) at an altitude of 18–22 miles during a one-month period to measure the X-rays produced by the electrons in hopes of solving some of the mysteries of the radiation belts that surround Earth. The balloons work in conjunction with the **Van Allen probes**, which are orbiting Earth to study the Van Allen belts and which will help us understand some of the sun's effects upon our planet. It is part of the **Living with a Star** program. Refer to appendix B for a list of its instruments.

BEAN, ALAN (March 15, 1932–). The fourth man to set foot on the **moon**, Bean was the **lunar module** pilot for ***Apollo*** *12* in November 1969, four months after **Neil Armstrong**'s initial landing. It almost didn't happen. Thirty-six seconds after liftoff, *Apollo 12* was hit by lightning. At the recommendation of NASA engineer John Aaron, Bean executed the commands that saved the mission. Bean was also the commander of **Skylab** 3, from July to September 1973.

Bean received a degree in aeronautical engineering from the University of Texas and had been commissioned in the **U.S. Navy** as a test pilot before he was selected as an **astronaut** candidate in 1963. During his NASA career, he participated in 11 world records in astronautics and space. After he left NASA in June 1981, he pursued his acclaimed painting career, most of which feature some aspect of space or the moon, and wrote five books.

BIG BIRD SPY SATELLITE. The Big Bird spy **satellites** were a series of high-resolution photographic reconnaissance satellites otherwise known as KH-9 or Keyhole-9. Nineteen successful launches from **Vandenberg Air Force Base** on **Titan** 3D rockets between 1971 and 1986 returned film (60 miles of film in each satellite) to Earth in reentry buckets that were quickly recovered by aircraft. NASA involvement was primarily attempting to study atmospheric density at high altitudes. Only declassified in September 2011, a KH-9 spacecraft was on display at the National Air and Space Museum for one day later that month, though it would go on public display at the National Museum of the **U.S. Air Force** the following year. Refer to appendix B for a list of typical instruments.

BIOSATELLITE PROJECT. Prior to human spaceflight, NASA's Biosatellite project—a set of three **satellites** named Biosatellite 1, 2, and 3—was launched to test the effects of space on biological systems between 1966 and 1969. Biosatellites 1 and 2 carried the first **animals in space**—fruit flies—as well as wheat seedlings, frog eggs, and bacteria, although Biosatellite 1 was not recovered. The three-day Biosatellite 2 project—before advanced computer technology—required the successful cooperation of 20 scientists, 120 technicians, 770 **contractors**, 180 foreign nationals, more than 200 military personnel to recover the satellite, and 200 NASA personnel to coordinate the mission—but it gave scientists the first real data, however limited, about life in space. Bonnie, the monkey on Biosatellite 3, was the subject of research on brain states, bodily functions, and behavioral performance, but stress on Bonnie ended the flight three weeks early, after less than nine days, and she died shortly after recovery. Her death is still the subject of controversy. See appendix B for a typical list of instruments carried on board each of the spacecraft.

BLUE GEMINI PROGRAM. The **U.S. Air Force** wanted to develop a flight program to train its military personnel in rendezvous, docking, and transfer procedures in space. To do so would have required cooperation from NASA regarding its **Gemini**-type spacecraft. Only a paper plan and never developed further, Blue Gemini was part of the **Man-in-Space program**'s Manned Orbital Development System (MODS) and would have entailed a military space station accessed via Gemini spacecraft. The plan called for NASA to complete six Gemini missions with Air Force personnel "learning the ropes." Air Force command officers had mixed feelings regarding the program; some felt it competed with their **Dyna-Soar** glider project while others thought it promised a quicker path to military space flight. NASA originally supported the project, but when Secretary of Defense Robert McNamara proposed a union of the Air Force project and NASA's Gemini program in 1962, NASA was not convinced of the merits of the merger. Meetings continued throughout most of January 1963, but in the end, nothing really changed and the Blue Gemini project was canceled.

BOEING. A strong and consistent **contractor** and partner with NASA, Boeing has worked with NASA since the early 1960s when Boeing collaborated with Xerox, Bell, AT&T, and Western Union to provide central communications for the **Apollo** program. More recently, modified Boeing 747s were also used to transport **space shuttles** from landing sites to launch sites (**Kennedy Space Center** primarily). In addition, Boeing and NASA partnered in late 2013 to conduct wind tests on a Boeing 757 in an attempt to decrease drag and thereby lower flight costs and/or increase velocity. Numerous other partnerships between the two organizations exist, including projects to sustain the **International Space Station**, new batteries and antennas, and many more, totaling more than $1.71 billion in 2013, not including Boeing's share of ownership in **United Launch Alliance** (co-owned with **Lockheed Martin**). On September 16, 2014, NASA announced that Boeing (together with **SpaceX**) were the finalists for the $6.8 billion project of the **Commercial Crew Program** to return human spaceflight to U.S. soil.

Boeing, however, especially in the 2000s, has had a long history of penalties, fines, and convictions for corporate actions it has taken with regard to numerous projects, contracts, and environmental concerns. Among those, two are related directly to NASA: In November 2000, Boeing and United Space Alliance were fined and gave up $1.2 million in unpaid invoices for overbilling NASA between 1986 and 1992. (In fairness, that work was performed by **Rockwell** International, which was later purchased by Boeing.) In June 2006, Boeing agreed to pay $615 million to settle charges that it used illegal practices to obtain contracts for launch services from the **U.S. Air Force** and NASA.

BOISJOLY, ROGER (April 25, 1938–January 6, 2012). An employee of Morton-Thiokol, Boisjoly, a mechanical engineer, alerted his company in late 1985 to the potential problems of the O-rings that were later found to have caused the ***Challenger*** accident. However, when NASA asked if there were any reasons the *Challenger* flight should be delayed, Morton Thiokol said that there were none, and NASA continued with the flight. Seventy-three seconds after liftoff, the spacecraft exploded after the O-rings failed. Boisjoly subsequently left Morton Thiokol and became a public speaker on ethics; he was awarded the Scientific Freedom and Responsibility Award in 1988. Morton-Thiokol split into Morton, taking the majority of the chemical business, and Thiokol in 1989 with Alliant Techsystems (ATK) acquiring Thiokol in 2001 and continues to provide essential equipment to NASA.

See also CHALLENGER TRAGEDY.

BOLDEN JR., MAJ. GEN. CHARLES F. (August 19, 1946–). Nominated by President Barack Obama, Bolden is the 12th and current administrator of NASA, taking office on July 17, 2009. A former **astronaut** between 1986 and 1994 (he took part in four missions with the **space shuttle**, two of them as commander, and was later named assistant deputy administrator in 1992), he had received a BS degree in electrical science in 1968 from the U.S. Naval Academy, at which time he was commissioned a second lieutenant in the Marine Corps. He became a naval aviator in 1970, seeing combat missions during the Vietnam War. After the war, he earned an MS degree in systems management from the University of Southern California in 1977 and worked at the Naval Air Test Center's Systems Engineering and Strike Aircraft Test Directorate, logging more than 6,000 hours of flight time, until he was chosen as an astronaut candidate in 1980. After his astronaut career, he rejoined the Marine Corps at the U.S. Naval Academy, from which he retired in 2003 with the rank of major general after a successful career. Between 2003 and 2009, he entered the private sector as a consultant and motivational speaker.

During his six-plus years at NASA (2009–), Bolden has overseen the retirement of the space shuttle after 30 years of missions, the landing of ***Curiosity*** on **Mars**, and launched a spacecraft to **Jupiter**. He has pushed the agency to develop the **Space Launch System** rocket and the **Orion** spacecraft that may eventually take an astronaut to Mars. **Commercial space transportation** coordination has grown and he has led the agency to develop the ability to launch astronauts from the United States by 2017, among many other accomplishments. He is, however, a realist and acknowledges that **budget** constraints must be considered.

Several years before Bolden was appointed administrator at NASA, he auditioned for and won the chance to be the virtual host at **Kennedy Space Center** Visitor Complex's Shuttle Launch Experience, where visitors receive a simulated first-hand experience of going into space on board a space shuttle. His was also the first human voice heard on Mars—not that anyone was there to hear it—when a transmission of his voice went to the **rover** and then beamed it back to Earth (the rover itself has no speakers and his voice was not broadcast).

BOOSTER. Booster generally refers to a booster rocket, which is the first stage of a multistage **rocket** that propels another rocket (or rockets) to space. It can also refer to a booster engine, which amplifies the thrust of a sustainer engine. The booster rocket is distinguishable from the spacecraft that is then deployed to complete its mission. Boosters drop back to Earth (generally over the ocean or other unoccupied territory) once their fuel is depleted. Some boosters—such as the **space shuttle**'s solid rocket boosters (SRBs)—are reusable once recovered, while others are expendable. Boosters are SRBs, liquid-fuel boosters, or hybrids. Once started, they cannot be throttled or stopped—they continue to burn propellant and produce thrust until all the propellant is expended. SRBs provide greater thrust than liquid-fueled rockets. Although accidents with SRBs are rare, they tend to be catastrophic because of the danger for explosion or fire. A Brazilian launch pad accident killed 21 employees in a single explosion in 2003. Liquid rocket boosters tend to be more expensive to design, test, and manufacture. The Soyuz maintains a liquid-fuel booster, which can more easily be terminated in the event of an emergency. There can be one or more boosters on a rocket.

BRILLIANT PEBBLES. Brilliant Pebbles is a nickname for the smaller, **satellite**-based missile interceptors that have a nonexploding (nonnuclear) warhead. Brilliant Pebbles were part of the **Strategic Defense Initiative** begun in 1986, but the program was cancelled in 1994. A year later, NASA used the technology in its **Clementine** deep space **probe**. A later attempt to resurrect Brilliant Pebbles failed.

BUDGET/FINANCING. NASA's proposed budget for 2015 can be found at www.nasa.gov/sites/default/files/files/508_2015_Budget_Estimates.pdf. A condensed version is found in appendix A. A breakdown of the presidential, House, and Senate versions compiled by the Space Foundation can be found at www.spacefoundation.org/sites/default/files/downloads/Update%204%20FY%202015%20NASA%20Budget.pdf.

When NASA was founded in 1958, its budget was set at $89 million for fiscal year 1959 (FY 1959; October 1, 1958–September 30, 1959). (All figures given in non-inflation-adjusted amounts unless otherwise noted.) Its total budget between its inception and through 1969 when it sent men to the **moon** was $34.8 billion, which included the **Mercury**, **Gemini**, and some **Apollo** programs as well as a number of other missions. At its peak during those years, NASA accounted for 4.44 percent of the national budget (the highest it has ever been, in 1966), and it averaged 2.15 percent of federal expenditures in that time. In contrast, NASA's FY 2015 budget is expected to be $17.5 billion, or 0.44 percent of the annual budget. President Obama also included supplemental NASA funding of $885 million in the Opportunity Growth and Security Initiative, and that amount is not included in the $17.5 billion, nor has it yet been approved by the U.S. Congress. (To put that budgeted $17.5 billion in perspective, it is roughly equivalent to the amount Americans spend on chocolate each year, or about half the amount we spend on tailgate parties every year, or the amount we spend on military appropriations each week.) NASA is, in short, expected to do more with less each year.

NASA maintains and promotes the well-known human and robotic space exploration programs and their related **educational and public outreach** programs, and those projects have in recent years been the natural focus of the public's interest in NASA's operations. However, NASA has a strong aeronautics and other science activity base on which it spends approximately $5 billion each year to provide support to our nation's aerospace industry and safety infrastructure, weather support and climate **satellites**, and other programs of national interest. Those missions cannot be eliminated without serious damage to national priorities and commercial interests.

Politics greatly affects NASA's budget, with some folks wanting more spent on NASA and some less. Since the U.S. Congress sets NASA's budget, whims in political winds can blow the budget in any direction. Public perception (depending on which poll you read) shows that anywhere from 39 percent of voters to 72 percent feel NASA spending is about right, with the remaining voters split on whether it is too much or too little.

Something else to keep in mind is that NASA's budget is not nearly as cut and dried as accountants and fiscal analysts might like. Take the early human spaceflight programs, for example, which at the time received all the funds they needed in their exploratory programs pretty much just for the asking. Numerous missions and projects provided the research on which the human spaceflight programs were based yet were not directly accounted for in the Mercury, Gemini, and Apollo programs—and at times were limited by their individual budgets. Space weather is another example: **astronauts**' lives depend (at least in part) on avoiding radiation areas in space that are identified by space science (generally heliotropic) missions, yet mission funding for solar studies is often at risk.

But NASA isn't a black hole in space into which all our tax dollars sink. NASA employs roughly 17,000 direct personnel and 10,000–30,000 **contractors** at any given time. Those employees and contractors—most of them in professional fields of engineering, aerospace, and specialized construction—then spend their money in local communities. In addition, the innovations and **spinoffs** that NASA generates create new job opportunities and are generally made available at no cost to the recipient. Cutting NASA's budget means eliminating at least some of those jobs.

At the same time, there is a feeling that NASA is not providing adequate oversight into how its money is spent as it relates to large corporate contracts and cost overruns. In 2013, NASA's Office of the **Inspector General** reported that NASA had been overpaying contractors to the tune of $66.4 million because the calculation of performance awards was "overly complex." Whether these are overpayments or simply contract performance awards paid early is still a matter of contention.

This mind-set of doing more with less has created significant problems within NASA. First, they—like every government program—cannot do everything they want to do, and they have a tendency to underestimate the costs of innovative, technological projects (such as the current **Space Launch System**) that they wish to promote. Second, reserving funds for later cancelled programs (such as **Constellation**) makes the amount available to other viable projects inaccessible, at least for a time. NASA is trying to privatize many of its contracts in the hopes of lowering costs but runs the risk of letting those corporate clients set its budget for them. Third, infrastructure costs such as those contributions NASA makes to our **aviation safety** programs need to be maintained or they will lose efficiency and increase risk no matter how well run they may be. Last, the same bureaucratic tendency to compartmentalize that led to the ***Challenger* tragedy** might also be applied to NASA's budget. Each individual cost may have its justification, but the budget in its entirety may lose its focus.

C

CALIPSO. *See* CLOUD-AEROSOL LIDAR AND INFRARED PATHFINDER SATELLITE OBSERVATION (CALIPSO).

CANCELLED MISSIONS. Occasionally—normally for either political or **budgetary** reasons—incomplete and unfinished projects and missions are cancelled by NASA. Other missions are cancelled when their mission has been completed to its designed limits, such as that of the **space shuttle**, when it reached 30 years old and it had become risky and not economical to maintain an outdated spacecraft. Some of the cancelled missions never make it past the proposal stage, while other cancellations take place before the contracts upon which they were based have expired, leaving NASA responsible for payments (sometimes in excess of $1 million) to **contractors** for projects that will never happen. The **X-33**, a proposed space shuttle replacement, is one such example. Fortunately, NASA can use a large portion of elements on one project for another—such as **Constellation/Orion**, or the mission is resurrected under a new name when the timing is more appropriate. Some of the more recent cancelled missions include the Aerial Regional-Scale Environmental Survey (ARES), Comet Rendezvous Asteroid Flyby (CRAF), **Jupiter** Icy Moons Orbiter (JMO), **Mars** Astrobiology Explorer-Catcher (MAX-C), Mars Telecommunications Orbiter (MTO), **Pluto Kuiper Express** (PLUTOKE, though see the **New Horizons** project), Space Interferometry Mission (SIM), and Terrestrial Planet Finder (TPF).

See also BLUE GEMINI PROGRAM; BRILLIANT PEBBLES.

CAPE CANAVERAL AIR FORCE STATION. Cape Canaveral is NASA's primary launch site on the East Coast, located near **Kennedy Space Center** and Patrick Air Force Base. For a brief period (1963 to 1973) it was known as **Cape Kennedy**, but the name change in the aftermath of President **John F. Kennedy**'s death proved confusing. Cape Canaveral predates NASA, having been used as a **U.S. Air Force** launch site for test missiles after World War II. The location is an almost ideal site for rocket launches:

Because it is closer to the equator than most of the United States, rockets departing the station receive a boost due to the Earth's rotation, and its proximity to the Atlantic Ocean provides safer launches and recoveries.

CAPE KENNEDY. In the aftermath of the death of President **John F. Kennedy**, the governor of Florida (at the request of President **Lyndon B. Johnson**) changed the name of **Cape Canaveral** to Cape Kennedy. The change was confusing to the public, and the name reverted to its historical one in 1973.

CARL SAGAN MEMORIAL STATION. *See* MARS PATHFINDER.

CASSINI-HUYGENS. A combined **Saturn** orbiter (Cassini) and a probe/lander (Huygens), Cassini-Huygens is an uncrewed **satellite** launched toward Saturn on October 15, 1997. Cassini was the first spacecraft to enter Saturn's orbit (July 1, 2004), and Huygens landed on Saturn's largest **moon**, Titan, on January 14, 2005. Cassini's latest data (April 3, 2014) showed a vast underground ocean of liquid water (essential for life as we know it) on Enceladus, one of Saturn's moons.

Managed by NASA's **Jet Propulsion Laboratory**, Cassini-Huygens is the combined effort of 16 European countries (primarily the **European Space Agency** and the **Italian Space Agency**) and the United States. In 2008, at the conclusion of its initial mission, the program was **extended** for two years as the Cassini Equinox Mission; in 2010 it was extended again, this time for six-plus years, as the Cassini Solstice Mission. In addition to its primary goals of investigating Saturn and its atmosphere and magnetosphere, Cassini has discovered seven new moons around the planet and investigated several storms on the planet, including a hurricane 8,000 miles across and another that is thought to be the "largest, hottest stratospheric vortex" ever detected.

In February 2015, Cassini came within 30,000 miles of Rhea, another of Saturn's moons, providing color images in high resolution; in May 2015 it visited Hyperion, another moon; and in August 2015 flew within 300 miles of the moon Dione. The spacecraft will descend to the surface of Saturn in September 2017 to avoid contaminating any of the moons that might someday be habitable by humans.

The mission was named for two 17th-century astronomers Giovanni Cassini and Christiaan Huygens. Refer to appendix B for a list of instruments.

CENTER FOR NANOSPACE TECHNOLOGIES. The Center for Nanospace Technologies was NASA's partner in hosting several conferences, the latest being NanoTech 2002. The center is a Texas-based private, nonprofit foundation geared toward exploring cutting-edge technology in aerospace, among other areas.

CENTRE NATIONAL D'ÉTUDES SPATIALES (CNES). CNES is the French space agency, with whom NASA has partnered in several missions. The most recent is a May 2014 agreement to complete implementation of the Surface Water and Ocean Topography (SWOT) mission, during which the two organizations will construct, launch, and operate a spacecraft for a survey of Earth's ocean surface in detail. The initial feasibility studies began in 2009, and the launch is scheduled for 2020. Other CNES projects include the **Jason-**1 survey of ocean altimetry, launched in 2001 to review ocean circulation, rising seas, and climate; Jason-1 lost contact in June 2013 and is on a long orbit away from Earth, not due to return for at least 1,000 years. Other missions are the **Polarization and Anisotropy of Reflectances for Atmospheric Sciences Coupled with Observations from a Lidar** (PARASOL) and **CALIPSO**, as well as Jason-2, which was launched in 2008 and is still in orbit. CNES works as a partner with the **European Space Agency**, and in a cooperative division of labor focuses its attention on access to space, civil applications of space, sustainable development, security and defense, and ongoing missions.

CHAFFEE, ROGER (February 15, 1935–January 27, 1967). Chaffee was one of three **astronauts** (along with **Gus Grissom** and **Ed White**) who died during the ***Apollo 1*** launch pad catastrophe at **Cape Kennedy**. Originally from Grand Rapids, Michigan, Chaffee graduated from Purdue University with a degree in aeronautical engineering. Following a career in the **U.S. Navy**, he was selected as part of Astronaut Group 3 in 1963, and became capsule commander for ***Gemini*** *4*. He was 31 at the time of his death.

CHALLENGER. The *Challenger* (OV-099) was the second orbiter (following ***Columbia***) to join the **space shuttle** fleet, and flew from 1983 until the ***Challenger*** **tragedy** in January 1986, when an explosion caused the death of the crew and destroyed the vehicle. *Challenger* was originally built to be a test vehicle for the nascent space shuttle program and spent almost a year being subjected to vibration tests that simulated launches and landings. When the original prototype (***Enterprise***) would have required more expensive refitting than did the *Challenger*, NASA elected to go with the *Challenger* as the second orbiter. During its first year and a half as an orbiter, *Challenger* flew 85 percent of all space shuttle missions, including **Sally Ride**'s first mission. It was also the first orbiter to host **astronauts** on an **extravehicular activity** with jetpacks and the first to pull a **satellite** out of orbit.

CHALLENGER CENTER FOR SPACE SCIENCE EDUCATION. A nonprofit, educational center with headquarters in Washington, D.C., the Challenger Center's mission is to increase scientific literacy in the public by offering students a chance to explore opportunities in science, technology, engineering, and mathematics. The center was created by the families of the **astronauts** who perished in the ***Challenger* tragedy** on January 28, 1986, and today there are more than 40 learning centers around the world. More than 30 partners (including NASA) provide support for the organization.

CHALLENGER* TRAGEDY.** On January 28, 1986, the **space shuttle *Challenger (STS-51-L) broke apart 73 seconds after liftoff at **Cape Canaveral**, Florida, killing all seven crew members. The liftoff was viewed by millions of Americans, including many schoolchildren, because of the involvement of Teacher-in-Space **Christa McAuliffe**. Subsequent testing revealed that an O-ring seal in the right-side solid rocket **booster** (SRB) failed, leading to a breach that allowed pressurized hot gas to reach the nearby pieces. A hot jet of gases from the SRB impinged on the LH2 portion of the external tank, which ultimately exploded. When the pieces separated because of aerodynamic forces, the pieces were strewn over the Atlantic Ocean, where they were later recovered from the ocean floor. Although some crew members survived the breakup of the spacecraft, the impact with the ocean surface would have caused their deaths. The accident led to a suspension of the shuttle program for almost three years as the **Rogers Commission** investigated the accident and NASA instituted changes in its procedures and assembly. Subsequent to the tragedy, NASA concentrated its payloads on security and scientific missions, instead of the private, contracted work it had earlier accepted, which at least in part caused it to push for the increase in frequency of shuttle missions that may have led them to approve an accelerated schedule.

The failed O-rings were not the sole factor in the accident, though they were the precipitating cause. Prelaunch conditions had changed from the initial January 22 date, when the previous mission's delays pushed the date to January 23 and then January 24. With bad weather forecast at the **Transoceanic Abort Landing** (TAL) site in Dakar the next day, NASA elected to use Casablanca as the TAL, but Casablanca was not suitable for night landings, so the date was moved to January 26. Then, bad weather at **Kennedy Space Center** moved the launch date to January 27. The launch was again delayed because of problems with an exterior access hatch, so the launch was moved to January 28. Extremely cold temperatures (for Florida) overnight on January 27 caused concern over the O-rings, and engineers from Morton-Thiokol, who used the O-rings in its SRBs, recommended a postponement, but discussions between NASA and Morton-Thiokol decided to go ahead with the launch. At the same time, ice (rare for Florida) was also a potential issue, but after an inspection by the Ice Team, the *Challenger* was given a thumbs-up.

Due to the cold temperatures, the O-rings lost their elasticity. When the SRB case pressurized under the pressure generated from the burning propellant, the O-ring failed to stay in contact with the "tang" in one of the tang and clevis joints in the SRB case. This allowed hot gases to flow past the first O-ring and ultimately burn past the second O-ring. After the second O-ring was breached, there was a direct path to the exterior of the SRB case.

Later, based on recovered equipment, examination showed that three of the four Personal Egress Air Pacs had been activated, indicating that at least some of the astronauts survived the breakup. However, loss of air supply combined with hitting the ocean surface at 207 mph (an impact of over 200 g) would have caused the death of any astronauts surviving thus far. Recovery efforts began within minutes of the accident, and continued until February 7. On March 7, divers discovered the crew compartment on the ocean floor, together with the seven astronauts; the crew members were returned to their families on April 29.

The seven astronauts killed on board were: Dick Scobee, commander, on his second shuttle mission; Michael J. Smith, pilot, on his first shuttle mission; Ronald McNair, mission specialist, also on his second shuttle mission; Ellison Onizuka, payload specialist with over 70 hours of time in space; Christa McAuliffe, a member of the **Teacher in Space Project**; Gregory Jarvis, payload specialist 2, also on his first space shuttle mission; and Judith Resnik, a mission specialist on her second space shuttle mission.

See also BOISJOLY, ROGER (April 25, 1938–January 6, 2012).

CHALLENGING MINI-SATELLITE PAYLOAD (CHAMP). CHAMP was a **satellite** launched into **low Earth orbit** on July 15, 2000, on board a Russian COSMOS launch vehicle that collected precise gravity and magnetic field observations in our atmosphere and ionosphere. CHAMP was part of a 10-year cooperative project with Germany geared toward mapping Earth's structure and core that also included **Gravity Recovery and Climate Experiment** (GRACE), which is still active although in its final stages as its orbit is deteriorating, and Gravity Field and Steady-State Ocean Circulation Explore (GOCE), which ended November 11, 2013. CHAMP's unique shape—described as looking like a large manta ray—allowed it to collect and model highly precise data showing changes in Earth's atmosphere over time.

CHANDRA X-RAY OBSERVATORY. Named after Subrahmanyan Chandrasekhar, an Indian American astronomer known to most as Chandra, the observatory was launched on July 23, 1999, and currently orbits about 86,500 miles above Earth. The Chandra X-ray Center in Cambridge, Massachusetts, a part of the Smithsonian's Astrophysical Observatory, operates the spacecraft and oversees the distribution of data to scientists around the world. Its immediate purpose was to detect high-energy X-ray emissions from those regions of the universe—exploded stars, galaxy clusters, and the matter surrounding black holes. Using X-rays, Chandra's primary observations have revealed new images of galaxies more than 250 million light years away from Earth, as well as information about the origin of stars. More than 10 years in the making, Chandra's resolution is eight times greater than any previous telescope. It is part of NASA's "Great Observatories" fleet that includes the **Hubble Space Telescope**, the **Spitzer Space Telescope**, and the deorbited **Compton Gamma Ray Observatory**. Refer to appendix B for a list of its instruments.

CHARITY. The bulk of NASA's charitable outreach is through the Combined Federal Campaign, which is a voluntary program for employees. However, over the years, in addition to its many educational programs and contributions, NASA has participated in assorted projects that could also be classified as charitable outreach. For example, NASA constructed a sterile space in 1977 for David, the young boy who came to be known as the "Bubble Boy," as well as a spacesuit for him that is now part of the Smithsonian's collection. In recent years, its contributions have been focused on science and technology education, more in keeping with its mission.

CHINA NATIONAL SPACE ADMINISTRATION (CNSA). CNSA is the chief body concerned with the People's Republic of **China**'s (PRC's) development of space exploration and research. It is comprised of General Planning, System Engineering, Science, Technology and Quality Control, and Foreign Affairs departments, and has a budget of approximately $1.3 billion. (Compare this to NASA's **budget** of $17.8 billion.) Like the United States, CNSA has both civilian and military partners. The PRC has plans to put a man on the **moon** by 2020, has its own space station (the *Tiangong-1*), and has developed its own crewed launch capabilities, something the United States no longer has up and ready. Will the former US/USSR **space race** be repeated between the United States and China? *Tiangong-2* is scheduled to go into orbit, and around 2022 its first space station will be completed.

The Chinese are extremely adept at reverse engineering; at the same time, innovation and experimentation require vision, dedication, and a significant budget, all of which are not always accessible in any country at any given time. Due to political exigencies, the PRC was not invited to take part in the 16-country **International Space Station** and hence created its own station. Perceived slights like these, as well as a whole host of other issues, have fostered a competition and distrust between the United States and China and focus around three areas of international rules and regulations: the **International Traffic in Arms Regulations**, space policy, and scientific cooperation.

The International Traffic in Arms Regulations (ITAR) covers space technology transfer and technical know-how, and relations between the United States and the PRC had pretty much complied with those guidelines. However, the 2011 U.S. budget, at the behest of the House Appropriations Committee, led to the prohibition of the White House Office of Science and Technology Policy (OSTP) and kept NASA from participating in any joint scientific activity with China. Specifically, the regulations state that federal funds cannot be used "to develop, design, plan, promulgate, implement or execute a bilateral policy, program, order, or contract of any kind to participate, collaborate, or coordinate bilaterally in any way with China or any Chinese-owned company." This has continued through 2015.

Space policy and treaty negotiations are the purview of President Obama's Article 2 responsibilities and covers anti-**satellite** weapons and exploration. These are usually high-level dialogues that don't refer specifically to the technical information covered by ITAR.

Scientific cooperation is the third area of focus, such as when countries host conferences to discuss, for example, particle physics, dark matter, or astronomy, as might be taught in general, high-level university courses. Normally, this was not held to be an exchange of technical know-how prohibited by ITAR. However, politics reared its head when NASA banned Chinese scientists in November 2013 from attending the annual **Kepler** space telescope conference to be held at **Ames Research Center**. However, many U.S. and foreign scientists planned to boycott the meeting in protest—despite the fact that the meeting on the Kepler space telescope program was one of the most important annual events for scientists in that field. Rep. Frank Wolf (R., Va.) then issued a letter to **Charles Bolden**, head of NASA, stating that the law did not prohibit multilateral conferences such as the Kepler conference, and Bolden subsequently blamed the overaggressive prohibition on "midlevel managers."

As of November 2014, satellites are no longer to be considered "munitions" and are not covered by ITAR but instead by Export Administration Regulations. In the meantime, CNSA's relations with the **European Space Agency** seem to be less troubled, at least on the surface. The two participated in the Double Star mission, beginning with the *Tan Ce-1* (which means *Explorer-1* in Chinese) satellite in 2003.

CHINA, PEOPLE'S REPUBLIC OF (PRC). Political forces play an important part in NASA–China relations. Led primarily by Rep. Frank Wolf (R-Va.), a 2011 U.S. law forbids NASA from cooperating on space programs with China or with Chinese-owned companies. However, China is one of the more than 80 nations participating in **International Space Station** activities, China has landed a **rover** on the **moon** as part of its *Chang'e 3* mission, and the country is working toward completing its own space station in 2018–2022 (just as the International Space Station is due to retire). Controversy erupted in late 2013 when NASA prohibited Chinese researchers from attending their **Kepler** Science Conference held at **Ames Research Laboratory**; U.S. and other scientists boycotted the meeting, and NASA later retracted its prohibition. NASA's fears may have been provoked by a potential spying violation in March 2013 by Bo Jiang, a researcher at **Langley Research Center** who was later exonerated. However, meetings between space-oriented officials of the two countries do occur, as they did in early 2014 at the meeting of the International Academy of Astronautics.

The PRC has several ambitious programs in the works. On September 4, 2014, China launched two **satellites** with one rocket: the Chuangxin-1-04 to collect and transmit data on hydrology, electric power, disaster relief, and weather, and another to conduct telecommunications experiments. The **China National Space Administration** lists its goals as exploring the moon, studying space science, setting up remote sensing systems, providing **commercial space transportation** services (such as Russia is doing), establishing its own satellite navigational system and its own telecommunications networks, and building a long-term Earth observation system. Its longer-term goals include uncrewed and crewed missions to **Mars** and establishing a lunar base with full-time personnel. Advocates such as **astronaut** Leroy Chiao encourage cooperation between the United States and China.

CHRISTOPHER C. KRAFT, JR. MISSION CONTROL CENTER. The Mission Control Center at **Johnson Space Center**, also known by its radio call sign, "**Houston**," manages flight control for America's human spaceflight program, currently involving **astronauts** aboard the **International Space Station** (ISS). It is comprised of several flight control rooms (FCRs) including FCR-1 (ISS support), FCR-2 (used to display the historical **Apollo** configuration), and the Red, Blue, and White FCRs that were used for training, the space station, and the **space shuttles**, respectively. The White FCR, formerly known as the Shuttle Flight Control Room, was renovated after the termination of the shuttle program to assist in NASA's current and future launch vehicles and spacecraft and is now called Mission Control Center for the 21st Century. Because of its location in a hurricane zone, backup facilities are located at **Kennedy Space Center**.

CITIZEN SCIENTISTS. Citizen scientists are those amateurs and nonprofessionals who participate in coordinated and networked projects. An early example was the Search for Extraterrestrial Intelligence (SETI) project, SETI@home, which used participants' home computers to expand its reach. NASA has more than a dozen citizen science programs, many of which can be used in the classroom, for home-schooling, or by interested adults, ranging from cloud observations to exploring a virtual **Mars**. Actual available projects can be found at http://science.nasa.gov/citizen-scientists/.

CLEMENTINE. Launched on January 25, 1994, Clementine was a spacecraft cosponsored by the Strategic Defense Initiative Organization and NASA as part of the Deep Space Program Science Experiment. An advance test of **Department of Defense** spacecraft, its primary objective was to evaluate imaging sensors and technologies from far (lunar) distances. Clementine orbited the **moon** for two months, generating thousands of images, and then headed toward Geographos (an Earth-crossing asteroid). Unfortunately, a malfunction caused the spacecraft to use up its fuel supply, and the mission ended early in July 1994. Refer to appendix B for a list of its instruments.

CLOUD-AEROSOL LIDAR AND INFRARED PATHFINDER SATELLITE OBSERVATION (CALIPSO). Launched on April 28, 2006, CALIPSO is the third **satellite** of the **A-Train** set of (then) five satellites ("**constellation**") that studies the effects of clouds and atmospheric particles on radiation, weather, air quality, and climate on Earth. It was launched as a joint mission with **CNES**, together with **CloudSat**, out of **Vandenberg Air Force Base** on a Delta II as a dual payload attach fitting mission. As a complement to CloudSat, the two provide a three-dimensional picture of how aerosols move through our atmosphere; among other findings, CALIPSO showed that—although atmospheric particles in the lowest part of our atmosphere are returned to Earth quickly by rain—aerosols in the highest parts of our atmosphere affect air quality hundreds of miles away by traveling long distances before falling back to Earth. The CALIPSO project was originally named PICASSO (Pathfinder Infrared Cloud-Aerosol Spaceborne Satellite Observations) but the Pablo Picasso heirs wanted too large a sum of money to allow use of the artist's name. Refer to appendix B for a list of its instruments.

See also AQUA; AURA; ORBITING CARBON OBSERVATORY-2 (OCO-2); POLARIZATION AND ANISOTROPY OF REFLECTANCES FOR ATMOSPHERIC SCIENCES COUPLED WITH OBSERVATIONS FROM A LIDAR (PARASOL).

CLOUDSAT. Launched on June 28, 2006, from **Vandenberg Air Force Base**, CloudSat—as its name suggests—studies the composition, vertical structure, and motion of clouds. Part of the **A-Train constellation**, CloudSat's reports improve weather and cloud modeling, leading to more accurate weather forecasts. In 2011, weak battery cells caused the spacecraft to be removed from the constellation; repairs were effected, and it was placed back in service in 2012—all without returning to Earth. Although initially funded for only 22 months, CloudSat continues to perform. Refer to appendix B for a list of its instruments.

CLUSTER II. Cluster (which was destroyed on takeoff in 1996) and its subsequent Cluster II (launched in summer 2000) is a **European Space Agency** mission and NASA is a participant in the venture. Cluster II is investigating the Earth's magnetic properties and interactions with **solar wind** until December 2016 (after extensions). Cluster represents a set of four identical **satellites** that fly in formation, allowing the measurement of magnetic fields and particles in three dimensions by studying small fluctuations in interplanetary space. One of its major accomplishments was the revelation in June 2014 of how a solar wind breaks through Earth's magnetic field. This has repercussions for the study of **space weather** as well as communication satellites. In July 2015, data from Cluster II revealed the origins of the "noisy waves" emanating from the geomagnetic equatorial region of Earth. Refer to appendix B for a list of its instruments.

CNES. *See* CENTRE NATIONAL D'ÉTUDES SPATIALES (CNES).

COBRAS/SAMBA. *See* PLANCK.

COLLINS, EILEEN (November 19, 1956–). Eileen Collins was the first female pilot and first female commander of a **space shuttle**. Between earning a bachelor's degree (Syracuse University, mathematics and economics) and two master's degree (operations research, Stanford University; space systems management, Webster University), she became a pilot in the Air Force and was the second female pilot to attend the **U.S. Air Force** Test Pilot School, eventually earning the rank of lieutenant colonel. In 1990, she was selected for the **astronaut** program. Collins piloted the space shuttle in 1995 and commanded STS-93 (July 1999), which deployed the **Chandra X-ray Observatory**. She again commanded STS-114 to the **International Space Station**, and was the first astronaut to fly the space shuttle in a complete 360-degree pitch maneuver in order for the astronauts to photograph the shuttle's underside to determine if there was damage to the shuttle that would be a problem on reentry. Since her retirement from NASA in 2006, she has covered shuttle landings and launches for CNN, made occasional appearances, and been on the board of USAA. She has received numerous awards, including the French Legion of Honor, and NASA's Outstanding Leadership and **Space Flight Medals**.

COLUMBIA. *Columbia* (OV-102) was the first orbiter in NASA's **space shuttle** series, beginning in 1981 until the ***Columbia*** **disaster** on February 1, 2003, when it disintegrated on reentry, killing all of its crew. It completed 28 missions in space, deploying eight **satellites** and completing 4,808 orbits around Earth. *Columbia* was the only orbiter to land at **White Sands Space Harbor**, which it did in March 1992 when floods at Edwards Air Force Base would not permit landing there.

Construction began on a prototype of a reusable spacecraft in June 1974, and that vehicle became the ***Enterprise*** (which never flew in space). Work on *Columbia* began a year later, and its first mission (STS-1), a two-person crew, was headed by Commander John Young, a veteran of **Gemini** and **Apollo** missions. The system of tiles covering the vehicle was new, and several test flights—focusing on deploying and repairing **satellites** and telescopes—were needed to assure NASA that the system would work. ***Challenger*** joined the fleet in 1982, and after *Columbia*'s next flight in 1983, *Columbia* was removed from the fleet for updating for the **Spacelab** mission and then for further upgrades for compatibility with ***Atlantis*** and ***Discovery***. After the ***Challenger*** **tragedy** in January 1986, space shuttle flights were halted until 1989, when *Columbia* took to space once again. *Columbia* was used as a science and laboratory research shuttle for a time, but by 1999 its focus was once more on telescope deployment and repair, including the **Chandra X-ray Observatory** that year and repair to the **Hubble Telescope** in 2002. *Columbia*'s 26th mission on July 23, 1999, included the first female commander of a U.S. spacecraft, Lt. Col. **Eileen Collins**. Its last flight was STS-107 in early 2003, a 16-day science mission, which broke up on reentry.

COLUMBIA **ACCIDENT INVESTIGATION BOARD (CAIB).** Established shortly after the ***Columbia*** **disaster** on February 1, 2003, the board published its findings in a 248-page report on August 26, 2003. Included in the report were 29 recommendations for future operations, grouped into 13 main categories dealing with physical, systems engineering, and organizational issues. Primary among the recommendations and later reviews by others is that, according to Robert Gregory, Scott Marcellino, and Seth Moyer, "Risk acceptance and normalization of deviance cannot be allowed to become an embedded organizational behavior. . . . Normalization of risk should not occur, ever," which had been recognized after the ***Challenger*** tragedy but was still an issue years later for ***Columbia***. The complete report, as well as the list of recommendations, can be found at http://www.nasa.gov/columbia/home/CAIB_Vol1.html.

All of the recommended changes were instituted by NASA during the two years following the accident, and the next shuttle mission was ***Discovery***, on July 26, 2005. Later reviews by outsiders tended to agree with the report, adding among many other recommendations that the use of Microsoft's PowerPoint—with its short summaries, bullet points, and possible glossing over of detail—would be better replaced by text presentations (whether Microsoft's Word or others) that could include text, tables, data graphics, images, and scientific notation, and reserving PowerPoint or similar programs for summarization of the textual material.

The board consisted of the following members: Major General John Barry, Brigadier General Duane Deal, board chairman Admiral (retired) Hal Gehman, Dr. James N. Hallock, Major General Kenneth W. Hess, G. Scott Hubbard, Dr. John Logsdon, Dr. Douglas D. Osheroff, Dr. **Sally Ride**, Roger E. Tetrault, Rear Admiral Stephen Turcotte, Steven Wallace, and Dr. Sheila Widnall. Dozens of additional investigators and support staff assisted the board throughout its two-plus years of investigations and hearings.

COLUMBIA **DISASTER.** On February 1, 2003, the **space shuttle *Columbia*** disintegrated on reentry after a science mission, killing all seven **astronauts** aboard. At takeoff, 16 days earlier, a piece of insulating foam (which is to reduce boiloff and prevent ice from forming on the cryogenic propellant tanks) from the left-side of the external tank had punctured the leading edge of *Columbia*'s left wing and damaged the heat shield. Unaware of the extent of the damage (a 6- to 10-inch hole) the debris had created, NASA management concluded that it would not be critical, as such debris-shedding had occurred in the past and caused no significant mishap. Low-resolution cameras could not reveal the exact location or the extent of damage. On-the-ground engineer requests for astronaut inspection of the hole were ignored. During the reentry process 16 days later, the spacecraft descended toward Earth at 1,870 mph (Mach 2.46). The speed of reentry forced superheated gases to penetrate the left wing, destroying the internal structure. As the flight control computers could not maintain the orbiter's stability, the cabin eventually lost pressure and the entire shuttle broke apart 66,000 feet above the Earth. As the cabin depressurized, the astronauts would have lost consciousness before the final breakup. Pieces of the shuttle were found over an 84-square-mile area in Texas, Louisiana, and Arkansas, and most of the debris is now housed at **Kennedy Space Center** in an unused office. DNA analysis was used to return astronaut remains to their families.

The seven crew members who died aboard this final mission were: Rick Husband, the commander, who was on his second mission for NASA; William C. McCool, the pilot, who was on his first shuttle flight; Michael P. Anderson, payload commander, on his second mission and with almost 600 hours of space flight under his belt; David M. Brown, mission specialist 1, also on his first mission; Kalpana Chawla, mission specialist 2 and the first Indian-American astronaut; Laurel Clark, mission specialist 4, on her first flight; and Ilan Ramon, payload specialist 1 and the first Israeli astronaut.

COLUMBIA SCIENTIFIC BALLOON FACILITY. A NASA facility in charge of providing launch logistics, coordination, command systems, and recovery services for unmanned high-altitude balloons. The facility was established in Boulder, Colorado, in 1961 under the auspices of the **National Science Foundation** and was later moved to Palestine, Texas, in 1963 and designated the National Scientific Balloon Facility (NSBF) in January 1973. In February 2006, the NSBF was renamed the Columbia Scientific Balloon Facility in honor of the crew of the **space shuttle *Columbia*** as the facility is located in the area where the debris from the crash came down to Earth.

COLUMBUS LABORATORY. Part of the **International Space Station**, the Columbus research laboratory module was contributed by the **European Space Agency**. It includes 10 individual interior modules, each of which can be used for a different experiment with independent controls for temperature and other controls, as well as four exterior platforms. Measuring 23×15 feet, it was launched in February 2008 and its control center is in Oberpfaffenhofen, Germany.

COMMAND AND SERVICE MODULE (CSM). The **Apollo** Program, which ran from 1961 to 1972, developed the Apollo spacecraft, which was composed of three parts: a combined command/service module and a **lunar module**. The command module (CM), which was the only part of the Apollo spacecraft to return to Earth intact, contained the pressurized main crew cabin large enough for three **astronauts** as well as the navigation and environmental control systems. The propulsion engine, which allowed the **manned orbital vehicle** (MOV) to enter and leave lunar orbit, was housed in the unpressurized service module, which was jettisoned from the CM prior to reentry into Earth's atmosphere.

COMMERCIAL CREW PROGRAM (CCP). The CCP is aimed at aiding the aerospace industry in developing certificated systems and programs that can safely transport humans to **low Earth orbit** and to the **International Space Station** (ISS). As of May 2014, **Space Exploration Technologies**

(SpaceX), **Boeing**, and Sierra Nevada Corporation Space Systems (SNC) had completed the first phase of NASA's contracted $30 million program to regenerate American spaceflight capabilities by 2017. The second phase of the process (Commercial Crew Transportation Capability, or CCCap) is open to all companies and results were announced on September 16, 2014, as part of NASA's goal to return human spaceflight to U.S. soil. Testing under this phase required at least one crewed flight test for verification purposes. This program is to encourage commercial and private enterprises in developing space transportation, rather than a governmental program available to commercial and private enterprises. It was awarded to Boeing and SpaceX, but in late September Sierra Nevada Corporation challenged that award with legal action. In January 2015, NASA revealed that it had concerns about the complexity of Sierra Nevada's Dream Chaser space plane and its readiness to fly astronauts to the ISS. However, Sierra Nevada was given an extended deadline until March 31, 2016, to finish its work under the Commercial Crew Integrated Capability (CCiCap) agreement.

COMMERCIAL ORBITAL TRANSPORTATION SERVICES (COTS). COTS was the first phase of a program to encourage **commercial space transportation** by inviting private companies to develop vehicles capable of making deliveries to the **International Space Station** (ISS). The program was concluded in November 2013 after two of the initial 20-plus applicants successfully designed, built, and launched capable vehicles: **SpaceX** and **Orbital Sciences**. (A third, Rocketplane-Kistler, was eliminated out of funding concerns.) NASA considers the program to have been a huge success, principally because the $800-million investment in COTS was significantly less than it would have been to develop its own (for example, the proposed $12 billion **Orion** spacecraft). NASA limited itself to the roles of technical consultant and lead investor, as well as potential customer (as private companies were free to contract with other customers, just as NASA was free to choose its **contractors**). As of the end of 2013, both SpaceX and Orbital Sciences had successfully delivered cargo via their COTS vehicles to the ISS.

See also COMMERCIAL RESUPPLY SERVICES (CRS).

COMMERCIAL RESUPPLY SERVICES (CRS). Working in partnership with the **Commercial Orbital Transportation Services** (COTS), which was a program to develop vehicles to deliver crew and cargo to the **International Space Station** (ISS), CRS is a contracted service to deliver cargo to the ISS. NASA has signed contracts with **SpaceX** (12 deliveries) and **Orbital Sciences** (8 deliveries) for the use of their COTS cargo vehicles. (Note that a separate program—the Commercial Crew Development program—covers crew rotation services. All three commercial programs are coordinated by the Commercial Crew and Cargo Program Office—C3PO). NASA is currently in the midst of its Commercial Resupply Services 2 solicitation. The selection period for its next set of contracts has been deferred, and additional missions assigned to SpaceX and Orbital Sciences in the meantime. Proposals for CRS-2 are expected from SpaceX, **Orbital ATK**, **Boeing**, Sierra Nevada, and **Lockheed Martin**.

Orbital's third resupply mission ended in failure on 28 October 2014 after it exploded after launch, and they have since contracted with **United Launch Alliance** for the delivery of the fourth resupply mission.

SpaceX's seventh CRS mission on 28 June 2015 also failed when it exploded after launch.

COMMERCIAL SPACE TRANSPORTATION. Commercial space transportation in the United States is regulated by the Office of Commercial Space Transportation (FAA/AST), a division of the Federal Aviation Administration. The office must approve any commercial rocket launch operations. As a governmental agency, NASA is not included in the AST's purview, but it does include any private companies with whom NASA contracts (such as **SpaceX** and **Orbital Sciences**). Because the effects of any damage or disaster would not necessarily be limited by governmental borders, international law takes into account the nationality of the launch operator and the location of the launch to determine who is responsible for any damage. For that reason, the AST has (as of summer 2014) only eight commercial spaceports in the United States: the **Mid-Atlantic Regional Spaceport** (Wallops Flight Facility, Virginia), Kodiak Launch Complex in Alaska, Spaceport Florida (**Cape Canaveral**), California Spaceport (**Vandenberg Air Force Base**), Cecil Field Spaceport in Jacksonville (Florida), Mojave Air and Space Port (California), Oklahoma Spaceport (Burns Flat), and Spaceport America (Las Cruces, New Mexico). However, many more spaceports have applications in process or pending.

COMMITTEE ON SCIENCE, SPACE, AND TECHNOLOGY. *See* HOUSE COMMITTEE ON SCIENCE AND ASTRONAUTICS.

COMMUNICATIONS/NAVIGATION OUTAGE FORECASTING SYSTEM (C/NOFS). C/NOFS is a military **satellite** launched by **Orbital Sciences** on April 16, 2008, to research the Earth's ionosphere. On board is NASA's **Coupled Ion-Neutral Dynamics Investigations** (CINDI) experiment. Selected data from CINDI is available through NASA's **Goddard Space Flight Center** and through the University of Texas at Dallas.

COMPTON GAMMA-RAY OBSERVATORY (CGRO). Launched on April 5, 1991, during STS-37 from the **space shuttle** ***Atlantis***, CGRO orbited the Earth 280 miles (450 km) above the surface. It was the second of NASA's Great Observatories, which also includes the **Hubble Space Telescope**, the **Chandra X-ray Observatory**, and the **Spitzer Space Telescope**. CGRO specialized in high-energy radiation (gamma rays), and carried larger and more sensitive telescopes than any previously orbiting the Earth. (It was also the heaviest payload at that point to be flown aboard a shuttle.) After a successful eight years—including the discovery of seven pulsars as well as the set of objects known as blazars—CGRO began having trouble with its gyroscopes in December 1999, despite anticipated operation until 2007. Because it was not designed for in-orbit repair, NASA conducted its first controlled de-orbit on CGRO, and it burned up over the Pacific Ocean on June 4, 2000. Refer to appendix B for a list of its instruments.

CONGRESSIONAL SPACE MEDAL OF HONOR. The Congressional Space Medal of Honor is a civilian award presented by the U.S. Congress. To date, it has been awarded to 28 **astronauts**: the crews of the **space shuttles** ***Challenger*** and ***Columbia*** who perished in their respective accidents; the crew of the ***Apollo 1*** fire who perished in their accident; Robert Crippen, the pilot of the first shuttle flight (and the most recent award, in 2006); John Young, commander of the first shuttle flight; William Shepherd, the first commander of the **International Space Station**; **Neil Armstrong**; **John Glenn**; Shannon Lucid, for the longest female spaceflight; Jim Lovell, commander of ***Apollo*** *13*; Thomas Stafford, commander of the **Apollo-Soyuz Test Project**; Pete Conrad, first U.S. commander of **Skylab**; Frank Borman, commander of the first lunar orbit; and **Alan Shepard**.

CONSTELLATION. (a) Project Constellation is the name for the 2004–2009 program intended to develop a group of **crew exploration vehicles** to return humans to the **moon** and to and from **Mars** by 2020, and it consisted of both the launch vehicle and the crew vehicle. The program was cancelled in 2010 by President Barack Obama, primarily due to **budget** constraints, and signaled a delay in any human spaceflights to the moon or other planets. Since that time, NASA has proposed using the **Orion Multipurpose Crew Vehicle** (for cargo) and **Ares** V (for the **boosters**) that (if modified) are still in development for other programs.

(b) Constellation also refers to a group of **satellites** that travels in formation, such as the **A-Train** and the **Morning Constellation**, linked to the same network and often reporting data simultaneously from various perspectives or instruments.

CONTRACT WORK/CONTRACTORS. As of mid-2014, NASA has more than 17,000 direct employees (see Appendix C). However, much of what NASA does is contracted out to other firms, supporting an estimated 60,000 additional personnel at a cost in excess of $8 billion, or about 40 percent of its annual **budget**. Of that total, **Jet Propulsion Laboratory**, the top contractor and a division of the California Institute of Technology, has about 5,000 employees and contractors, with an annual budget of $1.7 billion, virtually all of which is related to NASA. Compare this NASA's budget in 1992, where they spent approximately $13.4 million on contract work (out of a $13 billion budget), much of which was for large, state-of-the art, pushing-the-envelope projects. The percentage of contracted work has increased as private companies like **SpaceX** and **Orbital Sciences** take on more of the duties NASA used to accomplish, such as recent launches to transport supplies to the **International Space Station** (ISS). Once a reusable spacecraft is again available, deliveries to the ISS are expected to drop from $7 million to $5 million per round trip. NASA, however, came under fire in 2013 for paying out almost $70 million in bonuses to contracting firms out of monies intended to be used as incentives for finishing the contract early and well—but paid before the jobs were finished. Contract work for the individuals, however, is not necessarily a cushy job, as many of the individual contractors are also subject to furlough when government shutdowns occur, as they did in October 2013 during a budget dispute in Congress.

NASA also contracts with foreign countries, such as its agreement to pay **Russia**'s Federal Space Agency **Roscosmos** $424 million to train and launch six **astronauts** to the ISS on board the **Soyuz**. As of 2013, NASA's top 10 contractors in 2013 (before Orbital Sciences's wave of launches) were:

California Institute of Technology	$1.72 billion
Boeing Company	$1.69 billion
Lockheed Martin Corp.	$1.61 billion
SpaceX	$532 million
Raytheon Company	$459 million
Jacobs Engineering Group	$452 million
Northrop Grumman Corp.	$373 million
Stinger Ghaffarian Technologies	$305 million
United Launch Alliance	$298 million
Roscosmos	$285 million

(Note that **United Launch Alliance** is wholly owned by both **Boeing** and **Lockheed Martin**, in effect increasing their share of contract work.) NASA's contention is that private companies can do routine missions for a lower cost, freeing NASA to develop and create the vision to focus on identifying future projects, such as humans on **Mars**. Critics, however, such as the late **Neil Armstrong**, believe that private companies have (perhaps unintentionally) overpromised their capabilities and underestimated their costs. For example, SpaceX and Orbital Sciences launch spacecraft for a lot less money than NASA had spent on the **space shuttle**—but they also have a lower capacity and must therefore make more trips. In addition, private companies are benefiting from the development and research done by NASA and by NASA's **inventions**.

NASA itself acts as a contractor, primarily with other branches of the U.S. government (**National Oceanic and Atmospheric Administration** in particular) and the **Department of Defense** for military **satellite** launches.

CONSTRUCTION AND ENVIRONMENTAL COMPLIANCE AND RESTORATION (CECR). CECR is a budget category referring to the revitalization construction of facilities, the demotion of facilities, and the environmental compliance of such facilities, whether current or in the past. CECR is separated into two major categories: Construction of Facilities, which ensures that the facilities can be used or are upgraded to the necessary requirements for current and future projects, and Environmental Compliance and Restoration, which is the cleanup of chemicals released into the environment from past activities, to protect human health and preserve natural resources. For example, the upgrading of **Kennedy Space Center** so that it will be able to handle the **Space Launch System/Orion Multipurpose Crew Vehicle** will be funded under this category. In addition, the successful decommissioning of the **Plum Brook Reactor Facility** and the Santa Susana Field Laboratory cleanup are also part of this budget classification. The **budget** for CECR has remained fairly stable over the years since it was created in 2010.

CORONA SATELLITE PROJECT. The Corona **satellites** were intended for superficial photographic reconnaissance of the **Soviet Union**, the People's Republic of **China** (PRC), and other countries throughout the 1960s and into the 1970s, at the request of the **U.S. Air Force** and the Central Intelligence Agency's Directorate of Science and Technology. Also called Keyhole (and the satellites were given names such as KH-1, KH-2, etc.) based on the act of spying through a door's keyhole, the 144 satellites provided photographic evidence of what was happening on the ground in areas where airplane reconnaissance was not allowed. Essentially, the camera(s) on board the satellite took photographs from 75–100 miles above the Earth, and the film was dispatched back to Earth in a reentry capsule that protected the film from the elements and from heat, to either be caught in mid-air by aircraft with a hook or to land at sea and reclaimed by the **U.S. Navy**. The entire program was classified as top secret until 1992, and its photos have since been used by universities and research institutions to explore ancient living sites, tombs, Paleolithic archaeological sites, and Mesopotamian prehistoric traveling routes.

CORONAL MASS EJECTION (CME). CMEs, like **solar flares**, are associated with intense bursts of plasma on the sun, but CMEs are massive clouds of magnetic particles that travel out into space, taking from one to three days to reach Earth. CMEs are associated with solar flares but can also occur independently. A large CME contains millions—and sometimes even billions—of tons of matter and sends solar material out through the interplanetary medium. This massive cloud of magnetic particles affects Earth's magnetic fields, driving particles toward Earth's poles and helping create the Northern and Southern Lights. They also affect high-frequency radio waves and **Global Positioning System** coordinates, which can then be off by several yards. Electrical currents can overload systems during one of these events. Based on the data that scientists have accumulated, the **National Oceanic and Atmospheric Administration**'s **Space Weather** Prediction Center makes estimates about when the CMEs will arrive, allowing power companies, airlines, and even **astronauts** to prepare.

COSMIC BACKGROUND EXPLORER (COBE). Launched on November 18, 1989, from **Goddard Space Flight Center**, COBE measured infrared and microwave radiation present from shortly after the "Big Bang" origins of our universe, and many of its measurements concentrated on cosmic dust. Its discoveries revolutionized our understanding of the early cosmos at the time by measuring the oldest light in the universe. The **satellite**'s mission was terminated in late 1993 and its operation was turned over to NASA's **Wallops Flight Facility** for use as a test satellite. Refer to appendix B for a list of its instruments.

COSMIC HOT INTERSTELLAR PLASMA SPECTROMETER (CHIPS). Launched on January 12, 2003, from **Vandenberg Air Force Base** as a secondary payload aboard a **Delta**-II launch vehicle, CHIPS was a prototype for a smaller, cheaper **satellite**. Intended for a useful life of 18 months, CHIPS scanned the sky for diffuse nebulae and hot plasma for over five years, changing then-current theories of interstellar media. After two years in space, CHIPS was converted to a solar observatory for the remaining three years. CHIPS was the first spacecraft to use end-to-end satellite operations via the Internet (TCP/IP). Its operation was terminated on April 11, 2008. Refer to appendix B for a list of its instruments.

COSMONAUT. *See* ASTRONAUT.

COUPLED ION-NEUTRAL DYNAMICS INVESTIGATIONS (CINDI). Launched on April 16, 2008, CINDI is a project of the University of Texas at Dallas, Center for Space Sciences, aboard the **U.S. Air Force**'s **Communications/Navigation Outage Forecasting System satellite**. CINDI's purpose is to discover the relationship between neutral gas and charged particle motions by studying neutral atmospheric wind velocity (difficult in an always-moving satellite) and charged particle drift velocity, particularly in the ionosphere where such things can result in a disruption of navigational and communications signals. Instead, CINDI found that the ionosphere was not where it was supposed to be, and so the satellite transitioned lower (deboosted) in December 2008. Because the ionosphere is higher during the day as ions build up, and then lowers at night when ions disappear, this has given scientists the opportunity to study the connections between the interior dynamics of the sun and the response of the atmospheric environment on Earth. Originally scheduled for a mission of two years, CINDI continues to operate. See appendix B for a list of its instruments.

CREW EXPLORATION VEHICLE. *See* ORION MULTI-PURPOSE CREW VEHICLE (MPCV).

CRITICS. Over the years, NASA has received its share of criticism, from mishandling large amounts of money (prepayments to **contractors**, funding large—and quickly cancelled—projects) to hoaxes and fabrication (a faked **moon** landing, for example, as well as the cover-up of large pyramids and roads on **Mars**), from being too daring in its dreams and goals to being not nearly daring enough. Many of the complaints stem from a personal definition of risk: At what point, in either financial costs or risks of human life, is space exploration considered too expensive? A large part of the criticism directed toward NASA can be dismissed as unimportant (the faked moon landings), but comments made by otherwise eminent people in their field must be taken into account. Robert Zubrin, for example, an American aerospace engineer, proponent of Mars Direct, and founder of the Mars Society, believes that NASA has lost its focus, vision, and drive, too caught up in its own bureaucracy. Other critics base their beliefs on financing decisions: that the choice to use **Soyuz** instead of our own **space shuttle** for trips to the **International Space Station** (ISS) was short-sighted, provided no backup plan, and ultimately financially irresponsible; others meanwhile claimed that shuttle flights were too expensive. **Carl Sagan**, a strong proponent of robotic spacecraft, felt that NASA pushed the space shuttle and the ISS to the detriment of other missions. NASA's reputation and **budget** have fluctuated since the initial moon landings, and it is in the unenviable position of never knowing what is "just right" on any sort of continuum.

CROSS-AGENCY SUPPORT (CAS). NASA defines cross-agency support, a **budget** category, as a service that "provide[s] critical mission support activities that are necessary to ensure the efficient and effective operation and administration of the Agency but cannot be directly aligned to a specific program or project requirement." CAS is generally divided into two main categories: Center Management and Operations and Agency Management and Operations. The former refers to support given to any of the 10 field centers, while the latter ensures compliance with external and internal requirements as well as safety and mission success. There has been some talk of renaming CAS the "Safety, Security, and Mission Services" to more appropriately describe its responsibilities. CAS's budget has held fairly steady over the years, not allowing for inflation.

CUBESAT LAUNCH INITIATIVE (CSLI). CubeSats are small research "nanosatellites" that piggyback onto the payloads of other missions. CubeSats are approximately 4 inches cubed, weigh about 3 pounds, and can be combined in one, two, or six units. They are primarily used for educational purposes, and their research must be in line with both NASA's strategic plan and the **Education** Strategic Coordination Framework. For example, on the **SpaceX**-3 mission launched in February 2014, the Educational Launch of Nanosatellite (ELaNa) mission included five CubeSats. Students had designed, developed, and assembled the CubeSats, which then studied images of Earth in full color (University of Colorado at Boulder), long-term performance of consumer technologies in spacecraft (**Ames Research Center**), ChipSats (femtosatellites) from Cornell University, SporeSats studying effects of gravity on cells (Ames Research Center), and the reliability of a weather bus working with a global nanosat network (Taylor University and the University of Chile in Santiago). ELaNa-VIII in October 2014 was destroyed in Orbital ATK's launch failure. Calls for proposals are issued periodically (the latest in August 2014).

CURIOSITY. The **rover** *Curiosity* launched from **Cape Canaveral** on November 26, 2011, and arrived 350,000,000 miles away in the Gale Crater on **Mars** on August 6, 2012, eight months later, and only 1.5 miles from its intended target. Weighing almost 2,000 lbs., the rover is powered by a radioisotope thermoelectric generator fueled by plutonium-238 dioxide. As part of the **Mars Exploration Program** and coordinated from **Jet Propulsion Laboratory**, the six-wheeled *Curiosity* has four main goals: (1) to determine if life ever arose on Mars; (2) to characterize the climate of Mars, (3) to characterize the geology of Mars, and (4) to prepare for human exploration. To these ends, ***Curiosity*** carries a number of specific instruments (see appendix B, Mars Science Laboratory). Unlike previous Mars rovers such as ***Spirit*** and *Opportunity*, *Curiosity* was turned on before it touched down on the surface, giving so many viewers their first sight of Mars late on August 6 that computers were temporarily overwhelmed. *Curiosity* has been in operation greater than 700 sols (720+ days), and celebrated its Martian birthday, and its mission—originally scheduled for one Marian year (approximately 23 months)—has been **extended** indefinitely. The rover has gone up, over, and through small craters, rock beds, and sandy **regolith** (Martian "soil"), drilled down into the surface, and sampled the atmosphere. Among other things, it has found evidence of a dry lake bed that might have held water billions of years ago; has measured radiation in transit, which will help establish protection for future **astronauts**; found iron meteorites ("space rocks"); and taken a "selfie." So far, *Curiosity* has traveled about five miles, traveling sometimes only centimeters before stopping and reassessing.

In October 2015, *Curiosity* confirmed that there had been lakes of water on Mars approximately 3.8 to 3.3 billion years ago, just a short while after NASA revealed that water is still flowing on Mars's polar ice caps. A NASA Mars lander mission (**InSight**) in 2016 will explore the interior of Mars, and NASA is scheduled to participate in the **European Space Agency**'s ExoMars missions in 2016 and 2018.

CYGNUS. As part of NASA's **Commercial Orbital Transportation Services**, private corporations have contracted to deliver supplies to the **International Space Station** (ISS) in the wake of the termination of the **space shuttle** system. Developed by **Orbital Sciences** Corp. and launched on their Antares rocket from the **Mid-Atlantic Regional Spaceport**, Cygnus is an uncrewed resupply vehicle consisting of two parts: a pressurized cargo module (PCM) and a service module (SM). The PCM is built in Italy by Thales Alenia Space, with a delivery volume of 18 cubic meters (soon to be increased to 27 cubic meters). The SM is manufactured by **Orbital ATK** and is used to propel the spacecraft to maneuvering room with the ISS, after which the Canadarm2 robotic arm on the ISS catches the vehicle and berths it for approximately 30 days. The Cygnus is then loaded with trash and obsolete equipment, and is scheduled for destructive reentry over uninhabited parts of Earth. Its first mission delivered food and spare parts in September 2013; its second mission (the first of a series of eight **contracted** deliveries) in January 2014 carried research expansion supplies, crew provisions, and spare parts, as well as a dozen experiments from the **Student Spaceflight Experiments Program**. The third mission carried almost 3,300 lbs.—the heaviest to date—of assorted supplies, hardware, **CubeSats**, equipment, and tools. Its next delivery, the third to the ISS, exploded on October 28, 2014, shortly after liftoff.

D

DAWN. Coordinated by **Jet Propulsion Laboratory**, Dawn launched on September 27, 2007, on a 1.7-billion-mile journey. The first **satellite** to orbit an asteroid, Vesta, in the main asteroid belt between **Mars** and **Jupiter**, Dawn is now orbiting around Ceres. Both Vesta and Ceres are considered protoplanets, so information from Dawn will help us discover the origins of our solar system and planets, including Earth. In September 2014, Dawn lost thrust and entered safe mode when a component in the ion propulsion system was disabled by a high-energy particle of radiation. Temporary repairs were made, and Dawn came back online pending further repairs, which are complicated because radio signals take more than 53 minutes round trip. Refer to appendix B for a list of its instruments.

As part of an education and public outreach project, the names of more than 350,000 space fans are part of a memory chip carried by Dawn.

DEEP IMPACT. Launched in January 2005, the Deep Impact spacecraft (part of the **Deep Space Network**) sent a **probe** (made totally of copper, because scientists could identify copper's "signature" and ignore its spectrographic readings) into the path of the Tempel 1 comet on July 4 of that year. Photographing its journey until just three seconds before impact, the probe created a crater equivalent to five tons of dynamite, 2–14 stories deep. Scientists had expected a much flashier impact from the probe's crash, indicating a solid crust, but the probe's dusty flash revealed a more porous surface. Deep Impact's mission was then extended and its name changed to EPOXI (a combination of Extrasolar Planet Observations and Characterization [EPOCh] and Deep Impact Extended Investigation [DIXI]). Five years later, the spacecraft flew past the comet Hartley 2, and scientists have now concluded that short-period (smaller elliptical orbits, i.e., less than 200 years reoccurrence as seen from Earth) comets once formed closer to the sun than those with long periods. A second visit to the comet Tempel 1 by the **Stardust NExT** mission measured changes to the comet after the probe's impact and gave scientists more information about comet erosion. Communications were lost with the spacecraft in August 2013, and in September 2013 NASA ended the mission. Refer to appendix B for a list of instruments for Deep Impact and EPOXI. (Note: The Deep Impact mission was named well before the *Deep Impact* movie was released.)

DEEP SPACE 1 (DS1). DS1 was a **New Millennium** program spacecraft dedicated to testing a payload of advanced, high-risk technologies. Launched on October 28, 1998, the Deep Space 1 mission carried out a flyby of asteroid 9969 Braille, which was selected as the mission's science target. Its mission was **extended** twice to include an encounter with Comet Borrelly and additional engineering testing. Deep Space 1 was the first NASA spacecraft to use ion-powered rocketry as opposed to traditional chemical-powered rockets. DS1's engines were shut down in December 2001 after completion of its mission, and it continues to orbit the sun in case its services are needed at some future time.

DEEP SPACE 2 (DS2). DS2 was a **New Millennium** program set of **probes** that were on board the **Mars Polar Lander**. The two probes, nicknamed "Scott" and "Amundsen" after the first explorers to reach the Earth's South Pole, were designed to be the first spacecraft to penetrate below the surface of another planet. Ultimately, the mission was a failure after the lander failed to reestablish communications following the descent.

DEEP SPACE NETWORK (DSN). Jet Propulsion Laboratory operates DSN, a worldwide network of antennas that support interplanetary missions. Located in Goldstone, California (1958), in Madrid, Spain (1961), and near Canberra, Australia (1965)—all of which are approximately 120 degrees apart from each other, giving the antennas 360 degrees of coverage—the ground stations allow NASA, the **European Space Agency**, the **Japan Aerospace Exploration Agency**, and the **Indian Space Research Organization** to receive data and images from far-flung spacecraft and to monitor their progress, effect software repairs, send commands, and initiate course corrections. The antennas themselves measure from 111 feet in diameter to 230 feet in diameter. The Deep Space Network provides backup services to the **Near Earth Network** and the **Space Network**. Ongoing maintenance for the aging ground stations presents a problem for NASA because major repairs can take months during which the antenna is out of service, leaving a gap in communications.

DEFENSE ADVANCED RESEARCH PROJECTS AGENCY (DARPA). DARPA was created as a defense agency with priority on military applications, as opposed to the **Advanced Research Projects Agency** (ARPA), which did not have the military focus. However, by 1976, DARPA had incorporated virtually all aspects of ARPA. DARPA was briefly redesignated ARPA between 1993 and 1996, when it reverted back to DARPA. As the **Department of Defense**'s lead agency for technological development, its current focus is on electronic support to military operations, which includes computers, sensors, and communications instruments (which itself includes **satellites**).

NASA partners with DARPA in hosting competitions and conferences, grant projects such as 100-Year Starship, robotics projects, and research projects such as Global Hawk, as well as Phoenix (not to be confused with **Phoenix** Mars lander), which is geared toward recycling retired satellite parts.

DEFENSE SATELLITE COMMUNICATIONS SYSTEM (DSCS). The DSCS constellation (as well as DSCS II and DSCS III) of military communications **satellites** is a group of spacecraft launched between 1980 and 2003. These satellites are geared to provide high-volume secure voice and data communications to support military users. The satellites were launched by NASA from **Cape Canaveral** on behalf of the **Department of Defense**. The program followed the Initial Defense Satellite Communications System that began with the launch of the first eight satellites beginning in 1967. It is due to be replaced by the Wideband Global SATCOM system from **Boeing**.

DEFENSE SUPPORT PROGRAM (DSP). Established in 1970, DSP is a constellation of more than 19 **satellites** that provides an early warning system meant to detect the launch of long-range ballistic missiles and nuclear detonations. It is a military program, controlled from Scott Air Force Base, with stations around the world in Guam, Australia, and the United States. They were used most recently during the Gulf War to detect Iraqi Scud missile launches. Since NASA is a civilian program that occasionally does contract work for the military, its involvement with DSP is limited.

DELTA. Subject to many modifications over the years, Delta boosters (launch vehicles, **rockets**) were inherited by NASA from a **U.S. Air Force** military project carried out by the earlier **Advanced Research Project Agency** as the fourth in a series (Able, Baker, Charlie, Delta, Echo, . . .). Delta (also called Thor-Delta, because it was based on the PGM-17 Thor missile launcher) vehicles were used as the mainstay of NASA's medium **satellites** and small space-**probe** launches, beginning with the **Echo 1** launch of the first communications satellite on May 13, 1960, which was unsuccessful, followed by the Echo 1A payload a few months later. The **boosters** were finally discontinued when the **space shuttle** became the only launcher for satellites. However, after the **Challenger tragedy**, NASA began using the

Delta II in 1989, the Delta III in 1998, and the Delta IV heavy lift booster in 2002. More than 300 Delta launches make it the most heavily depended-upon booster, with a 95 percent success rate. Over the years, different numbering systems have been used depending on assembly, mission, and modifications. Delta II and Delta IV are still in use.

DELTA IV HEAVY BOOSTERS. The Delta IV Heavy (Delta 9250H), made by **United Launch Alliance** (a partnership of **Boeing** and **Lockheed Martin**) is the largest of the Delta IVs, and like all **booster rockets** it is expendable, meaning that once it carries its payload to the assigned point, it separates and returns to Earth, burning up as it returns. The Delta IV Heavy is the largest, highest capacity rocket currently in use. In addition to serving many of NASA's launches, it also delivers national security missions for the **National Reconnaissance Office** and the **U.S. Air Force** from **Cape Canaveral** and from **Vandenberg Air Force Base**. Its first launch in 2004 was a test payload and launch, and was not successful. Its second launch three years later, a **satellite**, completed as expected. Powered by two hydrogen-fueled common booster cores, all three cores operate at full thrust at lift off. After 44 seconds, only the boosters continue until they burn out and separate, at which time the core booster ramps back up. After a further 86 seconds, the second stage takes over until orbit is reached. Scheduled to lift off aboard a Delta IV Heavy was the **Orion Multipurpose Crew Vehicle** (unmanned) in December 2014, which successfully completed its test. The Delta IV Heavy is similar (though higher capacity) than **Russia**'s Angara and Proton rockets, the **European Space Agency**'s **Ariane** 5, and **SpaceX**'s **Falcon** Heavy (expected in 2015). At over 230 feet tall and having a mass of over 513,000 lbs., the booster has a payload capacity of approximately 60,000 lbs. to **low Earth orbit**.

DEMONSTRATION AND SCIENCE EXPERIMENTS (DSX). The **U.S. Air Force**'s Research Laboratory developed DSX to investigate technologies needed to advance our capability to operate spacecraft (civilian and military) in the harsh radiation environment of **medium Earth orbit**. Their focus is to provide the ability to operate space systems engaging in space surveillance, **satellite**-based communications, **GPS** navigation, and protection from **space weather**. NASA and the **Defense Advanced Research Projects Agency** are frequent partners of the missions, providing support for, for example, the **Space Environment Testbeds** program; DSX also benefited from data transmitted by NASA satellites, such as **Imager for Magnetopause-to-Aurora Global Exploration** (terminated 2005).

DEPARTMENT OF DEFENSE (DoD). The Department of Defense (DoD) is, by its very nature, more concerned with military applications such as missiles and spy satellites, whereas NASA is a separate civilian agency of the U.S. government reporting directly to the White House and more concerned with exploration and science (note that NASA launches for DoD are generally not open to the public). However, the two overlap significantly in several areas. For example, for the launch of a military **satellite**, the use of NASA's assets (whether launch pads, **boosters**, or personnel) may be more appropriate than its own facilities. The DoD's Synchronized Position, Hold, Engage, Reorient, Experimental Satellites-Chip Scale Atomic Clock (DOD SPHERES-CSAS), which is also part of a **Defense Advanced Research Projects Agency** program and has the potential for high-security UHF communication, is currently on board the **International Space Station**. At times, the reverse holds true and NASA uses military assets. Each contracts with the other to provide the agreed-upon service or transfers ownership of, for example, **Hubble**-like telescopes from the **National Reconnaissance Office** to NASA. From NASA's earliest days, there have been both cooperation and competition between the two organizations (NASA and the DoD), and the primary contention has been for **budget** funding. The pendulum in recent years has been markedly swinging toward DoD.

DESTINY LABORATORY. Delivered to the **International Space Station** on February 7, 2001, by the **space shuttle *Atlantis***, Destiny is, according to astronaut Tom Jones, "the guts of the Space Station's research and command and control capabilities. . . . [and we can do] quality science because of the arrival of the Lab." Operating on a modular basis with interchangeable racks, the laboratory consists of state-of-the-art research equipment, powerful cameras, and control of robotic arms, as well as a Minus Eighty Degree Laboratory Freezer for the ISS (MELFI), which is used to store materials as well as transport them to and from the space station. Constructed by **Boeing**, the laboratory is an aluminum cylinder 28 feet long and 14 feet in diameter, weighing 32,000 lbs. Replacing **Skylab**, which was vacated in February 1974—almost 30 years earlier—it became the United States' permanent orbital research station, where the bulk of U.S. space research in biotechnology, **medicine**, engineering, physics, material sciences, Earth sciences, and other fields takes place.

DISCOVERY. (a) *Discovery* was the third orbiter (OV-103) in the **space shuttle** series to become operational and the first to be retired. Discovery flew more missions than any other orbiter (39), from 1984 to 2011. It also launched the **Hubble Space Telescope**, two Hubble maintenance missions, several **probes** and **satellites**, and carried **John Glenn** into space on his last mission. *Discovery* is currently on display at the Smithsonian's National Air and Space Museum.

(b) Discovery is also a NASA program that involves small planetary missions. **Mars Pathfinder** and **Near-Earth Asteroid Rendezvous** were the first two Discovery projects, beginning in 1994. Other Discovery projects include **Deep Impact**, **MESSENGER**, and **Dawn**, as well as failed project CONTOUR, among others.

DISTINGUISHED SERVICE MEDAL. The Distinguished Service Medal is NASA's highest form of recognition, representing "a government employee who, by distinguished service, ability, or vision has personally contributed to NASA's advancement of United States' interests." More than 80 recipients, most of them **astronauts** and senior administrators, have received this award.

DLR. *See* GERMAN AEROSPACE CENTER (DLR).

DRAGON. Dragon, manufactured by **Space Exploration Technologies** (SpaceX), was the first privately owned, commercial spacecraft to deliver cargo to the **International Space Station**, which it did on May 25, 2012, and returned six days later. Previously, such deliveries and returns had been the purview of governments (the United States, **Russia**, **Japan**, and the **European Space Agency**); since then, **Orbital Sciences** (**Orbital ATK**) has also delivered cargo. The Dragon spacecraft can carry a payload of 13,228 lbs. (6,000 kg) into space, and return about half that amount. Eventually, the Dragon is also designed to carry personnel into space with its Dragon Version 2 spacecraft. It is launched in conjunction with the **Falcon** 9 launch vehicle.

DRYDEN FLIGHT RESEARCH CENTER (DFRC). Located at Edwards Air Force Base in southern California near the western Mojave Desert, DFRC originally was the Muroc Flight Test Unit run by the **National Advisory Committee for Aeronautics** and was used by the **U.S. Army** and Army Air Force. It was renamed in honor of **Hugh L. Dryden**, the former deputy administrator of NASA. Taking advantage of its nearly unlimited visibility, clear desert skies, and flat, 44-square-mile lakebed terrain, DFRC is NASA's leader in aeronautical research, including atmospheric flight research and operations. The facility supported the **space shuttle** and continues to support the **International Space Station**. In January 2014, on its third attempt through Congress, the facility was renamed the Neil A. **Armstrong Flight Research Center**. Other research centers include **Langley Research Center**, **Ames Research Center**, and **Glenn Research Center**.

DRYDEN, HUGH LATIMER (July 2, 1898–December 2, 1965). Director of the **National Advisory Committee for Aeronautics** (NACA) since 1947, Dryden moved with that organization into NASA's fledging organization in October 1958—indeed, he had helped create the new civilian space agency—at which time he became its deputy administrator. An accomplished engineer, writer, scientist, and administrator, Dryden laid the foundations of NASA by creating a Ten-Year Plan, setting the organization on its path. Despite early **failures** (seven of the 17 launches in 1959 failed), Dryden kept the nascent organization pointed toward the future. The politically charged atmosphere that was centered on ***Sputnik*** criticized NASA and its efforts, though NASA had only been in existence for two years at that point. Later, Dryden was credited by writer Tom Wolfe with suggesting to President **John F. Kennedy** that the opportunity to put a man on the **moon** would send a strong signal to the Soviets—and NASA moved forward from there. Dryden was the recipient of numerous awards, medals, and honorary doctorates. Dryden's career at NASA was the ideal culmination of his years of wind tunnel research, aeronautical engineering, and physics. During his time as deputy administrator, NASA worked feverishly to put a man on the moon (which occurred three years after Dryden's death) by developing capable and safe spacecraft and researching all the space-related sciences about which most scientists and engineers knew very little at the time. Dryden was personally involved in the early agreements with the **Soviet Union** about the peaceful use of space. NASA's **Dryden Flight Research Center** (now the **Armstrong Flight Research Center**) in Edwards, California, was named in his honor in 1976.

See also GLENNAN, T. KEITH (September 8, 1905–April 11, 1995).

DYNA-SOAR. *See* X-20.

E

EARTH OBSERVING-1 (EO-1). Part of the **New Millennium program** that was **cancelled** in 2009 due to **budget** cuts, EO-1 launched on November 21, 2000, with an initial one-year mission to provide advanced land-imaging systems and to test new software. Fourteen years later, although the **satellite** is still owned by NASA, it operates in conjunction with the U.S. Geological Survey (USGS) and it is still submitting weekly reports to the USGS. Refer to appendix B for a list of its instruments.

EARTH RADIATION BUDGET SATELLITE (ERBS). Launched on October 5, 1984, into **low Earth orbit** by the **space shuttle *Challenger***, the ERBS **satellite** studied the net flow of energy (shortwave radiation into the Earth's atmosphere, infrared long-wave radiation into space) as well as aerosols and gases. Intended to be a two-year project, ERBS provided scientific data for more than 20 years. Among other discoveries, ERBS provided data to show that chlorofluorocarbons (CFCs) were in part responsible for destroying the ozone layer, leading to the near elimination of CFCs in most countries. The Earth Radiation Budget Experiment consisted of data from ERBS as well as two **National Oceanic and Atmospheric Administration** satellites. ERBS was deactivated in October 2004 due to **budget** constraints, but it had been one of the longest-running spacecraft missions to date. Refer to appendix B for a list of its instruments.

EARTH SCIENCE PROJECT OFFICE (ESPO). The ESPO, located at **Ames Research Center**, manages and coordinates a variety of projects for NASA's **Science Mission Directorate**, including most recently ARISE (summer, 2014), AJAX (ongoing), and ATTREX (2011–2015), as well as many others. The majority of its programs focus on upper atmospheric research but also cover atmospheric chemistry and modeling analyses, the troposphere, the Airborne Science Program, and Earth Observing System satellites. Other programs and missions are **Fire and Smoke**, **Earth Observing** System, the **Orbiting Carbon Observatory-2**, and many others.

ECHO 1. The first NASA communications **satellite**, Echo 1 (in reality, Echo 1A after the original Echo 1 on board a **Delta** rocket failed to complete orbit) was in essence a giant metal balloon floating in orbit. It launched on August 12, 1960, and enabled voice communication between New Jersey and California by "bouncing" communications signals off its shiny metal surface. Measuring 100 feet in diameter, it was nicknamed "satelloon" by those working on the satellite. Its orbit eventually decayed and it burned up in our atmosphere on May 24, 1968, operating five years longer than expected. Echo 2, larger at 135 feet in diameter, was launched in 1964 and had a harder metal surface, but returned to Earth's atmosphere and burned up on descent in 1969. By that time, Telstar, an international communications satellite network launched by NASA, had taken over commercial communications.

EDUCATION AND PUBLIC OUTREACH (EPO). NASA's **Science Mission Directorate** has an elaborate program of educational websites and webcasts, research awards, exhibitions, partnerships with universities, research projects that involve students and teachers in NASA missions, public television broadcasts, and the **NASA TV** channel. Citizens as well as teachers and students (both young children and teens, as well as university researchers and teams) have programs geared specifically toward their age-appropriate interests. In addition, NASA actively searches for volunteers for review panels in specific subject matters to ensure an unbiased evaluation of its findings and studies, and partners with private organizations to provide programs in science, technology, engineering, and mathematics (STEM). Among the most successful programs have been the **Teacher in Space Project** (until the ***Challenger*** **tragedy**), the **Hubble Space Telescope** EPO program, and its Human Resources Program (collaborating with the University of Houston to provide a doctoral track in the Space Life Sciences program). Proposed budget cuts in 2013 (both due to the sequester and to annual budgetary issues) to eliminate NASA's educational STEM outreach raised the ire of many supporters.

EISENHOWER, DWIGHT D. (October 14, 1890–March 28, 1969). Though known primarily as the 34th president of the United States and as a World War II supreme commander, Eisenhower is also the president who signed the **National Aeronautics and Space Act** on July 29, 1958, which gave way to the creation of NASA later that year. As ***Sputnik 1*** and *Sputnik 2* orbited Earth, Eisenhower spoke on television in late 1957 of the importance of improving scientific and technological education. Though the U.S. military would launch its ***Explorer 1*** and *2* satellites in only a few weeks, the general feeling in the country was that the United States was lagging and that something had to be done. **Hugh Dryden**, director of the **National Advisory**

Committee on Aeronautics (NACA), published a paper proposing a civilian agency be created to deal with the space challenge created by *Sputnik*. A month later, the President's Science Advisory Committee also recommended the creation of a civil space program based on NACA, but one that could be expanded quickly to meet the **Soviet** threat. Looking for a way to show the world that the United States was not willing to play second fiddle to anyone, Eisenhower delivered a speech to Congress suggesting such an agency, and the House Select Committee on Astronautics and Space Exploration drafted such a bill. Eisenhower submitted it to Congress, it was signed by Eisenhower, and NASA officially began operations on October 1, 1958.

Eisenhower was a proponent of a smaller, more cohesive NASA. His successor, **John F. Kennedy**, greatly expanded NASA's properties, **budget**, and vision.

ENDEAVOUR. *Endeavour* (OV-105) is the fifth and last orbiter built for NASA's **space shuttle** program, beginning service in May 1992 and flying until it was retired in May 2011. *Endeavour* completed 25 missions, deploying three **satellites** and accomplishing one docking with **Mir** and 12 with the **International Space Station**. It was named as a result of a national competition of elementary and secondary schools. *Endeavour*'s flight STS-47 in November 1992 became the flight of the first married couple in space: Jan Davis and Mark Lee had married in secret, revealing their marriage only after it was too late to train a replacement. NASA has since made it an official policy that married couples are not allowed to fly together in space. *Endeavour*'s last flight was to the California Science Center in Los Angeles, where it remains on display. Note: Endeavor was also the call sign for the ***Apollo 15* command module**.

ENTERPRISE. *Enterprise* (OV-101), named after the *Starship Enterprise* from *Star Trek*, was the first orbiter to be built but never reached space. Although intended for retrofitting after test missions were completed, design modifications were too extensive and the ***Challenger*** was completed instead. It was partially deconstructed to allow use of the parts for other shuttles and went on an international tour, ending up at the Smithsonian's National Air and Space Museum. In April 2012 the *Enterprise* took its last flight and was delivered to the Intrepid Sea, Air, and Space Museum in New York City. After damage from Hurricane Sandy, the exhibit was reopened on July 10, 2013.

EPOXI. *See also* DEEP IMPACT.

EUCLID. Part of the Physics of the Cosmos program managed by **JPL**, Euclid is a **European Space Agency** program initiated in 2012. NASA became a partner in 2013, and the spacecraft is scheduled to launch in 2020 to map the geometry of the dark universe by investigating the distance-redshift relationship of galaxies and clusters. In essence, Euclid will be examining the universe of 10 billion years ago, when dark energy propelled the universe into expansion. It will consist of a 1.2-meter space telescope equipped with two cameras, one for photometry and one that is also a spectrometer. NASA will be contributing 16 infrared detectors and four spare detectors. The Euclid Consortium (1,000 members) will oversee development of the instruments, manage the science operations, and analyze the data. Refer to appendix B for a list of its instruments.

EUROPA MISSION. Managed by **Jet Propulsion Laboratory**, the Europa Mission (also known as the Europa Multiple-Flyby Mission, and previously as the Europa Clipper [2012–2015]) will be a detailed study of Jupiter's moon Europa in order to determine if the icy moon has the potential for life. As currently planned, the mission, which would launch sometime between 2022 and 2025, would include 45 flybys of Europa ranging from 16 miles to 1700 miles above the surface. Instruments on board will search for indications of water at or near the surface and in the moon's thin atmosphere, in response to **Galileo** (1990s) and the **Hubble Space Telescope**'s (2012) indications of water vapor above the south polar region. If the water plumes are in fact linked to a subsurface ocean, the study of the ecosystem will be invaluable in determining if life could potentially exist on the moon. Europa is about the size of the Earth's **moon**, and if a subsurface ocean does exist, could potentially hold more than twice as much water as Earth. Even if human settlement is not possible, with its salt water, rocky sea floor and a tidal environment, Europa might support simple organisms. The 2016 budget includes more than $30 million for this project, and the **European Space Agency** is a potential partner.

EUROPEAN SPACE AGENCY (ESA). Established in 1975, the European Space Agency predated the European Union (though it now has an office in Brussels and the EU is its single largest donor) and represents 20 member-states, with headquarters in Paris, France. Other European countries have (or are in the process of obtaining) agreements with ESA. The ESA is one of the four organizations that directly participate in the **International Space Station**, but it also independently launches uncrewed exploratory missions into space. The **Mars Express** and **Venus Express** missions have been particularly successful. Equivalent to the **Delta IV heavy booster** rocket, the **Ariane** 5 is their main launch vehicle. ESA has offices throughout Europe: Italy is home to the Earth observation missions, the Netherlands coordinate science missions, Mission Control is in Germany, the European Astronaut Center is in Germany, and the European Space Astronomy Center is in Spain. The ESA cooperates closely with **Russia** on scientific space research and through the purchase of **Soyuz** launchers that the ESA manufactures in French Guyana. ESA also worked with the People's Republic of **China** on the Double Star Mission.

Plans are underway to pursue a project with **Roscosmos**, and the ESA has joined NASA's **Orion** program, both of which are geared toward crewed spaceflight. ESA astronauts have been on **space shuttle** flights, and joint NASA–ESA projects have included **Spacelab**, **Cassini-Huygens**, **SOHO**, the **Hubble telescope**, the **James Webb Space telescope**, as well as many others, including a possible **Mars Sample Return Mission**.

EXCEPTIONAL SERVICE MEDAL. This award is presented to a government employee "for sustained performance that embodies multiple contributions which contribute to NASA projects, programs, or initiatives." Many, though not all, of the recipients are **astronauts**. In addition, mission specialists, payload specialists, astronauts, and civilians are eligible for the **Space Flight Medal**.

EXPLORATION PLANS. Subject to **budget** constraints and political whims, NASA's plans for the future of uncrewed scientific exploration have been remarkably straightforward and successful, but its plans for crewed missions much less so. Human spaceflight missions to **Mars** come and go, and a repeat trip to the **moon** is at this time not likely. Part of the disparity between crewed and uncrewed missions is the relative human danger involved, and the fallout that would occur should something disastrous happen. This has led a number of private organizations such as **SpaceX** and Mars One to consider human spaceflight to Mars as much more likely. Still, NASA continues to push for a safe and repeatable mission to Mars sometime after 2030. In preparation for that, NASA is planning to send **astronauts** to a captured, relocated asteroid before 2025 and has a number of uncrewed missions scheduled to perform the scientific basis of a later expedition to Mars. To that end, NASA will be replacing the **space shuttle** with the **Orion Multi-Purpose Crew Vehicle** and the Orion **Space Launch System** as a way to get astronauts returned safely to Earth after any deep space travels, and a trial uncrewed flight is scheduled for 2017. Space travel through radiation is a problem, so they are working on radiation-safe space suits; and propulsion systems are currently too costly and heavy, so they are working on prototypes for solar electricity and nuclear thermal propulsion. Astronauts will need to replace consumables (such as food), so NASA is working on ways for astronauts to farm or use on-site materials. The reported discovery of flowing water on Mars's polar ice caps in late 2015 may provide impetus for such plans.

Searching for planets with the potential for life is another of NASA's exploratory endeavors, and telescopes such as **Hubble**, **Kepler**, **Spitzer**, and the forthcoming **Webb Space Telescope** identify possibilities. In September 2014, NASA released information about a Neptune-sized planet (HAT-P-11b) with clear skies and water vapor, 120 light years away in the constellation Cygnus. Although the planet is unlikely to have life, the presence of confirmed water vapor was an important milestone.

Commercial space transportation will become more common, so NASA has expanded its **contracts** with aerospace private firms (such as SpaceX and **Orbital Sciences**) to develop safe and cost-effective travels, rather than focus internally on such expensive programs.

In the field of aeronautics, NASA is continuing to explore ways to make air travel safer, less congested, and with less strain on the environment by developing newer and quieter aircraft, more airport runway management software, and cleaner engines. Technologically, NASA is continuing to develop state-of-the art propulsions (solar sails, green propellants, cryogenic storage tanks for refueling in orbit), and robotic extensions.

See also ASTEROID REDIRECT MISSION (ARM).

EXPLORATION SYSTEMS MISSION DIRECTORATE. One of the two divisions that were combined to establish the **Human Exploration and Operations Directorate** (HEO) with the aim of expanding our human presence into the solar system. The Exploration Systems Mission Directorate is focused on:

- **International Space Station**, both expansion and missions
- **Commercial Space Transportation**, including the **Commercial Crew Program**
- Crew Vehicle and Launch System, including the **Orion** spacecraft, the **Space Launch System**, and a modernized **Kennedy Space Center**
- Human Health and Safety, which is focused on **astronaut** health, including optimal nutrition and the best technologies and methods to keep astronauts fit and healthy during space travel
- Analog Missions and Field Testing, including simulated missions such as Research and Technology Studies, projects such as **NASA Extreme Environment Mission Operations**, **rover** testing in Hawaii, **Mars** environment testing in the desert, and so forth
- Human Exploration Technology, which includes the Exploration Technology Development Program (to develop long-range technologies to enable human exploration at low risk), the Advanced In-Space Propul-

sion (to develop concepts for high-power electric and nuclear thermal propulsion systems), **Extravehicular Activity** Technology, and many more

EXPLORER 1. Launched on January 31, 1958, as part of the **International Geophysical Year**, *Explorer 1* (the first of more than 90 scientific spacecraft in the Explorer series) was the first U.S. **satellite** placed in orbit, quickly following on the heels of ***Sputnik*** *1* and *2*. **JPL** designed, built, and operated *Explorer I*, while the U.S. **Army Ballistic Missile Agency** used its Jupiter C rocket with a Juno I launch vehicle, developed by **Wernher von Braun**. Bullet shaped, *Explorer 1* was 80 inches long and weighed an estimated 31 lbs. *Explorer 1* was the first spacecraft to detect the **Van Allen radiation belts**, charged particles captured by Earth's magnetic field that surround our planet. The satellite's batteries died four months later and the mission was terminated.

EXPLORERS PROGRAM. The Explorer series of scientific satellites began with the launch of ***Explorer 1*** on January 31, 1958. The **National Advisory Committee for Aeronautics** (NASA's forerunner) gave the name Explorer as a generic term for its future explorations into space. The satellites—of which there were more than 90—were intended to study the atmosphere and ionosphere, the magnetosphere and interplanetary space, astronomical and astrophysical phenomena, and the Earth's surface and magnetic field. The series also includes 12 spacecraft included under the Medium-Class Explorers (defined as "not to exceed a cost of $180 million for development, launch services, and mission operations and data analysis, exclusive of the launch vehicle"; 10 Small Explorers (not to exceed $120 million for the same services); two University-Class Explorers (not to exceed $15 million); five Missions of Opportunity (non-NASA missions in which NASA participates at a cost not to exceed $55 million in noncash contributions), and an International Explorer, the **International Gamma-Ray Astrophysics Laboratory**. Many of the Explorer series' discoveries became the basis of further studies; for example, *Explorer 42* (launched December 12, 1970) discovered X-ray pulsars and the evidence of black holes. The Explorer series is NASA's longest running series, and many of them have lasted well beyond their expected termination dates. IMP 8 (*Explorer 50*) operated for more than 28 years of continuous service, but was retired in October 2001. As of August 2014, there are 13 operational Explorer missions; the two newest Explorer missions (**Nuclear Spectroscopic Telescope Array** [NuSTAR] and **Interface Region Imaging Spectrograph** [IRIS]) have been in space just over two years and one year, respectively. Refer to appendix B for a list of some of their instruments, particularly *Explorer 1*.

See also ADVANCED COMPOSITION EXPLORER (ACE); COMMUNICATIONS/NAVIGATION OUTAGE FORECASTING SYSTEM (C/NOFS); COUPLED ION-NEUTRAL DYNAMICS INVESTIGATIONS (CINDI); TIME HISTORY OF EVENTS AND MACROSCALE INTERACTIONS DURING SUBSTORMS (THEMIS).

EXTENDED MISSIONS. When a mission is proposed, accepted, designed, and budgeted for, it has a life expectancy planned out for it. Many times, the spacecraft exceeds the limits of the mission; ***Voyager*** *1* and *2* are a prime example, as they were designed for a five-year mission and are still operating—35 years later! NASA **budgets** for the initial mission, but subsequent funding depends on a multitude of factors, not all of them financial. If a key instrument is failing but the spacecraft is still viable, the spacecraft might be reprogrammed for a different purpose, its mission might be terminated, or it could be placed in a holding pattern, orbiting until its instruments can serve another purpose. Alternatively, such as with the **Hubble Space Telescope**, it might be repaired and extended due to public interest. Each extension is taken on its own merits, though budgetary concerns play an important role.

EXTRAVEHICULAR ACTIVITY (EVA). NASA defines an extravehicular activity as "any activity performed by a pressure-suited crewmember in unpressurized or space environments." This includes spacewalks where **astronauts** repair a **satellite** or telescope, perform maintenance on a spacecraft, or in an unpressurized module. EVAs can be planned (scheduled) or unplanned (unscheduled, but necessary to complete mission). There are limitations to an EVA, but these mainly deal with the limitations of the life support system: They are time bound (6–8.5 hours), or if the astronaut has sensory degradation (for example, limited vision to the sides), limited mobility and dexterity, and they can present hazards to the astronaut. Over the years, different spacesuits have provided differing degrees of function and form, allowing astronauts to complete their assigned tasks. Astronauts practice their dexterity and mobility in underwater tanks, simulating the gravity-free environment found during an EVA, and in the **Active Response Gravity Offload System** located at **Johnson Space Center**.

EXTREME ULTRAVIOLET EXPLORER (EUVE). Launched June 7, 1992, EUVE was the first space telescope designed for short-wave ultraviolet (4–76 nm) astronomy (ultraviolet wavelengths are generally absorbed by Earth's atmosphere and cannot be studied from Earth's surface). Ultraviolet studies help us determine how the universe has changed since its beginnings, giving us more information on how stars—and the planets that orbit them—are formed. Its Guest Observer Program used the four photometric imaging systems and a three-channel EUV spectrometer to conduct a sky survey. The combination of mirrors and filters maximized the telescope's sensitivity to EUV sources. EUVE operated for nine years before it reentered Earth's atmosphere in 2002. Refer to appendix B for a list of its instruments.

F

FAILURES. Failures are distinct from tragic accidents, although accidents can generally be traced to a failure of some sort. Some accidents are flukes of nature, and NASA must deal with these. The ***Challenger*** **tragedy** was both accident (cold weather) and failure (of management to heed warnings about predicted frozen O-rings, leading to fire); the ***Columbia*** **disaster** was both as well (accidental damage to the reinforced carbon-carbon wing leading edge and subsequent belief that it would not create a problem). Perhaps this is a fine line between events with a loss of human life versus those whose only loss is of mechanical structures and people's careers. In any event, the following are among the most costly of NASA's failures:

- Demonstration for Autonomous Rendezvous Technology (DART), which was a robotic **satellite** intended to take over a portion of satellite maintenance. Unfortunately, it instead collided into the failing satellite and depleted its fuel approximately four minutes later. It is now, as one observer noted, "a floating ice cube" in **low Earth orbit**.
- **Mars Climate Orbiter**, which failed because the **contractor**, **Lockheed Martin**, used Imperial units in its software, as opposed to the metric units specified by NASA. A radar altimeter that supplied altitude in meters reported back to the computer that was programmed to process the data as feet. Retro rockets were fired too early, fuel was consumed, and spacecraft crashed on the surface of **Mars**.
- **Mars Polar Lander**, which crashed on Mars on December 3, 1999, for reasons still unknown, along with its **Deep Space 2 probes**. It is speculated that an early firing of the engines caused it to land uncontrollably. Cost: $110+ million.
- **NOAA-19**, a **weather satellite**, which fell off its short platform in Lockheed Martin's factory because the bolts securing it into position had been removed. Cost of repair: $135 million.

- **Orbiting Carbon Observatory**, which failed to reach orbit because the Taurus XL launch vehicle payload fairing failed to separate during its initial attempt on February 24, 2009, at a cost of $280 million. A second OCO-2 was launched five years later.

Project cancellations could also be considered failures as such. For example, NASA's **Constellation** program had been funded for four years before it was cancelled, costing almost $9 billion. Not all of that is wasted, however, as parts of the program are reused in other capacities or for other missions. Over the years, an estimated $20 billion has been spent on cancelled projects (though, to keep this in perspective, the U.S. military has spent more than twice that amount on cancelled weapons programs).

FALCON. The Falcon rockets are a family of launch vehicles manufactured for **Space Exploration Technologies** and a Falcon was the first launch under NASA's program to privatize launch services under its **Commercial Orbital Transportation Services** project. Currently, SpaceX uses its Falcon 9 version, comparable to the **Delta IV** and Atlas V rockets, though it has a Falcon Heavy under development for larger deliveries. It is used in conjunction with the **Dragon** spacecraft.

FAR ULTRAVIOLET SPECTROSCOPIC EXPLORER (FUSE). Part of the **Origins Program**, FUSE launched on June 24, 1999, as a NASA astrophysics **satellite**/telescope to explore the universe using high-resolution spectroscopy in the ultraviolet region. Part of NASA's **education and public outreach**, the mission's lead partner was Johns Hopkins University, with collaboration from the University of Colorado at Boulder, the University of California at Berkeley, the Canadian Space Agency, the French Space Agency (**CNES**), and corporate partners. Its three-year mission was **extended** in 2002, and continued operation until July 2007 when the spacecraft lost its ability to hold its focus on points of interest. FUSE's measurements helped researchers answer questions about the beginnings of the universe, studied how chemicals disperse throughout the galaxies, and evaluated interstellar gas clouds. Among other results, FUSE showed that a large amount of water escaped from **Mars**, discovered far more deuterium in our galaxy than had been expected, and detected an atmosphere of hot gas surrounding the Milky Way. Over 400 scientific research papers used data from FUSE; once repaired and updated, the **Hubble Space Telescope** continued much of FUSE's research. Refer to appendix B for a list of its instruments.

FAST AURORAL SNAPSHOT EXPLORER (FAST). Second in NASA's **Small Explorer** series, FAST launched on August 21, 1996, to record and measure the plasma physics of the Northern and Southern Lights from an elliptical orbit, and to study the accelerated particles that cause the aurora. It continued operation until 2002 when one of its instruments began to fail, after which time it supported the **Time History of Events and Macroscale Interactions during Substorms** mission before it was terminated in May 2009. It continues to orbit the poles. See appendix B for a list of its instruments.

FERMI GAMMA-RAY SPACE TELESCOPE. More than a telescope, Fermi is a space observatory detecting and exploring the gamma rays (highest-energy form of light), invisible to our eyes. Launched on June 11, 2006, by NASA in partnership with the U.S. Department of Energy and contributions from Austria, France, Germany, Iceland, Italy, Japan, Spain, and Sweden, Fermi was scheduled for a five-year mission. In March 2010, for the first time, Fermi detected gamma rays emitted from stellar eruptions (novae), explosions on far-distant stars that release energy up to 100,000 the annual output of our sun and are believed to accompany the burst of new stellar-mass black holes. Three more similar detections over the next four years confirmed the energy source, and Fermi is operating in the **extended** phase of its mission until at least 2018. The National Academy of Sciences had ranked Fermi as a top-priority mission in light of the potential for new research and scientific collaboration, and the project has an extensive **education and public outreach** program. It was the first imaging gamma-ray observatory to survey the total sky and used Einstein's principle of $E = mc^2$ to convert gamma rays into matter to track cosmic origins. The 2011 Bruno Rossi prize was awarded to the Fermi Large Area Telescope team. Refer to appendix B for a list of its instruments.

FEYNMAN, RICHARD PHILLIPS (May 11, 1918–February 15, 1988). Feynman, an extremely accomplished and charismatic theoretical physicist, was on the board of the **Rogers Commission** that investigated the causes of the ***Challenger* tragedy**. His explanation and demonstration of the failure of the O-rings during a televised hearing (by dipping the material into his glass of ice water) cemented the public's perception of NASA and Morton Thiokol as risk-taking, bureaucratic mismanagers.

FIELD CENTERS. NASA has operations across the United States. These 10 field centers include: **Ames Research Center**, **Armstrong Flight Research Center**, **Glenn Research Center**, **Goddard Space Flight Center**, **Jet Propulsion Laboratory**, **Johnson Space Center**, **Kennedy Space Cen-**

ter, **Langley Research Center**, **Marshall Space Flight Center**, and **Stennis Space Center**. In addition, NASA conducts operations through the **Goddard Institute for Space Studies**, **Independent Verification and Validation Facility**, **Michoud Assembly Facility**, **NASA Headquarters**, **NASA Shared Services Center**, **Plum Brook Station**, **Wallops Flight Facility**, and **White Sands Test Facility**. NASA also maintains Field Center Educator Resource Centers for all 50 states, Puerto Rico, and the U.S. Virgin Islands that are co-located at one of the above facilities. Finally, NASA has interests in the Canberra Deep Space Communications Complex in Australia, the Goldstone Deep Space Communications Complex, and the Madrid Deep Space Communication Complex, all part of the **Deep Space Network** managed by JPL; the Infrared Telescope Facility in Hawaii, a **Near Earth Network** in Alaska, and **Space Network** Ground Terminals in Guam.

FIRE AND SMOKE. NASA maintains a network of **satellites** and aircraft that photograph fires and smoke as they are happening, and studies the impact of these on Earth's climate and ecosystems. As of September 2014, the following missions were taking part in at least some part of monitoring wildfires: **Arctic Research of the Composition of the Troposphere from Aircraft and Satellite**, **Aqua**'s Atmospheric Infrared Sounder Instrument, **Aura**'s tropospheric emission spectrometer and its microwave limb sounder, **CloudSat**, **CALIPSO**, **Earth-Observing-1**, the Fire Information for Resource Management System (FIRMS), the Moderate Resolution Imaging Spectroradiometer Rapid Response System, **GOES** satellites, **SERVIR**, and **Terra** (with its Moderate Resolution Imaging Spectroradiometer [MODIS], ASTER, MOPITT, and MISR instruments). NASA maintains an ever-changing map on its website of fires burning around the world that are detected by its MODIS instrument, updated every 10 days or so.

FIRST LADY ASTRONAUT TRAINEES. The first woman in space, **Valentina Tereshkova**, flew on June 16, 1963, two years after her compatriot **Yuri Gagarin**. The first American woman in space didn't fly until 20 years later when **Sally Ride** finally achieved that honor. Could NASA have elected to choose women astronauts before that time? Perhaps, though **Russian** and American societal climates regarding the genders were completely different than they are today. In early 1959, NASA's director of space medicine, Dr. William Randolph Lovelace, proposed the idea of female **astronauts** and argued that women were smaller and lighter, required less oxygen (and presumably food), and coped better with stress than did men. He recruited 24 women to participate in a program that became known as First Lady Astronaut Trainees, and included Jerrie Cobb. She underwent physical tests like those given to the **Mercury** astronauts, and ranked in the top 2 percent of all astronaut candidates of both genders, though she never had the opportunity to fly in space. When **Alan Shepard** achieved the honor of being the first American to fly in space on May 5, 1961 (three weeks after Gagarin), he could indeed have had more competition for the flight had Lovelace had his way.

See also ASTRONAUT TRAINING; GLENN, JOHN HERSCHEL JR. (July 18, 1921–).

FREEDOM 7. Named by **Alan Shepard**, the *Freedom 7* capsule carried Shepard into suborbital space on May 5, 1961. The number 7 was an homage to the **Mercury Seven**, as well as being model 7. Afterward, the number was appended to all **Mercury** missions in their honor.

FRIENDSHIP 7. *Friendship 7* was the name of the capsule in which **John Glenn** made his orbit around Earth. Its name derived from the friendship Glenn wanted to carry around the world.

G

GAGARIN, YURI (March 9, 1934–March 27, 1968). The first human to fly in outer space and the first to orbit the Earth on April 12, 1961, Gagarin (the "Columbus of the Cosmos") was a pilot and **cosmonaut** in the **Soviet Union**. One of only 20 candidates for the young Soviet space program, and by all accounts an intelligent, unassuming, and popular young man, Gagarin was chosen to fly aboard ***Vostok 1*** and became an instant celebrity. He traveled at an altitude of 187 miles above the Earth at 18,000 mph for 108 minutes. Reentry was controlled by a computer, and Gagarin ejected from the spacecraft and landed by parachute nearby. Following a worldwide tour (though not to the United States, as Cold War tensions would not have allowed it), Gagarin later returned to Star City, the cosmonaut training center, where he worked on designs for reusable spacecraft, eventually becoming the deputy training director there. On a training flight on March 27, 1968, as he prepared for his second mission in space, he and a flight instructor died in the crash of their MiG-15, and there is some controversy as to the cause of the crash, though it was widely believed to be either a combination of weather and equipment or a sudden sharp turn that caused the plane to stall and spin. In 2004, declassified reports revealed that a Su-15 jet flying in the area at low altitude might have broken the sound barrier, passing close to Gagarin and causing extensive turbulence, sending the plane into a spin. Gagarin was the recipient of many awards, medals, and honorary citizenships. A plaque on the **moon** (left by the crew of ***Apollo*** *15*) honors his and others' achievements.

On the 50th anniversary of Gagarin's historic flight, the crew of the **International Space Station** sported shirts with Gagarin's image; the Russian, American, and Italian crew wished the world "Happy Yuri's Night" and shot a film (*First Orbit*) tracing Gagarin's route.

GALAXY EVOLUTION EXPLORER (GALEX). Launched on April 28, 2003, GALEX is a space telescope searching ultraviolet ranges in space to determine the origins of our universe and the evolution of star formation. Unlike the **James Webb Space Telescope** and the **Hubble Space Telescope**

(which are in geosynchronous orbit around Earth), GALEX hovers just 433 miles above the surface of the Earth orbiting near the equator. Just 3 feet 8 inches wide on liftoff, GALEX extended its solar panels once it reached orbit to become 9 feet wide. Its ultraviolet detectors are so sensitive that all of the stars that we can see with our eyes are too bright for it to resolve, and so it examines the "dark" spaces in the sky. Its initial mission was for 29 months, but it continued operation for a further seven years until it was put on standby in February 2012 after **budget** cuts. Creative financing and programming resurrected it to continue operation through fund-raising campaigns and educational transfers. In January 2013, working with the European Southern Observatory's Very Large Telescope and NASA's **Spitzer Space Telescope**, GALEX found the largest spiral galaxy discovered to date (522,000 light years across from the end of one arm to the end of the opposite arm, or more than five times the size of our Milky Way). However, with budget cuts still a consideration, the spacecraft was turned off on June 28, 2013. It continues to orbit the Earth; see appendix B for a list of its instruments.

GALILEO. Launched on October 18, 1989, from the **space shuttle Atlantis**, Galileo (designed and built by **Jet Propulsion Laboratory**, with a set of experiments devised by **Carl Sagan**) arrived at **Jupiter** six years later. It circled Jupiter for more than seven years, making 35 complete orbits, evaluating its **moons**, and sending down a **probe** to sample the planet's atmosphere. On September 21, 2003, after it failed to operate, the spacecraft was sent spinning down into Jupiter at 30 miles per second (107,955 miles per hour).

Galileo had its beginnings in 1977 and was initially planned to launch from the ***Columbia*** in 1982, but delays after the ***Challenger*** **tragedy** and scheduling adjustments postponed its launch. Initially intended to be a two-year study of the Jovian system that would allow close passage near the largest moons, Galileo performed flybys (gravity assists) of both **Venus** and the Earth, studying them as it gained velocity. On its journey, Galileo also passed by two asteroids, 951 Gaspra in 1991 and 243 Ida and Dactyl in 1993 (Dactyl being Ida's small moon, 0.87 miles in diameter), and coming within 1,000 miles of Gaspra—near enough to take photographs. Then, in 1994, Galileo was the first spacecraft to record a comet crashing into a planet when Shoemaker-Levy 9 collided with Jupiter. In July 1995, the Galileo probe was released, subjected to 230 Gs on its descent to the planet, and recorded 58 minutes of data, stopping only when it reached 307°F; the probe found more turbulence and less helium than had been expected. In 1997, the mission was **extended** to allow the satellite to fly by Europa and Io (two moons with a high radiation exposure that would potentially damage Galileo's equipment). By late 2001, the radiation damage was irreparable and it was deorbited in 2003.

Galileo provided a wealth of data for researchers and planetary scientists. Among other things, Galileo detected ammonia clouds in the atmosphere; confirmed extensive volcanic activity on Io; indicated that liquid oceans (subsurface saltwaters) exist beneath the frozen surfaces of Europa, Ganymede, and Callisto; discovered Ganymede's magnetic field (the first moon found to have one); and much more. Galileo was not the first spacecraft to reach Jupiter—***Pioneers*** *10* and *11* and ***Voyagers*** *1* and *2* had all flown by the planet—but it was the first to make an extensive study of Jupiter. See appendix B for a list of its instruments.

Note: Galileo is also the name of a joint European Union/**European Space Agency GPS** network of 30 **satellites** expected to begin operations by 2016. Four experimental satellites have been deployed; unfortunately two are in an incorrect orbit due to a possible fuel freeze on the **Soyuz** rocket that carried them to space in August 2014 and there will undoubtedly be delays as a consequence.

GAMMA-RAY LARGE AREA SPACE TELESCOPE (GLAST). *See* FERMI GAMMA-RAY SPACE TELESCOPE.

GEMINI. The second of NASA's human spaceflight programs (following **Mercury**, preceding **Apollo**), Gemini explored the possibilities of docking with another vehicle in space as well as extending orbital flight time by enlarging the one-man Mercury capsules. Throughout its five-year history, ending in 1966, Gemini included 12 missions, 10 of them with a crew of two astronauts each, many of them making legendary "firsts." As a step up from Mercury, Gemini had longer spaceflights, successfully attempted rendezvous and docking in space between two spacecraft, performed controlled reentry and landing of the spacecraft, and studied the effects of weightlessness on crew members. Gemini missions were also referred to as Gemini-**Titan** in reference to the launch vehicle that carried them into space, and all were launched from **Cape Canaveral**. *Gemini III* (with **Gus Grissom** and John Young) was the first crewed Gemini flight; *Gemini IV* (with Jim McDivitt and **Ed White**) conducted the first **extravehicular activity** (EVA); *Gemini V* (with Gordon Cooper and Pete Conrad) included the first use of fuel cells for electrical power; *Gemini VI-A* (with **Wally Schirra** and Tom Stafford) and *VII* (with Frank Borman and Jim Lovell) conducted the first space rendezvous; *Gemini VIII* (with **Neil Armstrong** and David Scott) accomplished the first docking with another spacecraft; *Gemini XI* (with Pete Conrad and Dick Gordon) set a crewed Earth orbital record altitude (still standing as of August 2014) of 739.2 miles; and *Gemini XII* (with Jim Lovell and **Buzz Aldrin**) set an EVA record of 5 hours 30 minutes.

All of the capsules and trainers have found homes in museums or on display, with the exception of *Gemini I*, which disintegrated on reentry. See appendix B for a typical list of Gemini's instruments.

GENERAL ELECTRIC (GE). GE is a frequent **contractor** and collaborator with NASA. Much of the work of tracking the **Apollo** spacecraft was coordinated by GE, and GE designs for the **command module** were considered for the program (though in the end NASA went with its own design). GE continues to provide services to NASA, such as GE Ultrasound's equipment on board the **space shuttle *Atlantis*** for medical monitoring and testing, GE Aviation's wind tunnel research with NASA at **Glenn Research Center**, with NASA's Cloud Computing for Air Traffic Management program, and so forth. GE's diversification into a variety of fields and overseas operations has led to a lesser role with NASA.

GENESIS. Launched on August 8, 2001, Genesis's purpose was to discover the origins of the sun by studying the ions in **solar wind** (solar electromagnetic radiation and heat being two obstacles to actually traveling—and returning from—the sun). Conducting an orbit around Earth–sun **Lagrange point** 1 (L1), Genesis opened its collector arrays and began to sample space on December 3. For 884 days, Genesis made halo orbits around L1 through three types of solar wind—fast, slow, and **coronal mass ejections**—opening the appropriate collector array each time a particular wind appeared, for a total collection of about 10^{20} ions (0.4 mg). Then, for the first time since the **Apollo** program, NASA navigators needed to return a spacecraft to Earth from beyond Earth's orbit to a predetermining landing zone. In April 2004, the sample return case was sealed and the spacecraft began its return to Earth. Using the **Deep Space Network**, NASA tracked its return path. The sample return probe was ejected from the spacecraft but on its return to Earth a drogue parachute failed to open (having one of its switches installed backward). The case crashed in a Utah desert. Damage was not as severe as it could have been, given that the **probe** was traveling at about 193 mph, but it landed on muddy ground. Most of the sample tiles were still sealed on retrieval. Scientists are still studying its results, but one of the more recent findings is that the composition of the sun's chemistry is not the same as that of Earth (Earth having a different set of oxygen and nitrogen isotopes, compared to the sun). Scientists had assumed that when our solar system formed 4.6 billion years ago everything was created from the same roiling mess of dust and gas and that there would be more similarities at the basic level. Analysis is still underway, researchers having completed five of the 18 planned examinations. See appendix B for a list of its instruments.

GEOSTATIONARY OPERATIONAL ENVIRONMENTAL SATELLITES (GOES). GOES is a network of **geosynchronous** satellites operated by the **National Oceanic and Atmospheric Administration** (NOAA) for weather, but the technical guidance and project management is conducted by NASA's **Goddard Space Flight Center**. The first GOES **satellite** was launched in 1975. GOES satellites have a letter appended to their name before launch (when in NASA's control), but are later given numbers for NOAA operations—GOES-A became GOES-1 when it became operational, for example. Because they only have visible and infrared sounders, GOES satellites provide images of trends in atmospheric temperature and moisture distribution, giving weather forecasters more data for short-term forecasts. Their environmental instruments measure ocean and land temperatures, crop conditions, hazard detection, and so forth. GOES-R (the next in the series) is scheduled to launch in 2016, with GOES-S following a year or so later. See appendix B for a list of some of their instruments.

GEOSTATIONARY ORBIT. A **satellite** in geostationary orbit appears to an observer on Earth as if it is not moving. The spacecraft has an orbit that matches that of Earth, generally approximately 22,236 miles (**high Earth orbit**) when above the surface of the planet at the equator, and travels about 17,500 mph, equivalent to Earth's rotation speed. A geostationary orbit is a particular kind of circular **geosynchronous orbit**. Communications and meteorology satellites tend to operate in a geostationary orbit (also called geosynchronous when above the equator). More than 400 satellites are operating in geostationary orbit, some as close as 50 miles apart from each other.

GEOSYNCHRONOUS ORBIT. A satellite in geosynchronous orbit appears to an observer on Earth as if it moving around a far distant fixed point, unless it is orbiting the equator in which it appears not to be moving at all. The difference between a geostationary orbit and a geosynchronous one is that the geosynchronous satellite is not necessarily matching the orbit of the Earth at the equator. When it does not match the orbit at the equator, it is referred to as an inclined orbit. Geosynchronous satellites are permanently in view of their ground stations and are thus a common choice for 24-hour availability, such as communications and weather. The two terms are often used interchangeably, though they do not mean precisely the same thing. Approximately 600 geosynchronous satellites are in orbit, though not all of them are functional.

GEOTAIL. Launched on July 24, 1992, Geotail is a **satellite** developed in collaboration with Japan's Institute of Space and Astronautical Science (ISAS), who designed and built it to study Earth's magnetosphere, with the

Japan Aerospace Exploration Agency, and with NASA. Unlike most satellites orbiting Earth, Geotail ranges from 8 Earth radii to about 200 Earth radii (or about 31,000 to 800,000 miles from the center of the Earth). Data are transmitted to the Usuda Deep Space Center in Japan and relayed by NASA's **Deep Space Network**. Intended to have an operating period of about four years, Geotail flew far from Earth for its first two years and then returned to near Earth, where it has mainly operated in an elliptical orbit for 20 years. With **WIND**, **Polar**, **SOHO**, and **Cluster**, Geotail is part of the International Solar-Terrestrial Physics Science Initiative project. See appendix B for a list of its instruments.

GERMAN AEROSPACE CENTER (DLR). The Deutsches Zentrum für Luft- und Raumfahrt e.V., or DLR, is Germany's aerospace, energy, and transportation research center, located in Cologne and with offices throughout Germany. DLR and NASA operate the infrared telescope SOFIA (**Stratospheric Observatory for Infrared Astronomy**). Germany has sent **astronauts** on board the **space shuttles** and with **Spacelab**, has an agreement with NASA to study atmospheric effects of emissions of alternative fuels in jet engines, and the two have a joint project in the **Gravity Recovery and Climate Experiment** through the end of its on-orbit life, expected to be sometime in 2015.

GLENN, JOHN HERSCHEL JR. (July 18, 1921–). Best known for becoming the first American to orbit the Earth, Glenn had a hugely successful career both before and after his historic flight—as a test pilot, U.S. senator, professor, and speaker. With regard to his NASA career, he is the only surviving member of the **Mercury Seven** and the oldest person to fly in space.

Born in Cambridge, Ohio, he studied engineering at Muskingum College but his academic career was interrupted by Pearl Harbor. He earned his private pilot's license in 1941 and enlisted in the **U.S. Navy** as an aviation cadet, transferring to the Marine Corps for advanced training. After the war (twice returning to land with more than 250 holes in his plane), he became a flight instructor, then flew more than 90 missions during the Korean War, and became a test pilot. In 1957, he completed the first supersonic transcontinental flight. Returning to school (a science-based degree was required for **astronaut training**), he earned his degree and was accepted into the program. Launching from **Kennedy Space Center**, the ***Friendship 7*** with Glenn on board lifted off for its historic orbit around Earth at an altitude of 162 miles, traveling 17,500 mph; he circled the planet three times in a flight of almost five hours, and became an instant American hero, complete with medals and ticker-tape parades. Glenn resigned from NASA shortly after

President **John F. Kennedy**'s assassination (he was a good friend of the Kennedys), became a business executive, and then ran for U.S. senator from Ohio, a post he held for 24 years. While still serving as a senator, Glenn (at age 77) also flew on board **space shuttle *Discovery*** to study the effects of spaceflight on the elderly. In 1998, he founded the John Glenn Institute for Public Service and Public Policy at Ohio State University in order to encourage public service. Glenn has received numerous awards, medals, and honorary degrees, and has had buildings, highways, boats, landing platforms, and schools named after him.

A product of his times, Glenn testified before the House Space Committee in 1962 to exclude women from the **astronaut** program. (NASA at the time had no official policy excluding women, but they did have a requirement that astronauts be military test pilots—from which career women were excluded.) However, later in life he supported Judith Resnik in her bid to become an astronaut. He also opposed the "**space tourist**" program to the **International Space Station** as well as the termination of the space shuttle program.

GLENN RESEARCH CENTER (GRC). Named after Ohioan **John H. Glenn**, the first American to fly in space, the facility in Cleveland, Ohio, was originally part of the **National Advisory Committee for Aeronautics** when the latter was acquired by NASA in 1958. Prior to that time, it had been called the Aircraft Engine Research Laboratory, the Flight Propulsion Research Laboratory, and the Lewis Flight Propulsion Laboratory. After it was transferred to NASA, it became the Lewis Research Center (LeRC) and specialized in aircraft engine research. It was renamed the Glenn Research Center in 1999. Its facilities at **Plum Brook Station** are the only location capable of testing full-scale upper-stage launch vehicles and rocket engines under simulated high-altitude conditions. The research center also includes a space environment vacuum chamber, cryogenic test facilities, and a hypersonic wind tunnel. The electrical power system designed at GRC for **Space Station Freedom** is essentially the same used on the **International Space Station**. Despite being primarily a research facility, GRC was a key participant in the **Mars Pathfinder** mission in 1997 and will be managing the **Orion Crew Exploration Vehicle**'s service module as well as the **Space Launch Systems**' payload shroud.

GLENNAN, T. KEITH (September 8, 1905–April 11, 1995). Chosen by President Dwight D. Eisenhower to be the first administrator of NASA, Glennan presided over the new organization from August 19, 1958, through January 20, 1961, when he returned to Case Institute of Technology upon the transfer of the presidency to **John F. Kennedy**. Together with Deputy Administrator (and former director of the **National Advisory Committee for**

Aeronautics) Hugh Dryden, Glennan oversaw NASA, **Langley Memorial Aeronautical Laboratory, Ames Aeronautical Laboratory**, and **Lewis Flight Propulsion Laboratory**, and created the **Goddard Space Flight Center**. He worked with the **Advanced Research Project Agency**, acquired control of **Jet Propulsion Laboratory**, and oversaw the acquisition of the **Army Ballistic Missile Agency** (ABMA), which he renamed **Marshall Space Flight Center**. Early in his NASA career, he also hired German rocket engineer **Wernher von Braun** (then working at the ABMA) to work on space flight projects. In essence, he took a disparate group of related and networked organizations under a single broad umbrella, creating a national agency capable of sending humans to the **moon**—and at the same time tried desperately to keep NASA small enough that it would be a cohesive and scientific group. He was succeeded in his position by **James E. Webb**.

GLOBAL GEOSPACE SCIENCE (GGS). A NASA space program designed to study **solar wind** and its effect on the Earth's atmosphere and magnetosphere. The two **satellites** included in the program are **WIND** and **Polar**.

GLOBAL POSITIONING SYSTEM (GPS). A space-based **satellite** navigation system that provides precise location and time information anywhere on or near the Earth. It was developed in 1973 in a combined effort from several U.S. military departments to overcome the limitations of previous navigation systems and is maintained by the U.S. government.

GLOBAL PRECIPITATION MEASUREMENT (GPM). Launched by Japan on February 27, 2014, GPM was developed and tested at **Goddard Space Flight Center**. It is a state-of-the-art precipitation measurement **satellite** examining water and energy cycles on Earth and is part of an international **constellation** of such satellites, including the **Tropical Rainfall Measurement Mission**. The Core Observatory satellite is a collaboration between NASA and the **Japan Aerospace Exploration Agency**, and other GPM satellites were contributed by **India** and Europe, giving the network a global outlook every two to three hours. GPM operates in a non-sun-synchronous orbit from near the Antarctic Circle to the Arctic Circle 253 miles above the surface, in **low Earth orbit**. Ongoing data from the mission became available to researchers in September 2014. See appendix B for a list of its instruments.

GLORY. Glory, one of NASA's **Earth Science** research **satellites**, was scheduled to join the **A-train constellation** in February 2011 by helping scientists understand the aerosol contribution to climate change and record solar activities. However, the $424 million Glory failed to reach orbit after **Orbital Science**'s Taurus XL rocket's fairing system failed to fully open after launching from **Vandenberg Air Force Base**. The investigation into the **failure** is not available to the public because it contains details covered by the **International Traffic in Arms Regulations** as well as private proprietary information, but a summary was released in 2013. See appendix B for a list of its instruments.

GODDARD INSTITUTE FOR SPACE STUDIES (GISS). Established in 1961 for research in space science, GISS is a laboratory managed by the Earth Sciences Division of **Goddard Space Flight Center** and currently part of Columbia University's Earth Institute. GISS is located at Columbia University's campus in New York City. Its focus is on environmental global changes and trends, whether in our atmosphere, on land, or in the oceans, and research varies from natural cycles to man-made interruptions in an attempt to understand our planet's past and future environments.

GODDARD, ROBERT HUTCHINGS (October 5, 1882–August 10, 1945). Best known for building the first liquid-fueled rocket, Goddard was also a professor and physicist, an inventor and engineer, a theorist and researcher. He held 214 patents (more than half awarded posthumously). Though he died more than a decade before NASA was created, his contributions to spaceflight were immense, and the **Goddard Space Flight Center** in Greenbelt, Maryland (established May 1, 1959), bears his name. Interested in science from a very young age (he wrote "The Navigation of Space" while still in high school), Goddard studied physics and his first two patents, one for a solid-fuel rocket and one for a possible liquid-propelled rocket, were issued in 1914. Soon, he was working on fuel efficiency, converting more of fuel into thrust on his rockets. Further research continued, sometimes sponsored by the Smithsonian Institution, Clark University, Worcester Polytechnic, and the U.S. government (military rockets). In 1919, the Smithsonian published his *A Method of Reaching Extreme Altitude*, which received scathing reviews from newspapers, particularly the *New York Times* (which, to its credit, published a correction 49 years later). Goddard continued to study rocket science in the interwar years, and his test of a liquid-fueled rocket in 1926 proved groundbreaking, though little appreciated by either public or government. Three years later, his rockets carried the first scientific payload: a barometer and a camera. Goddard was probably the first scientist to have both the creativity and the practical experience to realize the potentials of spaceflight, though most of what he accomplished went unrecognized in his day. NASA presents the Robert H. Goddard Award for Science to individuals for outstanding work in science at Goddard Space Flight Center.

GODDARD SPACE FLIGHT CENTER (GSFC). Established May 1, 1959, GSFC is the first spaceflight center and one of NASA's 10 major **field centers**, and is located near Washington, D.C., in Greenbelt, Maryland. The facility had previously been known as the Beltsville Space Center but was renamed for **Robert H. Goddard**, a pioneer in early rocket studies. GSFC focuses on uncrewed scientific studies in Earth orbit for the space program and the related technologies. It employs approximately 3,000 people, and more than twice that number of supporting **contractors**. Some of the programs it operates are the **Earth Observing System**, **Explorer**, and the **Hubble Space Telescope**. As part of its **education and public outreach**, GSFC operates a visitor center open to the public as well as an Educator's Resource Center for teachers and community volunteers. Three of its facilities are located elsewhere: the **Goddard Institute for Space Studies** (New York City), the **Independent Verification and Validation Facility** (Fairmont, West Virginia), and **Wallops Flight Facility** (Wallops Island, Virginia). GSFC is also responsible for the **White Sands Complex** (Las Cruces, New Mexico).

GOES. *See also* GEOSTATIONARY OPERATIONAL ENVIRONMENTAL SATELLITES (GOES).

GPS. *See* GLOBAL POSITIONING SYSTEM (GPS).

GRAVITY PROBE-B (GP-B). Launched in 2004, GP-B was designed to test Einstein's general theory of relativity. Four extremely precise gyroscopes circling Earth 400 miles above the surface measured the theoretical warping of space and time around Earth, as well as the amount a spinning object (in this case, Earth) pulls space and time with it. By pointing at IM Pegasi (a star) while in polar orbit around Earth over the course of a year, GP-B confirmed Einstein's theories. Gravity **Probe**-A, almost 30 years earlier in 1976, had compared elapsed time in three identical clocks—two on the surface and a third traveling in a rocket—and had already tested some aspects of Einstein's theory. See appendix B for a list of its instruments.

GRAVITY RECOVERY AND CLIMATE EXPERIMENT (GRACE). A joint mission of NASA (**Jet Propulsion Laboratory**) and the **German Aerospace Center**, GRACE's two **satellites** (nicknamed "Tom" and "Jerry") launched in March 2002 from **Russia** to study the Earth's gravitational fields and thus its mass. They circle the Earth 15 times a day, traveling 127 miles apart, 310 miles above the surface, and mapping any changes in gravitational pull. As the first satellite passes over a region of stronger gravity (more mass), it pulls farther away from the second satellite, then slows. The second satellite repeats the procedure, getting closer to the first, and then reverts to its initial status. Each of these anomalies helps scientists determine what lies below the surface—how thick an ice shelf is, how deep and strong the ocean's currents are, and how magma flows deep in Earth's center, for example. The GRACE satellites were expected to last for five years, and are now in an **extended mission** but researchers anticipate they will reenter Earth's atmosphere in 2016 in a decaying orbit. A follow-up mission (GRACE-FO) is planned between NASA and Germany in 2017. See appendix B for a list of their instruments.

See also GRAVITY RECOVERY AND INTERIOR LABORATORY (GRAIL).

GRAVITY RECOVERY AND INTERIOR LABORATORY (GRAIL). Launched on September 10, 2011, the GRAIL mission stationed two **satellites** (GRAIL-A and GRAIL-B, nicknamed "Ebb" and "Flow" after a contest opened to students throughout the United States) in formation in the same orbit around the **moon**. As the two satellites pass over the lunar surface, changes in their relative velocity will indicate what is hidden below the surface by mapping the gravitational forces at work. (This is essentially what the **Gravity Recovery and Climate Experiment** has done for Earth.) The spacecraft also carried cameras for MoonKAM as part of its **education and public outreach** project. Twice each day, the spacecraft passed in view of the **Deep Space Network** and transmitted their data. After 88 days, with its mapping complete, the two satellites were powered down and crashed into the moon. See appendix B for a list of its instruments.

GREENHOUSE GASES OBSERVING SATELLITE (GOSAT). The GOSAT project is a joint mission between the National Institute for Environmental Studies, Japan's Ministry of the Environment, and the **Japan Aerospace Exploration Agency** (JAXA). Data from the satellite is shared with NASA. After the loss of the **Orbiting Carbon Observatory** (OCO), JAXA invited NASA's OCO team to contribute to data analysis. A new task force was formed by **Jet Propulsion Laboratory**, the California Institute of Technology, and Colorado State University, but was disbanded when the OCO-2 team was set up for its mission in 2014. GOSAT's mission is to evaluate the global distribution of carbon dioxide and to study how it varies depending on season, year, and location.

GRISSOM, VIRGIL IVAN "GUS" (April 3, 1926–January 27, 1967). One of the original **Mercury Seven astronauts**, Grissom was the second American (after **Alan Shepard**) to fly in space and the first American to fly in space twice. After serving briefly as a clerk in World War II, Grissom earned his mechanical engineering degree at Purdue University, reenlisted during the Korean War, and trained as a pilot, flying over 100 combat missions. He then served as a test pilot at Wright-Patterson Air Force Base. He was selected from a field of 110 military pilots as one of the seven **Mercury** astronauts in 1959, and on July 21, 1961, became the second American to fly in space during a suborbital flight of 15 minutes. The flight was intended to prove that Shepard's original flight was not a fluke, and it fully succeeded. American pride was immense after the bruising that had been given it by **Yuri Gagarin**'s flight just a few months earlier. Intensely private, Grissom had been known to don a floppy straw hat and sunglasses to avoid reporters. Three years later, he commanded ***Gemini*** *3* in its three-time orbit around Earth.

Grissom and fellow crew members **Ed White** and **Roger Chaffee** were killed during the ***Apollo*** *1/AS-204* fire during a prelaunch test exercise at Launch Pad 34 at **Cape Kennedy**. He was the recipient of many awards, medals, and commendations.

GROUND NETWORK (GN). *See* NEAR EARTH NETWORK (NEN).

H

HERSCHEL. A **European Space Agency** project, Herschel is a space observatory focusing on a study of the evolution of stars and galaxies. The telescope's primary mirror is 11.5 feet in diameter, more than 1.5 times larger than the **Hubble Space Telescope**, and it is capable of covering the entire range of far-infrared to microwave wavelengths. It was launched on May 14, 2009, from ESA's spaceport in French Guyana; NASA's **Jet Propulsion Laboratory** built the bolometers for Herschel's imaging receiver (40 times more sensitive than previous instruments). It orbits almost 1 million miles away, at Earth–sun **Lagrange point** 2 (L2). See appendix B for a list of its instruments.

HIGH EARTH ORBIT. High Earth orbit refers to the geocentric orbit of a spacecraft at least 22,236 miles above the Earth (about one tenth of the distance between the Earth and the **moon**). Many weather and some commercial **satellites** have a high Earth orbit, which creates a time lag (latency) when communicating with Earth. Because their rotation matches that of Earth, they tend to be geosynchronous, meaning that they seem stationary to a reference point on Earth. So many of them hover in a band along the equator, the area has gotten crowded. NASA and the Department of Defense have coordinated the **Orbital Debris** Office to track the more than 500,000 pieces of **space debris**, though most of the debris is significantly lower than high Earth orbit.

HIGH-ENERGY TRANSIENT EXPLORER-2 (HETE-2). Launched on October 9, 2000, HETE-2 was designed to study gamma-ray bursts—what they are, where they come from, why they occur. Because the **satellite** could relay information about where the gamma-ray bursts were, observers could direct ground-based instruments to study the bursts within seconds of their occurrence. They found that gamma-ray bursts are the explosive collision of two neutron stars or the merger of a neutron star and a black hole; that there is a connection between gamma-ray bursts and type Ic supernovae; and that "dark bursts" are not, in fact, completely dark. The international collaboration included a dozen organizations, including the **Centre National d'Études Spatiales**, the Massachusetts Institute of Technology, the Los Alamos National Laboratory, and the University of Chicago. HETE-2 was terminated prior to March 2008. An earlier mission, HETE-1, was a cooperative effort with Japan and France, but was lost on launch on November 4, 1996. See appendix B for a list of HETE-2's instruments.

HIGH SCHOOL STUDENTS UNITED WITH NASA TO CREATE HARDWARE (HUNCH). An **education and public outreach** project, HUNCH is a program that promotes middle school and high school student interest in science, technology, engineering, and mathematics (STEM). It partners **Johnson Space Center**, **Marshall Space Flight Center**, and **Langley Research Center** with students across the country. Students build hardware and prototypes for NASA, conduct experiments, and create soft goods with NASA-provided materials, equipment, and mentoring. The students get real-world experience, technical skills, and training in STEM fields. NASA has used some of these products on the **International Space Station** and it has the potential to encourage students to become engineers, scientists, and **astronauts**.

HINODE (SOLAR-B). NASA's **Marshall Space Flight Center** and **Goddard Space Flight Center** (together with multiple other international partners and collaborators) participated in the Hinode ("Sunrise") mission with Japan's Institute of Space and Astronautical Science (a division of the **Japanese Aerospace Exploration Agency**). Launched on September 23, 2006, Solar-B is a set of three solar telescopes studying the solar corona—how its transient energies affect **GPS** and communications **satellites**, for example—in a sun-synchronous (meaning they are pointed at the sun at all times) polar orbit. Its original three-year mission has been **extended**. See appendix B for a list of its instruments.

HOUSE COMMITTEE ON SCIENCE AND ASTRONAUTICS. After the creation of the **Select Committee on Astronautics and Space Exploration** shortly after ***Sputnik 1***'s surprise launch, the Select Committee wrote the **National Aeronautics and Space Act** that created both NASA and the permanent House Committee on Science and Astronautics. This was the first standing committee established in more than 11 years, and the first time since 1892 that Congress acted to create a standing committee in an entirely new field. Its purview includes both science and space—outer space, aeronautical research and development, and technology development—necessary "to maintain American preeminence in science and technology." In 1974, the committee's name was changed to the Committee on Science and Technology to reflect the role technology played in aeronautical research, and oversight was given to energy, the environment, our atmosphere, civil aviation, and the National Weather Service; three years later, the committee was also given jurisdiction over civilian nuclear research. The committee was extraordinarily lucky in its formative years because its chairmen had either extensive science and technological backgrounds or engineering careers before serving in Congress. The name was changed to the Committee on Science, Space, and Technology in 1977, became the Committee on Science in 1995, reverted to the Committee on Science and Technology in 2007, and is once again the Committee on Science, Space, and Technology as of 2011. In addition to name changes, political polarization combined with bureaucratic requirements, appointments and elections for political purposes only, and a lack of scientific background in recent years has caused the committee to become significantly less effective in promoting a science-based mission, despite its name.

A copy of the 2015 committee's Views and Estimates for budget consideration for NASA and other governmental agencies under its jurisdiction can be found at http://science.house.gov/sites/republicans.science.house.gov/files/documents/GOP%20Views%20and%20Estimates%20FY15.pdf.

HOUSE COMMITTEE ON SCIENCE, SPACE, AND TECHNOLOGY. *See* HOUSE COMMITTEE ON SCIENCE AND ASTRONAUTICS.

HOUSTON. The Capsule Communicator (CAPCOM) call sign that is used by the crew of all NASA manned space flights. The call sign was adopted after control of U.S. spaceflights moved to the Lyndon B. **Johnson Space Center** in the early 1960s, located in Houston, Texas. "Houston, we've had a problem" became a popular phrase after Jack Swigert on board ***Apollo 13*** uttered those words to Mission Control in Houston in April 1970 shortly before the mission was cancelled and the crew began evacuation procedures.

HUBBLE SPACE TELESCOPE (HST). Named for early 20th-century astronomer Edwin Hubble and often referred to as simply "Hubble," HST was launched into orbit by the **space shuttle *Discovery*** in 1990 and, after several repairs effected in part by public opinion after a threatened shutdown, will continue in operation until probably 2020. Its final repair mission was in 2009, and its semi-successor, the **James Webb Space Telescope**, will replace Hubble in 2018. It is thus far the only telescope designed for in-space repair. Hubble is a public telescope, and anyone can apply for time on the instrument; competition is fierce, however, and only 20 percent of applicants earn time on the telescope.

The telescope itself is a Ritchey-Chrétien design with two hyperbolic mirrors, the larger being 7.9 feet in diameter. The observatory also has four main instruments to observe the sky in the near ultraviolet, visible, and near infrared spectra. Its high-resolution images have led to numerous discoveries—for example, determining the rate at which the universe is expanding.

Use of the telescope was not without its problems after the initial launch. When images turned out to be not as clear as expected, scientists discovered that the large mirror had been ground incorrectly on its edges, resulting in a telescope that could not see faint objects—the main purpose of its mission—although it could be used for less-demanding tasks. The first service mission, therefore, corrected this problem, but not before NASA had to bear the brunt of much criticism and many jokes. The problem was eventually blamed on a contractor, Perkin-Elmer, though NASA came in for its share of faulty oversight. Many hours of **extravehicular activity** from the ***Endeavour*** in December 1993 finally corrected the mirror's optics. A second repair mission in February 1997 by *Discovery* replaced equipment, and a third in December 1999 again by *Discovery* replaced all six gyroscopes, the computer, and a thermal insulation blanket. The fourth service mission (called Service Mission 3B) by ***Columbia*** replaced the last of the original equipment as well as the solar arrays, giving Hubble 30 percent more power. One final service mission had been scheduled for February 2005. Unfortunately, one of the repercussions of the ***Columbia* disaster** in 2003 was the decision that all shuttle missions must be able to reach the **International Space Station** (ISS) in the event of troubles. Unfortunately, none of the space shuttles could reach both the ISS and Hubble on the same mission, so all future repairs on Hubble were cancelled. Public outcry was enormous—thanks to extensive **education and public outreach** to students and teachers over the years—especially considering that Hubble was a space telescope, something the wider public had not previously considered worthy of much attention. The National Academy of Sciences debated the issue and urged NASA in July to complete a final repair. Plans were completed for a robotic service mission, though not committed to. Finally, with a change in administration, approval was given in October 2006 and ***Atlantis*** was scheduled to make the repairs two years later. However, just weeks before *Atlantis*'s flight, Hubble's main data-handling unit became inoperable and a backup had to be brought online. The repair was delayed in order to include a replacement for the primary data-handling unit. In May 2009, *Atlantis* finally completed Service Mission 4, replacing and repairing many components. Although Hubble could fail at any time, it has at least continued until this time.

Some of the work it has completed or participated in has led to breakthrough discoveries. The colorized and composited photographs and images it makes possible are astounding. Thanks to Hubble, we know more about the origins of the universe, we have seen galaxies we never knew existed, we have watched as a comet (Comet Shoemaker-Levy 9) collided with **Jupiter**, and we know more about star formation and star-forming galaxies. Hubble has also raised new questions, such as: Is the rate of our universe expansion really accelerating?

Hubble's orbit will begin to decay at some point, and it will reenter Earth's atmosphere between 2019 and 2032 if nothing is done to prevent it. NASA had originally planned to retrieve Hubble with a space shuttle, but since the dismantling of the fleet that is no longer possible. See appendix B for a list of Hubble's instruments.

See also HERSCHEL.

HUMAN EXPLORATION AND OPERATIONS DIRECTORATE (HEO). HEO provides NASA's **low Earth orbit** and beyond crewed programs with administrative management and leadership. Created in 2011, it is a combination of two earlier divisions, the **Space Operations** Directorate and the **Exploration Systems Mission Directorate**, and represented a change of focus from near Earth operations to deep space (e.g., the **moon**, asteroids, **Mars**). To reach its goals, HEO will be responsible for overseeing the development of a new heavy-lift rocket, a capsule capable of carrying **astronauts** to and from an asteroid or Mars, and all the technology involved in such a mission. At the same time HEO was created, NASA is hoping to let private organizations (such as **Orbital Sciences** and **SpaceX**) take part in its **Commercial Crew Program** to low Earth orbit locations. HEO had an extensive **education and public outreach** program, especially for its **Space Launch System**.

Note: There is also a standing committee of the NASA Advisory Council called the Human Exploration and Operations Committee.

HUMAN SPACEFLIGHT PROGRAM. So far, NASA has participated in four human spaceflight programs—**Mercury**, **Gemini**, **Apollo**, and **Skylab**—in addition to the **International Space Station** and the **space shuttle**. Note that NASA now prefers the less gender-specific term "human spaceflight" instead of the earlier "manned spaceflight." NASA's human spaceflight program began with **astronaut Alan Shepard** on May 5, 1961, following **Yuri Gagarin** on ***Vostok 1*** on April 12, 1961. **China** became the third nation to achieve human spaceflight when Yang Liwei launched on October 15, 2003, although individuals from many nations had flown on board the space shuttle and International Space Station. The **India Space Research Organization** is studying human spaceflight, as are the **European Space Agency**, **Japan Aerospace Exploration Agency**, the Iranian Space Agency, and the Malaysian National Space Agency. Private organizations such as **SpaceX** and **Orbital Sciences** are gearing up to send humans into space.

Thus far, humans have only made it as far as the moon, though programs to **Mars** are in the planning stages by both nations (NASA's **Space Launch System**/Orion) and private organizations (Mars One, the Inspiration Mars Foundation). So far, the longest an individual has remained in space was that of Valeri Polyakov, who spent 437 consecutive days in space aboard **Mir**.

HURRICANES. NASA and the **National Oceanic and Atmospheric Administration** work together on a variety of missions to provide the latest in hurricane science, reporting, and forecasting. NASA's **Aqua**, Duo, **Global Precipitation Measurement**, **GOES** network, Nimbus 5, and **Terra**, as well as many others, provide hurricane and precipitation reports, which became particularly valuable during Hurricane Joaquin in October 2015. **Tropical Rainfall Measuring Mission** also provided data until July 2014 when its fuel became depleted. NASA's hurricane research was expanded in 2014 with the addition of two Global Hawk aircraft over the Atlantic Ocean. NASA maintains a Hurricane Research Team at **Goddard Space Flight Center**.

HYPERSPECTRAL IMAGER FOR CLIMATE SCIENCE (HySICS). HySICS is a high-altitude (123,000 feet above the surface of the Earth), helium-inflated balloon launched on August 18, 2014. By studying the amount of shortwave radiance reflected back toward space, researchers are better able to monitor and understand environmental changes on our planet. HySICS carried extremely precise instruments (spanning 350 to 2300 nm in wavelength) during its almost nine-hour flight, and the instruments were ejected from the balloon for a safe parachute landing near Fort Sumner, New Mexico. Analysis of the data is underway.

I

ICE, CLOUD, AND LAND ELEVATION SATELLITE (ICESat). ICESat and ICESat-2, launched on January 13, 2003, and scheduled for launch in 2017, respectively, measure the ice sheet mass balance and cloud and aerosol heights. ICESat was a single-instrument **satellite** composed of three lasers scheduled to operate for five years, but problems with corrosion on the diodes reduced its operational life to less than a year. ICESat's schedule was changed to reduce its observations, extending the mission, and it focused on the ice sheets of Greenland and Antarctica. It reentered the atmosphere in August 2010. **Operation IceBridge** (via aircraft) is being used as a transition before ICESat-2 is launched in 2017.

IMAGER FOR MAGNETOPAUSE-TO-AURORA GLOBAL EXPLORATION (IMAGE). Also known as ***Explorer*** *78*, IMAGE was decommissioned in December 2005 after it failed to respond to commands. It had been the first spacecraft dedicated to observing Earth's magnetosphere and had been operational for more than five years.

INDEPENDENT VERIFICATION AND VALIDATION FACILITY (IV&V). Located in Fairmont, West Virginia, NASA's IV&V facility is focused on safety and cost-effectiveness for mission-critical software. It was founded in 1993 as a result of the *Report of the Presidential Commission* on the **space shuttle *Challenger*** accident and the National Research Council. It operates under the purview of **Goddard Space Flight Center** and employs almost 300 people. As part of its **education and public outreach** Educator Resource Center, IV&V provides resources and training for 1,000–2,000 educators each year on science, technology, engineering, and mathematics subjects.

INDIA. Although a relatively recent addition to extraterrestrial flights, India has several projects in cooperation with NASA as well as independently managed and launched spacecraft. NASA's **Deep Space Network** is supporting Mangalyaan, India's **Mars** Orbiter Mission **satellite**, which launched on November 5, 2013, and is now orbiting Mars. NASA's **Mars Atmosphere and Volatile Evolution** mission followed two weeks later for the same purpose (though they are testing different aspects of Mars's atmosphere). The **Indian Space Research Organization** (ISRO) is participating in both of NASA's Deformation, Ecosystem Structure and Dynamics of Ice project scheduled for 2021 and **Ice, Cloud, and Land Elevation Satellite**-2 mission scheduled for 2017. Before its operations were terminated, NASA's **Tropical Rainfall Measuring Mission** was tracking cyclones and thunderstorms in India, giving forecasters there advance notice of potential storms. The Synthetic Aperture Radar program is a joint project of NASA and ISRO to co-develop and launch a dual-frequency synthetic aperture radar satellite.

INDIAN SPACE RESEARCH ORGANIZATION (ISRO). ISRO was founded in 1969 to enhance national space technology and coordinate **India**'s entry into space, initially with two **satellite** systems. It has expanded considerably since then with numerous communications and Earth observation satellites. ISRO sent its first spacecraft to the **moon** on October 22, 2008, followed by its **Mars** Orbiter Mission that reached Mars's orbit in late September 2014. It remains a frequent partner with other countries in space observation and research, and is working toward sending three **astronauts** into space in **low Earth orbit**. ISRO's headquarters are at Antariksh Bhavan in Bengaluru.

INSPECTOR GENERAL. NASA's Inspector General's office provides independent, objective analysis of NASA's activities, and reports to NASA, the U.S. Congress, and the public with regard to NASA's programs and operations. It has four main divisions: (1) The Office of Audits conducts independent audits to improve effectiveness and to identify waste and mismanagement in NASA's programs, projects, operations, and **contractor** activities; (2) The Office of Investigations investigates any charges of crime, fraud, abuse, or any such activity that might hinder NASA's activities in any way; (3) The Office of Counsel provides legal advice and assistance regarding the Inspector General's review of NASA's operations; and (4) The Office of Management and Planning provides financial, personnel, administrative, and information technology services to support Inspector General staff.

The Inspector General's office does not limit itself to dry analysis of past projects and expenses. For example, in September 2014, they issued a report on NASA's **Near Earth Object** program, finding fault with many of its assumptions and progress.

There are Inspector General staff members at each of the 10 **field centers** and at NASA headquarters, amounting to approximately 200 individuals. The current inspector general is Paul K. Martin, a former deputy inspector general for the U.S. Department of Justice.

See also INTERNATIONAL TRAFFIC IN ARMS REGULATIONS (ITAR).

INSTRUMENTS. Each of NASA's missions, whether a one-goal mission to sample atmospheric particles over the South Pole or a complex one to send humans to the moon, has customized instruments on board. Some instruments, such as guidance cameras or basic visible light and near-infrared imagers, may themselves be common to dozens of missions but still require custom installation. Others are one-of-a-kind, state-of-the-art instruments never before used and intended for only one purpose. See appendix B for a listing of the instruments used on current missions and the purposes for which they are intended.

INTEGRATED SYSTEMS RESEARCH PROGRAM (ISRP). ISRP is part of NASA's **Aeronautics Research Mission Directorate**. As such, ISRP is more concerned with national applications of mobility, capacity, safety, security, energy, and the environment of land-based aircraft using its Next Generation Air Transportation System. Many of the technologies involved have application to or are derived from space programs.

INTELSAT. Intelsats are communications **satellites** that operate in **constellations** and were owned and managed by the International Telecommunications Satellite Organization (ITSO), an international cooperative group from 1964 through 2001 (in 1973 its name was changed to Intelsat). In 2001, ITSO became a private company. Spacecraft operations are controlled through ground stations in Maryland, Hawaii, Italy, Germany, Australia, and the People's Republic of **China**. Intelsat charges satellite usage fees to anyone wanting to use its satellites and currently has over 100 members, serving more than 149 countries. The abundance of satellites allows rapid rerouting in the event of one satellite's failure. ITSO's first satellite, Intelsat 1 ("Early Bird") was launched in 1965, and weighed just 86 lbs.; Intelsat's latest fleet, launched in 2012, included satellites weighing in excess of 13,000 lbs. The vast majority of Intelsats operate in **geosynchronous orbit**.

The **space shuttle *Endeavour***, on its first flight, was contracted to repair Intelsat 6 in May 1992. Crew members captured the satellite in three **extravehicular activities** (EVAs), the first- and second-longest EVAs at that time, equipped it with a new perigee kick motor, and successfully released it back into orbit.

INTERNATIONAL COOPERATION. NASA works directly with many countries around the world, and indirectly (through organizations such as the **European Space Agency**) with many more. **International Space Station** partners include Brazil, Canada, Russia, and Japan (directly), and Austria, Belgium, the Czech Republic, Denmark, Finland, France, Germany, Greece, Ireland, Italy, Luxembourg, the Netherlands, Norway, Poland, Portugal, Romania, Spain, Sweden, Switzerland, and the United Kingdom indirectly. Other projects have probably fewer partners, though no less important: **SSERVI**'s partners include organizations from Republic of Korea, Saudi Arabia, and Israel, among others; and NASA Astrobiology includes Mexico and Colombia, as well as others. Over the years, NASA has participated in more than 2,000 cooperative agreements, almost all of which have an international connection. Six of its major programs have had costs of more than $1 billion, and with significant international participation: the ISS, the Earth Observing System, the Advanced X-ray Astrophysics Facility, the **Cassini** mission, the **Hubble Space Telescope**, and the **Galileo** mission.

See also CHINA, PEOPLE'S REPUBLIC OF (PRC); INDIA.

INTERIOR EXPLORATION USING SEISMIC INVESTIGATIONS, GEODESY, AND HEAT TRANSPORT (InSight). InSight is a **Mars** mission that will investigate what lies under the surface, unlike the **rovers** ***Spirit***, ***Opportunity***, and ***Curiosity*** that explore the surface—volcanoes, soil, and exposed geology. Hoping to find indications of the evolutionary processes of planetary structure, Mars provides a more stable record than does Earth since it has been (as far as we know) less geologically active. Contracted through **Lockheed Martin** Space Systems, testing for its lander began in May 2015. Initially developed through NASA's **Discovery** program, InSight is scheduled for launch in March 2016, with an arrival date in September 2016. It will be a joint operation with the **Centre National d'Études Spatiales** (France) for approximately two years and will be using equipment tested on the Mars **Phoenix** mission in 2007 that explored the polar ice caps. Other partners on InSight are the **German Aerospace Center**, the United Kingdom Space Agency, and the Swiss Space Office (through the **European Space Agency**). It has an international science team, and will be managed by **Jet Propulsion Laboratory**, launched from **Vandenberg Air Force Base**. Two **CubeSats** will be on board. See appendix B for a list of its instruments.

INTERNATIONAL GAMMA-RAY ASTROPHYSICS LABORATORY (INTEGRAL). Launched on October 17, 2002, from **Russia**, INTEGRAL is a **European Space Agency** mission in collaboration with Russia and the United States. INTEGRAL is a sensitive gamma-ray observatory that in late August 2014 demonstrated (for the first time) that dead stars ("white dwarfs") can reignite and explode as supernovae after interaction with a companion star. This had long been suspected. Its mission has been **extended** several times and continues to provide research data. See appendix B for a list of its instruments.

INTERNATIONAL GEOPHYSICAL YEAR (IGY). An international scientific exchange from July 1, 1957, through December 31, 1958, IGY drew collaborators from 67 countries, including the United States and the **Soviet Union**, then in the midst of the Cold War. ***Sputnik*** *1* and *2* were launched during the year followed by ***Explorer 1*** from NASA on behalf of the United States, **Van Allen radiation belts** were discovered, and solar radiation was found to be dangerous for human spaceflight.

The genesis of the IGY had come some seven years earlier when a group of scientists proposed the idea in lieu of the International Polar Years that had been held twice earlier. After the 1932 Polar Year, much of the documentation and scientific analyses were lost during the ensuing World War II, so this time the organizers created World Data Centers to archive data and each host country agreed to share data openly with all participating nations (which excluded mainland **China**, protesting the inclusion of Taiwan). Organization and funding was coordinated by a committee of the International Council of Scientific Unions, an independent federation of international scientific unions, to ensure that the cooperation was apolitical and free of nationalism. IGY led to new scientific research as well as the establishment of the Antarctic as a scientific location for peaceful purposes.

INTERNATIONAL SOLAR-TERRESTRIAL PHYSICS (ISTP) PROGRAM. ISTP is an international cooperative research program with NASA, the **European Space Agency**, and the Institute of Space and Astronautical Science of **Japan Aerospace Exploration Agency**. Its goal is to understand the Earth–sun system so that scientists can predict how the Earth's magnetosphere and atmosphere adapt to the **solar wind**. Its first **satellite** is called **Wind**, which launched on November 1, 1994. **Polar**, its sister spacecraft, was launched in February 1996, but the mission was terminated in April 2008, and the spacecraft remains in orbit.

INTERNATIONAL SPACE STATION (ISS). The ISS is the largest and most expensive ($150 billion) cooperative scientific and engineering project in space, and took more than 13 years and 37 trips into space to assemble (and it is not yet complete). About the size of a football field and weighing 450 tons, it orbits the Earth approximately 240 miles above the surface with a permanent habitation of rotating crews. Governed by a series of intergovernmental treaties and agreements, it is the ninth space station to be inhabited by crews from Earth (preceded by the **Russian** incarnations of Salyut ["Salute"], Almaz ["Diamond"], and Mir ["Peace"] and from the United States, **Skylab**).

Although it had its genesis long before President Ronald Reagan's State of the Union address in 1984 (the **Soviet Union** had linked two **Soyuz** together in 1969, forming the basis of a station), that speech set a 10-year goal to create a space station and gave the impetus to push the pieces together for a more collaborative approach. Its original goal was to be a laboratory, observatory, and factory in **low Earth orbit**, and to be a staging area for future missions to the **moon**, **Mars**, and beyond. In 2010, additional goals of commerce, diplomacy, and **education** were added. Each of the governmental partners owns the equipment that it has provided and retains jurisdiction in its particular components and modules.

Construction of the ISS began in November 1998 when Russian modules docked together; subsequent modules were delivered by the **space shuttles** until those ended in 2011, including 159 additional components. Zarya, launched on that initial rocket, did not have life support, but did provide propulsion, attitude control, communications, and electrical power. NASA's module Unity was attached to Zarya two weeks later by the crew from the space shuttle ***Endeavour*** during an **extravehicular activity** (EVA) and included pressurized mating adaptors. In July 2000, Zvezda launched and robotically docked to the Zarya-Unity spacecraft, adding sleeping quarters, sanitation facilities, life support equipment, and communications equipment. The ISS was ready for occupation. Its first crew (Expedition 1) arrived in November 2000 on a **Soyuz** spacecraft: Yuri Gidzenko and Sergei Krikalev from Russia and William Shepherd from the United States. The ISS expanded during the next two years with additions ferried by the space shuttles and the Russian deliveries, but was temporarily halted after the space shuttle ***Columbia* disaster** in 2003 and some modules had to be scrapped. As of June 2011, the ISS had 15 pressurized modules and the integrated truss structure (the central backbone providing mountings for radiators, solar arrays, and so forth). Still to be delivered are the multipurpose Laboratory Module from **Russia** and some external components.

The pressurized modules include the following:

- Zarya ("Sunrise"), which provided the base during the initial stage of assembly and the first two solar arms, launched by Russia in 1998.
- Unity, which included life support to work and living areas of the station, launched by the United States in 1998 two weeks after Zarya in order to connect with it.
- Zvezda ("Star"), which provided life support and living quarters, launched from Russia on July 12, 2000. The first crew arrived a few months later.

- **Destiny**, which provided research facilities, launched in February 2001. In 2005, the U.S. Congress designated it as a "national laboratory" in order to maximize its use for governmental agencies and academic and private institution, and in 2011 NASA selected the Center for the Advancement of Science in Space to manage the research laboratory.
- Quest, which is the primary airlock and hosts spacewalks and EVAs, launched on July 14, 2001. On the night before an EVA, crew members will sleep in this module, breathing a low nitrogen mixture in order to avoid decompression sickness in the low-pressure space suits.
- Pirs and Poisk ("Pier" and "Search") are Russian airlock modules, and were added in September 2001.
- Harmony, a node module that would later act as a connection point for other components, was added in October 2007.
- **Columbus**, the primary research laboratory for European payloads, was launched for the **European Space Agency** by NASA in February 2008.
- **Kibō** ("Hope") was the largest module and carried out space medicine, biology, and other research, was carried by space shuttles to the ISS on behalf of Japan in three components, the last in July 2009.
- Tranquility, another node module with expansion capability, also added life support systems to recycle wastewater and supplement oxygen production, was added in February 2010.
- Cupola, a seven-window observatory, is pointed toward Earth. It was initially a NASA/**Boeing** project but was cancelled due to **budget** restrictions. NASA and the ESA came to an agreement, and it was added in February 2010.
- Rassvet ("Dawn") is used for cargo storage and a docking port. It is the only Russian-owned module launched by NASA.
- Leonardo is a storage module attached to the Unity node. (Raffaello, Donatello, and Leonardo were made for NASA in Italy.) In exchange for their construction, the United States allotted Italy more research time on board the ISS.

Because the ISS is close to Earth and included in its magnetic field, it is partially protected from radiation in space. Occasional **solar flares**, however, can be a health hazard to crew members, so there are more heavily shielded parts of the station in Destiny and Zvezda designed to lessen these hazards.

Being close to Earth is not necessarily a good thing. The ISS's orbit decays about 1.5 miles each year due to atmospheric drag and must be "boosted" up to its preferred altitude periodically. In addition, **space debris** is an occasional hazard and the space station must be "steered" away from debris. When experts at **Vandenberg Air Force Base**, who maintain a list of known pieces of debris, determine a potential collision or even a close approach, they notify the controllers at **Johnson Space Center** (JSC). Then it's up to the attitude determination and control officer at JSC to figure out how to move the ISS. Zvezda is equipped with thrusters and the ISS has four gyroscopes that can change its altitude. More than a day before an expected approach, NASA will begin coordinating with partners in Moscow to begin avoidance maneuvers, leaving the astronauts to go about their normal business. In the event of an imminent, unpredicted collision, the crew would take shelter in the Soyuz escape spacecraft (which they have had to do three times).

The ISS has two sets of incompatible docking rings—one for Russian craft and the other for the (now decommissioned) space shuttles. These were designed for the arrival of heavy vehicles, but the next generation sets of vehicles are smaller and lighter. A 10-ton spacecraft would simply recoil from the rings because it does not have enough momentum to engage the locks. The ESA is experimenting with a new international standard, working with the Sierra Nevada Corporation on its Dream Chaser for testing.

Expedition 45 is the most recent team to arrive, joining Expedition 44. Most expeditions last approximately six months, though U.S. astronaut Scott Kelly and Russian cosmonaut Mikhail Korniyenko will remain on board from Expedition 44 as part of their yearlong stay aboard the ISS.

Critics of the ISS maintain that it is too expensive and too dangerous to continue. NASA **cancelled** its Centrifuge Accommodations Module in 2005 (built by **Japan Aerospace Exploration Agency** for NASA) due to budget cuts, so experiments are limited to those that do not require specialized equipment. The ISS's U.S. budget has been extended through 2020, and in November 2014 the heads of all the partner agencies of the ISS issued a joint statement continuing its operation through 2024, but its future beyond that point is uncertain. Some parts of it (particularly the Russian Nauka and Rassvet) are designed to be reused in their planned space station OPSEK. U.S. portions will need to be (a) returned to Earth with a random reentry such as that performed by Skylab in 1979 in Australia, or (b) a controlled de-orbit to a remote area.

See also ANIMALS IN SPACE; INTERNATIONAL COOPERATION.

INTERNATIONAL SUN–EARTH EXPLORER (ISEE). The ISEE was a joint NASA/**European Space Agency** program to study the effects of **solar wind** on the Earth's magnetosphere. Three spacecraft were involved: a "mother-daughter" pair (ISEE-1 [also known as ***Explorer*** *56*] and ISEE-2) and a heliocentric satellite, ISEE-3, all carrying complementary instruments. (ISEE-3 was later renamed International Cometary Explorer [ICE].) The first two were launched together in October 1977, completed their studies of plasmas, energetic particles, waves, and fields, and reentered Earth's atmosphere in 1987. ICE was launched on August 12, 1978, and passed through the plasma tail of Comet Giacobini-Zinner on September 11, 1985, measuring particles, fields, and waves. ICE was the first spacecraft to be placed in orbit around the Earth–sun **Lagrange point** 1. Its mission was **extended** to study **coronal mass ejections** of the sun, and it was terminated in May 1997, though it continued to orbit. In May 2014, communication was reestablished with an anticipated agenda for the **satellite**. Its thrusters failed, however, probably due to a lack of nitrogen pressurant in the fuel tanks. Its future is still uncertain.

INTERNATIONAL TRAFFIC IN ARMS REGULATIONS (ITAR). Among other responsibilities, ITAR controls the transfer of military and space-related technology. NASA's Office of the **Inspector General** (OIG) is expected to investigate any complaints made regarding potential illegal activities of any NASA office, employee, or officer. In 2009, complaints were made to the OIG that foreign nationals at **Ames Research Center** were accessing documents subject to ITAR control, setting off a four-year investigation that included the Federal Bureau of Investigation and the Department of Homeland Security. The report was not made public because it contained privileged information subject to the Privacy Act. Although there was considerable evidence to show that employees and managers had acted unwisely regarding the Divert Attitude and Control Assembly, which was Raytheon Company's small, rocket-propelled steering system for a now-defunct vehicle, no criminal charges were filed. In 2015, the Pentagon confirmed that foreign nationals did have access to the technology, but could not determine if such data was shared with other countries or individuals.

Separate but similar incidents occurred at **Goddard Space Flight Center** in August 2012 and at **Langley Research Center** in March 2013.

See also CHINA, PEOPLE'S REPUBLIC OF (PRC); GLORY.

INTERSTELLAR BOUNDARY EXPLORER (IBEX). Launched on October 19, 2008, by **Orbital Sciences**, IBEX is part of NASA's **Small Explorer** program and is mapping the boundaries of our solar system. By studying the **solar wind** and the interstellar medium at the edge of our solar system, IBEX's particle telescopes determine the edges of the bubble of our heliosphere (the area of our solar system around our sun)—and, to their surprise, found that our solar system has a tail, just like a comet, though that of the solar system is more like the shape of a four-leaf clover. In addition, IBEX had found a narrow strip of "energetic neutral atom" emission that shifts over time, which had also not been predicted. No one yet knows where these emissions are coming from. IBEX flies in an elliptical, **high Earth orbit** (to avoid our magnetosphere) ranging from 39,000 miles from the center of Earth to 175,000 miles.

INVENTIONS. When NASA was created in 1958 by the **National Aeronautics and Space Act**, the law stipulated that its research and technological advancements should benefit the public. Since that time, everything from ballpoint pens to Tang to Velcro (none of which is true) has been credited to NASA. NASA partners with a variety of private businesses, contracts with many more, and develops products in house, and they have filed more than 6300 patents for various products. NASA's Technology Transfer Program connects NASA to private industry, and its journal *Spinoff* highlights many of these related products. The list of inventions that have caught on with the public includes:

- Cochlear implants—created by an engineer at **Kennedy Space Center** in collaboration with BioStim
- Cordless tools—already developed by Black & Decker as battery-powered tools, NASA improved the technology that then spread to many more tools, including the Dustbuster
- Ear thermometers—developed by Diatek in a partnership with **Jet Propulsion Laboratory**, based on infrared heat sensors
- Freeze-dried food—developed for the **Apollo** missions
- Insulin pumps—developed by **Goddard Space Flight Center** to help **astronauts** monitor their vital signs, the insulin pumps were a spinoff of their medical equipment
- Invisible braces—made of translucent polycrystalline alumina, developed by NASA Advanced Ceramics Research and Ceradyne to protect equipment in space, combined with Unitek, who designed dental braces
- Memory foam—to help cushion spacecraft seats to reduce impact on landing
- Scratch-resistant eyeglass lenses—developed to protect astronaut helmet visors, and then used by Foster-Grant
- Shoe insoles—developed by both KangaROOS and AVIA using NASA technology developed for boots on the Apollo mission
- Smoke detectors—developed with Honeywell Corporation for use on **Skylab**
- High-resolution, navigable photographs—developed for use with ***Spirit*** and ***Opportunity*** at **Ames Research Center**, now produced and used by Gigapan (Carnegie Mellon) and TagOramic (Major League Baseball)
- Advanced eye-surgery techniques—using Goddard Space Flight Center's research into optics for the **James Webb Space Telescope**, WaveFront Science incorporated the algorithms for its Complete Ophthalmic Analysis System

INTERFACE REGION IMAGING SPECTROGRAPH (IRIS). IRIS is a Small **Explorer** mission exploring the area between the photosphere and solar corona and is still a mystery to scientists. IRIS hopes to track the flow, energy, and plasma through the chronosphere and the transition region by using spectrometry. Solar material moves, gathers energy, and heats up as it travel away from the sun, but researchers are at a loss to understand the sun's lower atmosphere. Understanding this might help explain the cause of solar disturbances such as **coronal mass ejections** (CMEs) that cause **space weather** with the potential to disrupt human technology and affect radiation. Built by **Lockheed Martin** and launched on June 27, 2013, from **Vandenberg Air Force Base**, IRIS is expected to last for at least two years. IRIS sent images of thin structures that had never been seen before and in May 2014 captured the images of a CME erupting at 1.5 million miles an hour in detail (and of material at 53,500°F. Analysis is still underway. In May 2015, IRIS celebrated its 10,000th orbit around the sun.

IONOSPHERIC CONNECTION EXPLORER (ICON). Scheduled to launch in June 2017, ICON is part of the **Explorers Program**. Headed for the boundary region between Earth and space, it will examine the area where ionized plasma (ionosphere) and neutral gases (from our atmosphere) form a boundary. The ionosphere, because it is connected to **space weather**, varies considerably, and ICON's purpose is to investigate the transfer of energy and momentum from Earth's atmosphere to space, and what effects the **solar wind** and magnetosphere have on our system. See appendix B for a list of instruments that ICON is planned to carry on board.

ISRAEL SPACE AGENCY (ISA). Founded in 1983, the ISA is part of Israel's Ministry of Science and Technology. The ISA partners with NASA on a number of experiments, both crewed and uncrewed, and had an **astronaut**, Ilan Ramon, on board the **space shuttle *Columbia*** when it exploded. On board the space shuttle ***Endeavour*** (STS-47) in 1992 were 230 Oriental hornets from Israel and astronauts studied the gravitational aspects of building combs in space. Israel became a member of the NASA Center for **Moon** Research in 2010. This was part of a follow-on discussion after a trip by NASA administrator **Charles Bolden** earlier that year to Israel. Israel also maintains a data center as part of NASA's Earth Observing System.

See also COLUMBIA DISASTER.

ITALIAN SPACE AGENCY. *See* AGENZIA SPAZIALE ITALIANA (ASI).

J

J-2X. The J-2X was the rocket engine planned for use on NASA's Project **Constellation**, but the construction was threatened with **cancellation** in 2013 (effective after testing in 2014) due to **budget** cuts. It would have burned cryogenic liquid hydrogen and liquid oxygen propellants and would have been heavier than earlier rocket engines. The J-2X was originally based on the J-2 engine used on **Saturn** rockets for the **Apollo** program, but weight considerations required a new design. The development of the engine, manufactured by Pratt and Whitney Rocketdyne, survived the budget battles and is intended for use in the upper stages of the **Space Launch System**, which will begin uncrewed missions in 2017.

JAMES WEBB SPACE TELESCOPE (JWST). JWST (or "Webb") is a large, space-based observatory that will study infrared wavelengths, expected to be launched in 2018. It was originally called the "Next Generation Space Telescope" but was renamed in 2002 after former NASA administrator **James Webb**. It will be similar to the **Hubble Space Telescope**, which is expected to begin failing sometime after 2020, though Webb will cover longer wavelengths and have greater sensitivity than Hubble. Longer wavelengths allow Webb to see further back in time (about a quarter of a billion years after the Big Bang), to when the first dust clouds from which the stars and planets were created. Webb will be searching for Earth-like planets (those that have enough water to have oceans) and will be exploring dark matter and the acceleration of the universe. NASA is partnering with the **European Space Agency**, the Canadian Space Agency, and the scientific community for this mission and will have a 21.3-foot (diameter) mirror capable of examining wavelengths from 0.6 micrometers to 28.5 micrometers. (By comparison, Hubble has a 7.9-foot mirrored telescope and studies wavelengths of 0.2 to 1.7 micrometers, while **Herschel** has an 11.5-foot mirror and studies wavelengths from 55 to 670 micrometers.)

Webb is designed for a mission of at least 5.5 years but, like Hubble, could last significantly longer. Unlike Hubble, it is not designed to be repaired from space, but will be capable of working at temperatures as low as –370°F (50 K). It will be positioned at the Earth–sun **Lagrange point** 2 (L2).

Ambitious plans for Webb tend to overshadow its enormous, over-**budget** cost, currently anticipated to be $8.7 billion, still more than three years away from launch. See appendix B for a list of its instruments.

JAPAN AEROSPACE EXPLORATION AGENCY (JAXA). Founded in October 2003 with the merger of three of Japan's space agencies, JAXA is an occasional partner with NASA on a variety of projects, the largest of which is the **International Space Station** (ISS). The first Japanese **astronaut** flew into space on the **space shuttle *Endeavour*** in September 1992, and since then almost a dozen more individuals have participated in shuttle and ISS missions. Other joint projects have included the **Tropical Rainfall Measuring Mission**, the **Global Precipitation Measurement** (GPM), and the **Aqua** Earth observation satellite. The two organizations have also held joint anime contests in conjunction with the GPM mission.

JAPANESE EXPERIMENT MODULE. *See* KIBŌ STATION.

JASON. Jason-1 (launched in 2001), Jason-2 (launched in 2008), and Jason-3 (scheduled for launch in June 2015 but postponed after the **SpaceX** launch failure) are follow-on projects of the **TOPEX/Poseidon** satellite mission of 1992, and all are collaborations with the **Centre National d'Études Spatiales**; the last two also include the European Organisation for the Exploitation of Meteorological Satellites as a partner. All four **satellites** are concerned with measuring the height of the ocean surface as well as its circulation patterns, which gives researchers information about potential climate changes and sea levels in the future. See appendix B for a typical list of instruments.

JET PROPULSION LABORATORY (JPL). When a group of amateur rocket enthusiasts and California Institute of Technology (Caltech) students were relegated to an isolated area near the mountains, they hardly knew they were starting one of the foremost space research and development facilities worldwide dedicated to the robot exploration of space. JPL continues to be managed for NASA by Caltech under a contract renewed every five years. Unlike the other **field centers**, JPL employees work for Caltech rather than the federal government; NASA, in turn, contracts with Caltech for JPL services. This is a similar arrangement as the Department of Energy labs, which are also known as Federally Funded Research Laboratories.

JPL primarily developed rocket and missile technology for the **U.S. Army** and was therefore named by them in that capacity in 1944. Less than three months after ***Sputnik*** *1*'s surprising flight, JPL created ***Explorer 1***, which went on to discover the **Van Allen radiation belts**. When NASA was founded in 1958, control of JPL was transferred from the army to NASA, and its focus became that of the payloads that the rockets would carry. Throughout the 1960s through 1980s, JPL concentrated on major space missions: **Pioneer**, Explorer, **Mariner** (particularly *Mariner 9*, which became the first **satellite** to orbit another planet), **Ranger** (most of which failed), **Surveyor** (which landed instruments on the **moon** in six successful missions), **Voyager** (which became the longest-running operation ever, launched in 1977 and still traveling in space), and the beginnings of the **Deep Space Network**, because communications and tracking were essential for all their missions. JPL also developed closer-to-home explorations of our atmosphere, oceans, and land, and in 1977 began experimenting with what became the **Hubble Space Telescope**. **Budget** cuts beginning in the 1980s forced JPL to work "smarter, faster, cheaper" with mixed results. **Mars Pathfinder** was a popular, successful program assigned by NASA to JPL, **Stardust** successfully returned samples from space in 2006, and **Genesis** collected particles of **solar wind** at Earth–sun **Lagrange point** 1. The **Deep Space** Program was part of the 1990s missions, as was **Cassini-Huygens**. Current projects that focus on **Mars** include the Mars Science Laboratory (including the ***Curiosity*** **rover**), **Mars Exploration Rover Opportunity**, and the **Mars Reconnaissance Orbiter**; other projects include **Dawn**, **Juno** (en route to Jupiter), the **Nuclear Spectroscopic Telescope Array**, and the **Spitzer Space Telescope**.

JPL employs an estimated 5,000 people with several thousand more **contractors**. It is free to conduct business with other institutions, such as with Hollywood movies and with the Department of Homeland Security. As part of its **education and public outreach**, JPL created the NASA Museum Alliance in 2003 to provide exhibition materials to museums, planetariums, and similar institutions, especially with regard to Mars rovers ***Spirit*** and ***Opportunity***. They also offer workshops and materials for teachers in science, technology, engineering, and mathematics.

JOHNSON, LYNDON BAINES (August 27, 1908–January 22, 1973). The 36th president of the United States, Johnson was an avid supporter of the **National Aeronautics and Space Act** signed into law in 1958 while he was in the U.S. Senate. As vice-president under President **John F. Kennedy**, he was instrumental in the selection of Houston/Clear Lake as the home of the new Manned Spacecraft Center in 1961 (and which was renamed **Johnson Space Center** on February 17, 1973, in his honor). After the Kennedy assassination, Johnson honored the late president by changing the name of **Cape**

Canaveral to **Cape Kennedy**. During Johnson's administration, NASA continued to grow, conducting the **Gemini** program, developing the **Saturn V** rocket, and beginning the **Apollo** program through *Apollo 8*.

JOHNSON SPACE CENTER (JSC). Established shortly after NASA's founding in 1958, the Manned Spacecraft Center began its operations in 1961 in the wake of the ***Sputnik 1*** launch and as President **John F. Kennedy** created the national goal of a man on the **moon**. It is the primary location for **astronaut training**, human flight research, and mission control for human spaceflights. Located in Clear Lake, Texas, on land donated by Rice University, it was renamed in 1973 to honor former president **Lyndon B. Johnson**, a Texas native. In addition to coordinating and managing the **Apollo** program, JSC also managed the **space shuttle** program. More than 3,200 people work at JSC, plus an additional 15,000 **contractors**. JSC also conducts research related to human spaceflight with regard to agriculture, communications, electronics, energy, medicine, and transportation.

Covering more than 1,600 acres, JSC's facilities include the **Mission Control Center**, which is the operations control room for crewed flights (changed from the historic Apollo Control Room), Space Center Houston, its visitor center, offers tours, "lunch with an astronaut," educational exhibits, and a wide variety of former spacecraft and mockups. It is managed by the Manned Spaceflight Educational Foundation, a nonprofit organization.

JPL. *See* JET PROPULSION LABORATORY (JPL).

JUNO. Juno launched from **Cape Canaveral** on August 5, 2011, on a voyage to **Jupiter**; it is expected to arrive in July 2016 after traveling an estimated 1.7 billion miles. It is the first mission to Jupiter that uses solar panels instead of radioisotope thermoelectric generators (such as those used by ***Voyagers 1*** and *2*, **Cassini-Huygens**, and **Galileo**). Juno will then orbit and map Jupiter's gravity field, magnetic field, and magnetosphere, and try to determine how much water is in Jupiter's atmosphere as well as the composition of its clouds. Because Jupiter is so large, it is hoped that its mass allowed it to hold on to its original composition, giving us a clue as to our solar system's history. Because of the radiation belts surrounding Jupiter (which can interfere with the electronics and solar panels), Juno has 0.4-inch thick titanium walls. Juno will take from 11 to 14 days to complete each polar orbit. It will remain in orbit around Jupiter for approximately a year, after which is will enter Jupiter's atmosphere to avoid hitting any of Jupiter's moons. The spacecraft carries three tiny aluminum LEGO figures: Galileo, the Roman god Jupiter, and his wife Juno.

This mission is the second of NASA's **New Frontiers Program**; the first was the Pluto **New Horizons** mission, launched in January 2006, which visited Pluto in 2015. See appendix B for a list of its instruments.

JUPITER. The fifth planet from the sun, Jupiter is the largest planet in our solar system and is referred to as a "gas giant." Jupiter's mass is more than 2.5 times that of all the other planets and their moons combined. It has at least 67 moons; its largest moon, Ganymede, is larger than the planet **Mercury**. Covered with clouds (primarily ammonia crystals), its most notable feature is the Great Red Spot, a whirling mass of clouds larger than Earth. Like **Saturn**, it has rings, although they are not nearly as striking as those of Saturn.

The first spacecraft to get close to Jupiter was the ***Pioneer** 10* space probe in 1973. That was followed a year later by *Pioneer 11* in 1974, which got even closer (21,000 miles). In 1979, both ***Voyager** 1* and *2* performed flybys on their way to the edge of the solar system. Ulysses followed in 1992, and **Cassini** (on its way to Saturn) flew by for some of the highest resolution photos of the planet, but **Galileo** was the first (and so far only) spacecraft to begin a deep exploration of the planet and the first to orbit it. **Juno** is expected to reach Jupiter in July 2016. The **European Space Agency** plans a mission scheduled to launch in 2022.

K

KÁRMÁN LINE. The Kármán line is the dividing line between Earth's atmosphere and outer space, 62 miles (100 km) above sea level on Earth. It was named after Theodore von Kármán, an engineer and physicist, who was the first to calculate that somewhere near this line the atmosphere is too thin to support flight.

KENNEDY, JOHN FITZGERALD (May 29, 1917–November 22, 1963). Kennedy, the 35th president of the United States, is commonly credited with spurring the country to put a man on the **moon** before the end of the 1960s. In a speech before a joint session of Congress on May 25, 1961, less than a month after **Alan Shepard**'s flight into space and less than two months after **Yuri Gagarin** had completed his historic flight, Kennedy encouraged the nation to stand behind NASA and lead the way. The country was still burning from the realization that the **Soviet Union** had not only sent ***Sputnik 1*** into space first, it had also sent the first man into space; the **space race** was a one-sided game. Kennedy needed a more dramatic goal. Undoubtedly, politics played a big part of his decision, if only to distract the public from other unfortunate events such as the Bay of Pigs fiasco. Unlike his predecessor, **Dwight D. Eisenhower**, Kennedy wanted a bigger, faster, more powerful NASA. Project **Mercury**, already in progress, became part of this effort, and **Gemini** and **Apollo** were designed to comply with Kennedy's—and the nation's—goal of landing a man on the moon. He repeated his plan, giving more details, on September 12, 1962, at Rice University in Houston, calling his inspirational speech "We choose to go to the moon." Although Kennedy died before his mission was accomplished, he succeeded in ramping up NASA's efforts and created a dream that inspired millions of Americans.

There were critics to Kennedy's plan and to NASA's efforts. The cost of the program was, yes, astronomical. The Apollo project was estimated to eat up $3.7 billion of NASA's total $5.7 billion budget. Some questioned the motives behind such a project—scientific, military, or egocentric? Some thought that NASA should work with the Soviet Union instead of in competition. In any event, Kennedy's dream was accomplished when the *Apollo 11* crew landed on the moon on July 20, 1969.

KENNEDY SPACE CENTER. In the aftermath of ***Sputnik 1***'s flight and U.S. determination to advance and improve its space program, NASA selected the oceanside area midway between Jacksonville and Miami at **Cape Canaveral** to be the site of its Launch Operations Center in July 1962. When President **John F. Kennedy** was assassinated a little more than a year later, the 140,000-acre center (and national wildlife refuge) was renamed the Kennedy Space Center in his honor. It has been the departure gate for every crewed U.S. mission from **Mercury** to **Gemini** to **Apollo** and the **space shuttles**, and for hundreds of spacecraft.

Cape Canaveral had been site of the **U.S. Army**'s Joint Long Range Proving Grounds established in 1949 to test missiles under development for several reasons: it was virtually undeveloped, so there was little danger to residents; it had a year-round climate suitable for operations; rockets could be launched (and recovered, if needed) over water; and a long string of islands allowed easier tracking of missiles. The **U.S. Air Force** established its Patrick Air Force Base nearby for the same reasons. Shortly after *Sputnik 1*'s flight on October 4, 1957, the **U.S. Navy** launched its ***Vanguard*** satellite on December 6, 1957, from the launch pad at Cape Canaveral but the effort proved unsuccessful. Not to be outdone, the Army launched its ***Explorer I*** satellite on January 31, 1958, built by **Jet Propulsion Laboratory**, and the United States was in space.

The U.S. Congress established NASA in late 1958, and it absorbed Project Vanguard personnel as all three branches of the military continued to use the launch site for their individual purposes. When President Kennedy declared his intention to put a man on the **moon**, the Cape Canaveral complex was ideal for its purposes. Renamed the Kennedy Space Center in 1963, the site grew as more and more was demanded of it for both civilian and military purposes. It absorbed the personnel from the Manned Spacecraft Center in Houston, leaving **Johnson Space Center** to concentrate on mission control and operations.

The **Ares** I-X launch from KSC on October 28, 2009, was the first uncrewed launch since **Skylab** in 1973. Uncrewed launches continue with **SpaceX**, **Orbital Sciences**, and **United Launch Alliance**—all private corporations. Exploration Flight Test 1 successfully concluded on December 5, 2014.

Today, approximately 8,000 people work at KSC, of whom an estimated 6,000 are **contractors** (this is down from the 13,000 people who were employed there during the operation of the space shuttles). The Visitor Complex employs 700 people and hosts 1.5 million visitors each year. Attractions of the Visitor Complex include a "rocket garden," the Apollo/Saturn V center, the **Astronaut** Hall of Fame, and a variety of exhibits, **educational and public outreach** programs, and theaters, as well as views of launch pad operations.

KEPLER. Launched on March 7, 2009, to orbit the sun, Kepler is a space observatory named after Johannes Kepler, a 17th-century astronomer. Kepler searches our region of the Milky Way for Earth-like planets (those in the habitable zone) and found more than 3,500 candidates through the use of its photometer, watching for signs of a planet passing in front of its star. The first planet in the habitable zone (where water is likely, based on distance from the star), is exoplanet Kepler-186f orbiting Kepler-186, a red dwarf about 500 light-years from Earth in the Cygnus constellation. Kepler's primary telescopic mirror is 4.6 feet in diameter, at the time the largest mirror located outside Earth's orbit. Kepler operated for almost five years before its two failed reaction wheels limited it to a field of view. Rather than dispose of the telescope, NASA solicited the science community for proposals to reuse the equipment, and the K2 program was approved for its **extended mission**. K2 focuses on data that aids researchers in understanding planet formation, stellar structure and evolution, and extragalactic science. K2 solicits targets through its **citizen-scientist** Guest Observer Program. Note: The Johannes Kepler ATV was the cargo resupply vessel to the **International Space Station** launched by the **European Space Agency** in 2011; it self-destructed on reentry to Earth's atmosphere, as planned. See appendix B for a list of the Kepler Observatory's instruments.

KESSLER SYNDROME. Also known as the Kessler effect or collisional cascading, it is a scenario proposed by the NASA scientist Donald J. Kessler in 1978 in which the density of objects in **low Earth orbit** (LEO) reaches a critical point where collisions between objects could cause a chain reaction that would increase the likelihood of further collisions. One implication is that the distribution of debris in orbit could render space exploration, and even the use of **satellites**, impractical for future generations.

See also SPACE DEBRIS.

KIBŌ STATION. The Japanese Experiment Module, known by its nickname Kibō ("Hope"), is the science and research module of the **International Space Station** (ISS) developed by the **Japan Aerospace Exploration Agency**. Launched by NASA on **space shuttle** missions STS-123, STS-124, and STS-127 (the ***Endeavour***, ***Discovery***, and *Endeavour*, respectively), it is the largest module on board the ISS. Its research focuses on space **medicine**, biology, biotechnology, and communications.

KILLIAN, JAMES RHYNE (July 24, 1904–January 29, 1988). Killian, born in Blacksburg, South Carolina, was the science advisor to President **Dwight D. Eisenhower** from 1957 to 1959, as well as the president of the Massachusetts Institute of Technology. As the chairperson of the **President's Science Advisory Committee**, he encouraged Eisenhower in the formation of NASA in 1958 and in developing a national curriculum of science and technology.

KISTIAKOWSKY, GEORGE B. (November 11, 1918–December 1, 1982). As a member of the **President's Science Advisory Committee** to President **Dwight D. Eisenhower**, Kistiakowsky, who had been born in Ukraine but later moved to the United States, succeeded **James R. Killian** as chairman. A former chemist and member of the Manhattan Project, Kistiakowsky was an advocate of banning nuclear weapons and was in part responsible for the civilian focus of NASA.

L

LAGRANGE POINTS. Also known as Lagrangian points, L-points, and libration points, Lagrange points are locations in space between two celestial bodies where the gravitational influences from each interact to become a stable place from which to make observations. They are named for the Italian-French mathematician Joseph-Louis Lagrange, who posited their existence in 1764. From our position on Earth, for example, the closer an object is to our sun, the faster it will move in its orbit, leaving Earth, as it were, in the dust—only to catch up with us some time later. However, if a **satellite** were placed where it still feels some of Earth's gravity—about one-hundredth of the distance to the sun—the pull from the sun is lessened and becomes equal to the pull from the Earth, so the satellite stays in the same location to us relative to the sun. That location where the pulls are equal is known as L1. **SOHO**, a solar observatory and a partnership of NASA and the **European Space Agency**, is positioned there, as are **WIND** and the **International Sun Earth Explorer 3**. L2 is located on the far side of Earth, away from the sun, and this is where the **James Webb Space Telescope** is positioned, out of Earth's shadow. L3 is on the far side of the sun and cannot be seen from Earth. L4 and L5 are both locations resistant to gravitational perturbations in the **moon**'s orbit, ahead of and behind the moon as it orbits Earth, and are truly stable. Asteroids, dust, and other debris tend to accumulate there.

LANDSAT. The Landsat program has provided the longest continuous space-based record of Earth's land masses. Since July 23, 1972, when Landsat 1 was launched, Landsat has archived images of the Earth, showing the changing landscapes and structures. Known at the time as an Earth Resources Technology **Satellite**, Landsat 1 was built on a **weather satellite** platform and operated until 1978. Landsats 2 through 8 continued, with Landsat 8 launching on February 11, 2013, by **United Launch Alliance**, each iteration with more equipment than the one before, with varying levels of launch and mission success. Landsat 8 (previously called the Landsat Data Continuity Mission) was a significant improvement over Landsat 7, and its images are being used in agriculture, education, business, and science. Over the years, management of the satellites switched from NASA to the **National Oceanic and Atmospheric Administration** to the U.S. Geological Survey (USGS); they were privatized for a period near the turn of the century, and then returned to USGS management in partnership with NASA. See appendix B for a list of some of its instruments.

LANDSAT DATA CONTINUITY MISSION (LDCM). *See* LANDSAT.

LANGLEY MEMORIAL AERONAUTICAL LABORATORY. *See* LANGLEY RESEARCH CENTER (LaRC).

LANGLEY RESEARCH CENTER (LaRC). LaRC is the oldest of NASA's **field centers**, established in 1917 by NASA's predecessor, the **National Advisory Committee for Aeronautics** (NACA). Named for aviation groundbreaker Samuel Pierpont Langley and located in Hampton, Virginia, LaRC's original research centered on aerodynamics (particularly wind tunnels and aviation performance) and rocket research. Langley Memorial Aeronautical Laboratory was dedicated on June 11, 1920, and created its first wind tunnel that same year, just 16 years after the Wright brothers' first flight; the building was later transferred to the **U.S. Air Force** in 1977. After the creation of NASA and the transfer from NACA to NASA in 1958, LaRC pioneered research leading to supersonic flight, simulation of lunar gravity, and the **Viking** program for **Mars** exploration. LaRC conceived and initially managed Project **Mercury**, including training for the **Mercury Seven** before that program moved to the **Manned Spacecraft Center**. Scientists at LaRC were also the lead proponents of the **Lunar Orbit Rendezvous** plan eventually selected for the **Apollo** mission. In recent years, it has become known for its electron beam freeform and plastic fabrication in aeronautics, for carbon nanotubes and boron nitride nanotubes, and for the design of the **Lunar Landing Research Vehicles** and the **Mars Exploration Rover** spacecraft. LaRC participated in the **Viking** mission, the **Long Duration Exposure Facility**, the **Stratospheric Aerosol and Gas Experiment**, **CALIPSO**, **Ares** I-X, the **Orion** Pad Abort 1, inflatable aeroshells, the X-43, and shell buckling knockdown factor research. It is also a leading partner in NASA's Earth Sciences studies. As part of its **education and public outreach**, LaRC has numerous student and intern opportunities, it participates in Virginia's Aerospace Science and Technology Scholars Program, and it recognizes Boy Scouts and Girl Scouts who achieve the highest rankings of their groups, the Eagle Rank and Gold Award, respectively.

LARGE BINOCULAR TELESCOPE (LBT). The LBT is an optical telescope for astronomy located on Mount Graham in southeastern Arizona. Part of NASA's **Origins Program**, the LBT is currently one of the world's most advanced optical telescopes using two 27-ft (8.4 m) wide mirrors.

LASTER GEODYNAMICS SATELLITE (LAGEOS). The LAGEOS satellites were designed for the use of high-precision laser ranging for accurate measurement of our planet's shape and its tectonic plate movements. LAGEOS-1 launched in 1976, and LAGEOS-2 was built by the **Italian Space Agency** and launched in 1992. LAGEOS-3, an **internationally cooperative** project with France, Germany, the United Kingdom, Italy, Spain, and the United States, is intended to measure the gravitational magnetic dipole moment, as predicted by Einstein's theory of general relativity. The spacecraft are designed to last for many decades. Both LAGEOS-1 and -2 are brass spheres (60 cm) covered with 422 fused silica glass retroreflectors and 4 retroreflectors made from germanium. Without any on-board sensors or electronics, they simply travel in an extremely stable orbit 3,700 miles (5,900 km) above the Earth, and are used primarily for use by ground-tracking stations to determine locations. LAGEOS-1, which is not expected to enter the Earth's atmosphere for another 8.4 million years, contains a plaque designed by **Carl Sagan**.

LAUNCH CONTROL CENTER (LCC). LCC, located on Merritt Island, Florida, is a four-story control tower responsible for launch operations at **Kennedy Space Center**. All crewed spaceflights from the United States are launched from Kennedy and therefore by the LCC, ever since ***Apollo** 8* was launched on December 21, 1968. The launch processing system has become increasingly automated, with computer programs monitoring and recording prelaunch performance of electrical and mechanical systems monitored from the firing rooms of the LCC. What used to take a control room of many individuals and computers can now be done by just a few highly trained professionals. When the booster clears the launch tower, control is transferred to **Mission Control Center** at **Johnson Space Center**.

LAUNCH SERVICES PROGRAM (LSP). Located at **Kennedy Space Center**, **Vandenberg Air Force Base**, **Wallops Flight Facility**, and the Kodiak Launch Complex in Alaska, as well as the Reagan Test Site at the Kwajalein Atoll in the Marshall Islands, "Earth's Bridge to Space" LSP is the coordinator and supervisor of all launch operations and countdown management. LSP oversees the expendable launch vehicle services directly from commercial providers, which allows NASA to host commercial, private launches from organizations such as **SpaceX** and **Orbital Sciences**, as well as military launches. Working closely with NASA, LSP's **contractors** allow NASA to concentrate on those programs in which it has a direct involvement and provides access to a wide variety of services for its programs.

LEWIS FLIGHT PROPULSION LABORATORY. *See* GLENN RESEARCH CENTER (GRC).

LITERATURE AND DRAMA. Hundreds of novels have been written that include some form of NASA in them, many of them by former employees or **contractors** of NASA, with varying degrees of accuracy and readability. One of the more recent is *The Martian*, by Andy Weir, a technically adept and realistic account of space missions for which the author studied orbital mechanics, astronomy, and the history of human spaceflight; the film debuted in October 2015. The *NASA Trilogy* (late 1990s) by Stephen Baxter is an alternative history/hard science fiction set that ends up being critical of NASA. Tom Wolfe's *The Right Stuff* (1979) and James Michener's *Space* (1983) are acclaimed histories of events that include NASA, while Stephen Harrigan's *Challenger Park* (2006) and Dan Brown's *Deception Point* (2002) are more recent suspense novels. Films that prominently feature NASA include the children's film *Planet 51* (2009), *Apollo 13* (1995), *Mission to Mars* (2000), *October Sky* (1999), *RocketMan* (1997), *Space Cowboys* (2000), and *The Right Stuff* (1983). Brief allusions are made in numerous other films such as 2014's *Interstellar*.

LIVING WITH A STAR (LWS). LWS is a comprehensive set of missions exploring our sun, how it interacts with Earth and within our solar system, and what effects these will have on humanity as we learn to travel in space or remain on Earth. The Heliophysics Science Program has two branches: Solar Terrestrial **Probes** and LWS. LWS is concerned with the sciences necessary to understand the sun (why do its flares heat up as they get farther from the surface of the sun?) and space environment (**space weather**) that affect humans and our society. Some of the future missions included within LWS are the Solar Orbiter Collaboration (estimated launch date 2017, a collaboration with the **European Space Agency**), **Solar Probe Plus** (estimated launch date 2018), **Space Environment Testbeds** (instruments on a Demonstration and Science Experiments mission sponsored by the **U.S. Air Force**, scheduled mid-2016). Current missions are the **Solar Dynamics Observatory** (launched February 11, 2010), the **Van Allen probes** (launched August 30, 2012), and the **Balloon Array for Radiation-Belt Relativistic Electron Losses** (December 2013/January 2014). LWS is managed by **Goddard Space Flight Center**.

LOCKHEED MARTIN. One of NASA's top three contractors (with California Institute of Technology/**Jet Propulsion Laboratory** and **Boeing**), Lockheed Martin is a worldwide aerospace company and partners with NASA on a multitude of projects. It is based near Washington, D.C., and is the top U.S. federal **contractor** when its defense and security divisions are included with that of NASA. It was founded in 1995 when Lockheed Corporation merged with Martin Marietta. The Lockheed Martin Space Systems (formerly Lockheed Missiles) is the division most closely related with NASA, primarily in Earth observation and exploration **satellites** as well as crewed spaceflight. Projects have involved the following, among others:

- **Geostationary Operational Environmental satellite**
- **Gravity Probe B**
- **Gravity Recovery and Interior Laboratory**
- **Hubble Space Telescope**
- **Imager for Magnetopause-to-Aurora Global Exploration**
- **Juno**
- **Landsat**
- **Lunar Prospector**
- **Mars Atmosphere and Volatile Evolution**
- **Mars Climate Orbiter**
- **Mars Global Surveyor**
- **Mars Odyssey**
- **Mars Reconnaissance Orbiter**
- **Orion crew exploration vehicle**
- **Phoenix Lander**

NASA and Lockheed Martin have also teamed up to provide educational opportunities for students, such as their Exploration Design Challenge, and the company has signed a Space Act Agreement to support **education and public outreach** activities.

Given the amount of business between NASA and Lockheed Martin, there has been surprisingly little friction between the two organizations, but when projects are this large, the mistakes tend to be expensive. In the 1990s, the loss of NASA's Mars Climate Orbiter was caused by incorrect figures provided by Lockheed Martin, at a cost of $125 million. In 2001, Lockheed Martin paid the U.S. government $7.1 million after its predecessor, Lockheed Engineering Science Corporation, submitted false claims to NASA. Lockheed Martin's failure to secure the **NOAA-19 weather satellite** to its base caused the satellite to crash to the floor during routine maintenance, for which the company reimbursed NASA for $30 million of the estimated $125 million repair.

LONG DURATION EXPOSURE FACILITY (LDEF). Constructed at **Langley Research Center**, the LDEF was built to study the harsh environmental conditions in space, particularly extreme temperature changes, radiation, and impacts from **space debris**. The **space shuttle *Challenger*** delivered it to **low Earth orbit** in 1984, and it was retrieved almost six years later by ***Columbia***. More than a half-dozen countries participated in 57 experiments on subjects ranging from optical fibers and thermal systems to the survivability of tomato seeds and bacterial spores.

LOW EARTH ORBIT (LEO). Low Earth orbit refers to the area between 99 miles and 1200 miles (160 and 2000 km) above the surface of the Earth, with an orbital period of between 88 minutes and 127 minutes, respectively. **Satellites** and spacecraft that zip around the Earth tend to be in LEO. The **Long Duration Exposure Facility**, which launched in 1984, spent 69 months in LEO testing the effects of space on different systems and materials. The majority of satellites as well as all space stations operate in LEO, and with the exception of lunar flights, all human spaceflights have been in LEO. Most Earth observation satellites (spy satellites) are also in LEO, and work in networks (**constellations**) to provide continuous coverage. Instruments such as the **Hubble Space Telescope** that at some point need maintenance need to be in LEO in order to effect repairs. Because there is some atmospheric drag at LEO heights, objects tend to slow down over time and their orbit decays, requiring their programs to be adjusted or terminated and allowed to disintegrate. A significant portion of **space debris** is crowded between 450 and 500 miles above Earth.

LUNAR ATMOSPHERE AND DUST ENVIRONMENT EXPLORER (LADEE). LADEE, as the name implies, investigated the dusty atmosphere of the **moon**. It was launched on September 7, 2013, with a dust detector, a mass spectrometer, and a UV spectrometer; LADEE also included a test of two-way laser communications. Seven months later, at the end of its extended mission, the $280 million **satellite** (including its development, launch, and other costs) was intentionally crashed into the far side of the moon on April 17, 2014. Because a meteor shower occurred at the same time as the crash, analysis of dust levels has been complicated, but researchers have discovered helium, argon, and neon in the moon's exosphere; a complete analysis is still underway. See appendix B for a list of its instruments.

LUNAR CRATER OBSERVATION AND SENSING SATELLITE (LCROSS). An uncrewed (robotic) spacecraft designed to explore the possibility of water-ice on the **moon**, LCROSS was launched on June 18, 2009, together with the **Lunar Reconnaissance Orbiter** (LRO). Its upper-stage

Centaur vehicle was sent crashing into the moon's crater Cabeus, near its south pole; four minutes later, the spacecraft followed, analyzing the plume of dust and debris, which revealed the presence of water molecules. The plume traveled 10 miles above the crater, exposing it to sunlight for the first time and revealing volatile content, which suggests an active water cycle in the moon's permanent shadows The water-ice had been predicted by the **Clementine** and Lunar Prospector missions but had not been confirmed.

A similar proposal, A119, involving a nuclear explosion, had been proposed in 1958 but the existence of the project was not revealed until the late 1990s when a researcher, in examining the life of **Carl Sagan** for a biography, stumbled across the documents. The project had been cancelled by the **U.S. Air Force** in 1959 while still in the planning stages, probably due to both negative public attention and the radiation risk to future lunar exploration.

See appendix B for a list of LCROSS's instruments.

LUNAR EXCURSION MODULE. *See* LUNAR MODULE (LM).

LUNAR LANDING RESEARCH VEHICLE (LLRV). Known as the "Flying Bedstead" because of its shape, the LLRVs were used to simulate the Lunar Excursion Module used by **Apollo astronauts** on the **moon**. The LLRVs were mounted vertically in a gimbal, and the throttled engine supported five-sixths of its weight to simulate the reduced gravity of the moon. Smaller thrusters gave the pilot control of pitch, yaw, and roll. Two of the LLRVs were built to NASA's satisfaction in 1964, which led to a contract with Bell Aerosystems to purchase three additional Lunar Landing Training Vehicles (LLTVs) in 1966. Three of the five vehicles (one LLRV and two LLTVs) were crashed by astronaut-pilots in learning to maneuver and control them; the pilot ejection seat ensured that there were no fatalities. **Neil Armstrong** credited his training on LLRV #1, from which he had to use the ejection seat at an altitude of about 200 feet (60 m), to his successful landing of the **Lunar Module** just over a year later.

LUNAR MODULE (LM). Known originally as the Lunar Excursion Module, the **Apollo** LM was that part of the orbiting Apollo spacecraft that carried two men to and from the moon's surface, beginning with *Apollo 11* (**Neil Armstrong** and **Buzz Aldrin**) and ending six voyages later with *Apollo 17* between 1969 and 1972. (A seventh LEM was used by the crew of *Apollo 13* for life support and propulsion when its service module was disabled on the way to the moon after an oxygen tank explosion; the crew survived and was rescued from the South Pacific Ocean. The *Apollo 13* film portrays a largely realistic view of the mission, despite some Hollywood liberties.) Modifications on earlier versions allowed the LM on Apollo missions 15–17 to have greater payload capacity (up to 268 lbs. of moon rocks) and the ability to stay longer on the lunar surface, up to 75 hours. A Lunar Roving Vehicle ("moon buggy"), built by **Boeing**, was carried by the LMs but left behind on the moon's surface after each of the three missions.

Once initial problems during manufacture and testing were overcome, the LMs proved reliable and enduring. Each had an overhead hatch that docked with the **command module** and a forward EVA hatch to provide access to the moon. The 16 thrusters, propulsion system, and hypergolic propellant tanks allowed the LM to rendezvous with the command and service module in orbit around the moon. *Apollo 11* carried an Early Apollo Scientific Experiment Package (EASEP) containing several experiments that were left on the lunar surface for subsequent collection and for data transmission, powered by two solar panels. The EASEP stopped functioning on August 27, 1969. It was a forerunner of the Apollo Lunar Surface Experiments Package (ALSEP) on subsequent missions, all of which were terminated on September 30, 1977, due to **budget** and power restrictions as well as space restrictions on Earth. Fifteen LMs were manufactured, at least in part; the last was scrapped, and two were never completed. Of the remaining 12, seven were launched to the moon on crewed missions, two were tested on the moon, two were tested on Earth, and a final one was put on display at the National Air and Space Museum in Washington, D.C.

LUNAR ORBIT RENDEZVOUS (LOR). An LOR is the method NASA eventually chose in July 1962 to land **astronauts** on the **moon**, which entailed a main spacecraft from which a lunar module would depart, land on the moon, and then return. Other options were a spacecraft that would itself depart from Earth, travel to the moon, and return directly from the moon's surface, or an Earth orbit rendezvous where astronauts would fuel while in Earth orbit, travel to the moon, and return. There was significant controversy at the time over this decision. An Earth-orbit rendezvous was believed to be less risky because if anything failed the astronauts would simply return to Earth; in addition, docking with a lunar orbiting spacecraft carried its own risks. At the same time, the construction of a single rocket, such as the proposed Nova that many at NASA preferred, would require an very large rocket, which would be extremely expensive and perhaps not even possible given the fuel consumption and size required in time to meet President **John F. Kennedy**'s goal of putting a man on the moon within the decade. Scientists at **Langley Research Center**, and in particular Dr. John Houbolt, advocated strongly for the LOR, and eventually succeeded.

LUNAR QUEST PROGRAM. The Lunar Quest Program was a set of missions designed to explore and analyze the **moon**. Its missions included the **Lunar Reconnaissance Orbiter**, the **Lunar Atmosphere Dust Environment Explorer**, the Robotic Lunar Lander, the Lunar Mapping and Modeling Project, and others. It was coordinated by **Marshall Space Flight Center** and responsible for the development of instruments, for research and analysis, and for oversight of lunar missions. **Budget** restrictions in 2012 cancelled or compromised many of the programs, effective in 2014.

LUNAR RECONNAISSANCE ORBITER (LRO). Launched with the **Lunar Crater Observation and Sensing Satellite** on June 18, 2009, LRO spent three years in a low orbit (31 miles) around the **moon** mapping its surface. It was the first U.S. mission to the moon in more than 10 years. It is currently operating under an **extended mission** after the completion of its primary mission, and is returning data such as day-night temperature maps, a global geodetic grid, high-resolution color imaging, and a map of the moon's ultraviolet albedo.

During a lunar eclipse—when temperatures drop and sunlight is temporarily gone, such as what happened during late September 2015—the LRO is placed in hibernation mode for the duration of the eclipse, with the exception of a radiometer. LRO's mission is funded through October 2016. Refer to appendix B for a list of its instruments.

M

MAGELLAN. The Magellan spacecraft was launched by NASA on May 4, 1989, to map the surface of **Venus**. The spacecraft arrived at Venus on August 10, 1990, and collected radar images of more than 98 percent of the surface. Additionally, the images that Magellan collected were 10 times more detailed than earlier **Soviet** missions (i.e., *Venera 15* and *16*). After its successful mission, it was directed to fall into the planet's atmosphere on October 11, 1994, to take measurements of the atmosphere as well as analyze the performance of how the spacecraft performed in the descent. See appendix B for a list of Magellan's instruments.

MAGNETOSPHERIC MULTISCALE (MMS) MISSION. MMS is a NASA **solar terrestrial probe** space mission designed to study Earth's magnetosphere using four identically instrumented spacecraft grouped in a tetrahedral shape, launched on 12 March 2015. Based on the **European Space Agency**'s **Cluster II** mission, which also studied Earth's magnetosphere, MMS has greater spatial and temporal resolution, allowing scientists to observe the three-dimensional structure of magnetic reconnection, or how the sun's and Earth's magnetic fields connect and disconnect. Additionally, MMS sensors measure energetic particle acceleration and turbulence, processes that occur in all astrophysical plasma systems and play an important role in the development of **space weather**. MMS entered its science phase on 1 September 2015. See appendix B for a list of its instruments.

MAN IN SPACE SOONEST PROGRAM (MISS). During the Cold War, the **U.S. Air Force** established the Man in Space (MIS) program, but MIS was replaced a month later by the accelerated MISS program to put a man into space before the **Soviet Union** did. The project was **cancelled** on August 1, 1958, but was later replaced by NASA's Project **Mercury**. Of the nine **astronaut** candidates who were selected for the MISS program, only two would actually go into space: **Neil Armstrong**, the first man to walk on the **moon**, and Joseph Walker, a test pilot, in July and August 1963 in the **X-15**.

MANNED ORBITAL VEHICLES (MOV). MOV refers to any orbiting vehicle with the capacity to sustain a human presence in space. The first crewed orbital vehicle was the **Soviet Union**'s ***Vostok*** *3KA*, a one-crew spherical spacecraft, of which the *Vostok 1* was the first to carry **Yuri Gagarin** into space on April 9, 1961. NASA's Project **Mercury** followed closely behind the Soviet efforts with their **Mercury-Redstone** 3 mission, which launched the ***Freedom*** *7* capsule that carried **Alan Shepard** into space on May 5, 1961, to become the second person ever to travel in space. The MOV used in the Mercury missions (1958–1963) consisted of a module that was large enough to fit room for 120 controls as well as a single crew member. Next came Project **Gemini**, equipped with an Orbit Attitude and Maneuvering System that, unlike Mercury, allowed the spacecraft to adjust its position in space in all three axes in addition to attitude control. Gemini also increased the size of the spacecraft's module to allow room for two crew members.

Other MOVs include **Skylab** and NASA's Space Transportation System, known more commonly as the **space shuttle**. Since the space shuttle retirement in 2011, NASA has reached out to private companies and provided funding for designing uncrewed and crewed spacecraft for the United States. Several companies are currently developing these spacecraft, including **SpaceX**, which has designed the **Dragon** V2, and **Boeing**'s CST-100 spacecraft, both capable of transporting up to seven people into **low Earth orbit**.

MANNED ORBITING LABORATORY (MOL). Originally referred to as the Manned Orbital Laboratory, MOL was a **U.S. Air Force** effort for crewed military space operations. The program started in the spring of 1964 after the **cancellation** of the **Boeing** X20 **Dyna-Soar** program, evolving from the earlier air force **Blue Gemini** program. The MOL was designed as a cylindrical laboratory with a **Gemini** capsule attached that would carry a crew of two **astronauts**. After being launched into orbit by a **Titan** IIIC **booster**, the crew would enter the laboratory through the modified capsule and conduct experiments and reconnaissance for two to four weeks. At the end of each crew's mission, the capsule would detach and reenter the Earth's atmosphere for landing, leaving behind the laboratory for future reuse. Three groups of armed forces officers were trained to be crew members; however, the program was cancelled in June 1969 as a result of the success and cost-effectiveness of uncrewed reconnaissance **satellites**. After cancellation, several of the systems and experiments were included in later space missions, including NASA's **Skylab**.

MANNED SPACECRAFT CENTER. *See* JOHNSON SPACE CENTER (JSC).

MANNED SPACE FLIGHT NETWORK (MSFN). The MSFN is a set of tracking stations that began operating in July 1961 to support the **Mercury** Program and later the **Gemini**, **Apollo**, and **Skylab** space programs. There were two other NASA space communication networks at the time, the **Spacecraft Tracking and Data Acquisition Network** (STADAN) for tracking unmanned satellites in **low Earth orbit**, and the **Deep Space Network** (DSN) for tracking more distant unmanned missions. After the end of Skylab, the MSFN and STADAN were merged to form the **Spaceflight Tracking and Data Network** (STDN). STDN was in turn replaced by the **satellite**-based **Tracking and Data Relay Satellite System** (TDRSS) during the **space shuttle** program, beginning in 2009.

MARINER. Mariner refers to the series of 10 missions designed by NASA from 1962 to 1973 to be the first spacecraft to other planets, specifically **Venus** and **Mars**. The spacecraft made accurate measurements of interplanetary space, took photographs of the surfaces of **Mercury**, Venus, and Mars, and studied their atmospheric and chemical compositions. *Mariner 2* and *5* visited Venus, *Mariner 4*, *6*, *7*, and *9* visited Mars, and *Mariner 10* visited Mercury. Of the 10 missions, only seven were successfully carried out, with *Mariner 1*, *3*, and *8* malfunctioning and never completing their missions. *Mariner 2*, launched in 1962, is still orbiting the sun. See appendix B for a list of some of the instruments.

See also MARS EXPLORATION PROGRAM (MEP).

MARS. The fourth planet from the sun, Mars is a cold, dry, rocky planet that is approximately half the diameter of Earth. Its reddish appearance is due to the abundance of iron oxide found on its surface. Mars's rotational period and seasonal cycles are very similar to Earth's with a 24.66-hour day and an axial tilt of 25.19 degrees. The geological features found on Mars are also quite similar to ones found on Earth, including valleys, deserts, volcanoes, and polar ice caps. Mars has two **moons**, Phobos and Deimos, which are very small and are likely asteroids that were captured by Mars's gravity well.

Second to the moon, Mars is the most explored object in the solar system. Over 40 missions have been planned since the early 1960s, although over half of them ended in **failure**. Throughout the 1960s and 1970s, the United States and the **Soviet Union** both focused heavily on sending spacecraft to Mars. The first successful mission to Mars, ***Mariner*** *4*, arrived at Mars on July 14, 1965, and provided 22 close-up photographs of the planet's surface. During the **Viking** program, NASA was able to successfully land the first lander, *Viking 1*, on the surface of Mars and collect data on the surface composition and chemistry. Years later, in 1997, the **Mars Pathfinder** mission landed on Mars and released the first **rover** to operate on the planet's surface, ***Sojourner***. The success of the Mars Pathfinder mission would later lead to the development of the **Mars Exploration Rover** mission that landed two more rovers on the surface, ***Spirit*** and ***Opportunity***, of which *Opportunity* is still functioning. Based on a similar design, the ***Curiosity*** rover employed a unique sky crane device that lowered the rover to the surface of Mars to find evidence of past habitability.

After the failure of the **Mars Observer** orbiter, NASA redesigned the Martian surface mapping mission and launched the **Mars Global Surveyor** in 1996. The Mars Global Surveyor successfully reached the planet in 1997 and was able to study and map the entire surface. In 2001, the **Mars Odyssey** orbiter arrived at Mars to survey the surface for the past or present evidence of liquid water, of which it detected an abundance of subsurface water in the form of ice. In September 2015, NASA revealed that photos from the **Mars Reconnaissance Orbiter** indicated intermittent flowing water on the polar ice caps. Several other successful orbiter missions conducted over the years include the **Mars Express**, **Mars Atmosphere and Volatile Exploration** (MAVEN), and Mars Orbiter Mission from India.

Currently, NASA's primary objective for the exploration of Mars includes searching for evidence of ancient life. Two active scientific rovers are on Mars (*Opportunity* and *Curiosity*) as well as five orbiters surveying the planet (Mars Odyssey, Mars Express, Mars Reconnaissance Orbiter, Mars Orbiter Mission, and MAVEN). At least five future missions from several countries are due to explore Mars by 2020, including the **InSight** lander and the **Mars 2020** rover.

MARS 2020 MISSION PLANS. In 2012, NASA announced plans to send a robotic science **rover** to study **Mars** in 2020. Following the successful landing of the **Mars Science Laboratory** (MSL) ***Curiosity*** rover, the 2020 mission will focus on exploring the potential for life on Mars as well as addressing the challenges for future human expeditions there. Design specifications will closely follow that of the MSL, including the landing system and rover build. One of the unique features of the rover will be an electrolysis experiment for oxygen conversion. Another feature still in development is an inflatable heat shield to protect any descending spacecraft from reentry heat issues.

See also MARS EXPLORATION PROGRAM (MEP).

MARS ATMOSPHERE AND VOLATILE EVOLUTION (MAVEN). MAVEN is a NASA space **probe** that arrived in **Mars**'s orbit on September 21, 2014, to study the planet's atmosphere. Specifically, MAVEN is examining how the Martian atmosphere has changed over time, as it was once presumably much thicker and capable of supporting flowing water on more of the surface. In order to determine this change, the probe examines how the upper atmosphere interacts with elements such as **solar wind** and measures the current rates of gas escaping into space. The probe is placed into an extremely elliptical orbit, coming as close as 93 miles above the planet's surface and as far as 3,900 miles away.

MAVEN was chosen to represent the Mars Scout 2013 mission as part of the **Mars Scout Program**, which was discontinued in 2010. The probe was set for launch on November 18, 2013, which took advantage of a natural alignment between Earth and Mars every 26 months. However, a government shutdown on October 1, 2013, threatened the projected launch window, which would have resulted in waiting 26 months for another alignment. In light of this situation, emergency funding was authorized to resume the spacecraft's preparation for an on-time launch. MAVEN is managed by **Goddard Space Flight Center**. See appendix B for a list of its instruments.

MARS CLIMATE ORBITER. Originally known as the Mars Surveyor '98 Orbiter, the Mars Climate Orbiter was a space **probe** sent to **Mars** by NASA to study Martian climate for an entire Martian year (approximately 687 Earth days). The probe launched in 1998 with two instruments that would provide detailed information about the interaction between the Martian atmosphere and the surface of the planet. However, due to an error in its computer software programming, the orbiter came too close to the planet and ultimately disintegrated in the upper atmosphere on September 23, 1999. See appendix B for a list of its instruments.

See also FAILURES; LOCKHEED MARTIN.

MARS EXPLORATION PROGRAM (MEP). Formed in 1993, the Mars Exploration Program is a longtime NASA effort to explore the climate, natural resources, and potential for life on **Mars**. There are four main goals of the MEP, which center on understanding the potential for life on Mars: (1) "Follow the Water," or determine if life ever arose on Mars, (2) characterize the climate of Mars, (3) characterize the geology of Mars, and (4) prepare for the human exploration of Mars. Previous missions have focused on the "Follow the Water" goal, including the **Mars Odyssey**, **Mars Exploration Rovers**, **Mars Reconnaissance Orbiter**, and the **Mars Phoenix Lander**. Currently, MEP is transitioning from the "Follow the Water" goal to a combination of characterizing the climate and geology of Mars with the **Mars Science Laboratory**'s ***Curiosity* rover**. Future missions will place more emphasis on seeking signs of life as well as preparing for human exploration of the planet.

MARS EXPLORATION ROVER MISSION. A current robotic space mission involving two **rovers**, MER-A (***Spirit***) and MER-B (***Opportunity***), that are exploring the Martian surface and geology as part of the **Mars Exploration Program**. The mission's scientific goal was to search for and characterize a wide range of rocks and soils that hold clues to past water activity on **Mars**. The MER-A and MER-B rovers were launched on June 10, 2003, and July 3, 2003, respectively, and landed on Mars in January 2004. Although the *Spirit* rover became stuck in sand in 2009 and unresponsive in 2010, the *Opportunity* rover is still carrying out its scientific mission. On January 24, 2014, NASA stated that current studies by the ***Curiosity*** and *Opportunity* rovers will now be searching for evidence of ancient life on Mars. The MER rovers were designed, built, and are being operated by the **Jet Propulsion Laboratory** for NASA. The cost of the initial 90-day mission was an estimated $820 million, but the duration has now extended 11 years, making it considerably more cost efficient. Although some of the equipment and instruments on *Opportunity* are slowing down due to age, the rover is still sending valuable data. See appendix B for a list of their instruments.

MARS EXPRESS. The first **European Space Agency** (ESA) mission to another planet, the spacecraft's goal was to study the geology, atmosphere, surface environment, and potential for life on **Mars**. The mission was comprised of two parts: the Mars Express Orbiter, which is still currently in operation taking scientific measurements and high-resolution imagery, and the Beagle 2, a lander from Great Britain that lost contact before it landed on the surface of the planet. Among the orbiter's accomplishments are providing subsurface measurements of Mars that allowed for the discovery of underground water-ice deposits as well as closer observations of Phobos, the larger of Mars's two **moons**. The mission has been **extended** several times; in December 2014 it was extended once again to continue operations until 2018. See appendix B for a list of its instruments.

MARS GLOBAL SURVEYOR (MGS). The Mars Global Surveyor was a NASA spacecraft sent to **Mars** to explore the planet's atmosphere and surface. MGS orbited Mars from the North to the South Pole and collected data on how seasons affect the Martian landscape. One of the more intriguing discoveries made by MGS was evidence that sources of liquid water are currently present at or near the surface of Mars in gullies and past delta flows. The orbiter arrived at Mars in September 1997 and lasted until November 2006, a much longer duration than expected due to innovative fuel efficiency engineering. See appendix B for a list of its instruments.

See also MARS EXPLORATION PROGRAM (MEP).

MARS OBSERVER. After the success of the ***Viking*** *1* and *2* missions, NASA decided to launch a spacecraft to **Mars** to study the planet's geology, magnetic field, and climate. The orbiter, which launched on September 25, 1992, was based on Earth-orbiting communications **satellites** and was scheduled to enter into Mars's orbit on August 24, 1993. However, contact with the spacecraft was lost three days prior to performing the orbital insertion maneuver and was not heard from again. Investigations later suggest that a fuel line ruptured during fuel tank pressurization, which would have caused the Mars Orbiter to spin uncontrollably. Although the Mars Orbiter did not complete any of its intended goals, it was able to provide information on the interplanetary voyage that would be crucial to future missions to Mars. See appendix B for a list of its instruments.

See also MARS EXPLORATION PROGRAM (MEP).

MARS ODYSSEY. Named in tribute to the film *2001: A Space Odyssey*, the 2001 Mars Odyssey is a spacecraft orbiting **Mars** searching for evidence of past/present water and volcanic activity on the planet. In order for the spacecraft to orbit Mars and reserve fuel, it used a technique called "aerobraking," which used the Martian atmosphere to naturally and gradually slow down its orbit. Additionally, the spacecraft acts as a relay for communications between several **rovers** on the planet, including the **Mars Exploration Rovers** ***Spirit*** and ***Opportunity***, the **Mars Science Laboratory**'s ***Curiosity*** rover, and the **Phoenix** lander. Launched on April 7, 2001, the 2001 Mars Odyssey **satellite** is currently operational and holds the record for the longest running spacecraft in orbit around another planet. See appendix B for a list of its instruments.

MARS PATHFINDER. Consisting of a lander, named the **Carl Sagan Memorial Station**, and a small wheeled robotic **rover**, ***Sojourner***, Mars Pathfinder is a NASA spacecraft that landed on **Mars** in 1997. It was designed as a proof-of-concept to show that a lander and free-ranging robotic rover could be sent to Mars efficiently and still be cost-effective. Both the lander and rover outlived their intended lifespans by several factors and returned thousands of photos of the surface. Additional scientific measurements made by Pathfinder, including chemical analyses of rocks and data on Martian climate, suggested that Mars once had a warm and wet past. See appendix B for a list of its instruments.

MARS POLAR LANDER. Also known as the Mars Surveyor '98 Lander, the Mars Polar Lander was a 640-lb. (290 kg) robotic spacecraft lander launched by NASA on January 3, 1999, to study the soil and climate of Mars's southern pole near Planum Australe as part of the Mars Surveyor '98 mission. On board the Mars Polar Lander spacecraft were two surface-penetrator **probes**, called **Deep Space 2**, which were designed to study Martian soil composition. On December 3, 1999, however, after the descent phase was expected to be complete, the lander failed to reestablish communication with Earth. It was determined that the most likely cause of the accident was premature termination of the engine firing before the lander touched down, resulting in it striking the surface at a high velocity. See appendix B for a list of its instruments.

MARS RECONNAISSANCE ORBITER (MRO). Launched in 2005, the Mars Reconnaissance Orbiter is a **Jet Propulsion Laboratory**/NASA spacecraft designed to search for evidence that water existed on **Mars** for a long period of time in accordance with the **Mars Exploration Program**'s goal to "Follow the Water." MRO is currently operational and is equipped with many scientific instruments designed to analyze the surface in great detail by taking high-resolution photographs and studying mineral composition as well as atmospheric conditions. Its telecommunications systems were designed to serve as a reliable link between Earth and future Mars spacecraft due to its energy efficiency and high performance. The orbiter carries the largest camera ever flown on a planetary mission, allowing the spacecraft to identify potential obstacles that could jeopardize current and future landers and **rovers**. See appendix B for a list of its instruments.

MARS SCIENCE LABORATORY (MSL). A NASA robotic space probe sent to **Mars** and managed by the **Jet Propulsion Laboratory**, MSL is notable for landing the largest **rover** sent to the planet, ***Curiosity***, on August 6, 2012. Because of *Curiosity*'s large size—standing seven feet tall and weighing almost one ton—MSL engineers developed a unique sky crane landing system in which the *Curiosity* rover was lowered to the surface by a rocket-powered descent stage. Its objectives include studying Mars's habitability by examining its climate and geology as well as collecting data for a future **human spaceflight** mission to Mars. The *Curiosity* rover is equipped with several sophisticated scientific instruments that allow it to drill rocks, scoop soil, and focus laser beams to directly analyze chemical compositions of various samples. The landing site, Gale Crater, was chosen because it provided an area with a wide variety of possible scientific objectives. Additionally, at the center of the crater lies Aeolis Mons, or Mount Sharp, which could provide data on several different periods in Mars's history; *Curiosity* reached the base of Mount Sharp in summer 2014. Each Earth year on the anniversary of its landing, *Curiosity* sings "Happy Birthday" to itself, programmed by scientists at **Goddard Space Flight Center**.

The Mars Science Laboratory Entry, Descent, and Landing Instrument (MEDLI) set was designed and developed by **Langley Research Center** and **Ames Research Center**, and consisted of seven integrated sensor plugs and seven pressure sensors placed on the heatshield of the spacecraft. MEDLI provides data to enable scientists to study environmental factors that might affect future landings on Mars. See appendix B for a list of MSL Aeroshell instruments.

MARS SCOUT PROGRAM. A NASA initiative with the goal of sending a series of small, inexpensive robotic missions to **Mars**, the Mars Scout Program selects from proposals made by the scientific community. The purpose of requiring that each mission be low-cost was to assess the feasibility of complementing major missions being planned as part of NASA's **Mars Exploration Program** or those under development by foreign space agencies. The **Phoenix** lander and **Mars Atmosphere and Volatile Evolution** orbiter were selected and developed prior to the program's retirement in 2010.

MARS SURVEYOR '98 LANDER. *See* MARS POLAR LANDER.

MARS SURVEYOR '98 PROGRAM. A NASA mission to study Martian weather, climate, and atmosphere, the program was comprised of two spacecraft launched separately: the **Mars Climate Orbiter** and the **Mars Polar Lander**. The objective of the two missions was to understand the reservoirs, behavior, and atmospheric role of volatiles in the Martian atmosphere and to search for evidence of long-term and periodic climate changes. Neither spacecraft successfully carried out its mission, with the Mars Climate Orbiter burning up in the Martian atmosphere and the Mars Polar Lander's engines malfunctioning during descent.

See also MARS ATMOSPHERE AND VOLATILE EVOLUTION (MAVEN).

MARSHALL SPACE FLIGHT CENTER (MSFC). Formally known as the George C. Marshall Space Flight Center, it is the U.S. government's civilian rocketry and spacecraft propulsion research center located in Huntsville, Alabama. Founded on July 1, 1960, MSFC has a long history of providing rocket systems that have powered U.S. missions into space. Initially, its focus during the 1960s was largely on developing the **Saturn** family of launch vehicles that were necessary for the **Apollo** program that ultimately took Americans to the **moon**. MSFC has since been the lead center for the propulsion behind the **space shuttle** as well as the expected **Space Launch System** and design and construction for several NASA programs, including **Spacelab**, the **Hubble Telescope**, and the **International Space Station**

(ISS). The center also contains the Huntsville Operations Support Center, which supports payload and experiment activities for the ISS monitors rocket launches from **Cape Canaveral** Air Force Station when an MSFC payload is on board, and is the lead center for advanced X-ray astronomy and microgravity research.

MCAULIFFE, CHRISTA (September 2, 1948–January 28, 1986). McAuliffe was an American teacher from Concord, New Hampshire, and one of the seven crew members killed in the **space shuttle *Challenger* tragedy**.

In 1985, she was selected from more than 11,000 applicants to participate in the NASA **Teacher in Space Project** and was scheduled to become the first teacher to travel into space. As a member of mission STS-51-L, she was planning to conduct experiments and teach two lessons from space shuttle ***Challenger***. On January 28, 1986, the shuttle broke apart shortly after launch. After her death, schools and scholarships were named in her honor and she was posthumously awarded the **Congressional Space Medal of Honor** in 2004. She was survived by her husband and two children.

MEDICINE IN SPACE. The term "space medicine" was first used by Hubertus Strughold, the first and only professor of space medicine at the School of Aviation Medicine at Randolph Air Force Base, Texas, in 1948. The objective of space medicine is to analyze how well and for how long people can survive the extreme conditions in space, and how fast they can adapt to the Earth's environment after returning from their mission. The risks associated with traveling in space are extensive, with factors such as zero gravity, radiation, close quarters, and artificial environments affecting **astronauts** both physically and mentally.

Some of the ways that astronauts use to lessen the risks associated with space travel include exercise, a well-balanced diet, and medication. Radiation exposure is currently the biggest threat to astronaut health and is countered by limiting the amount of time astronauts spend in space to an average of three months. To prevent muscle atrophy and weakening of the immune system, astronauts on the **International Space Station** work out about two hours a day, using treadmills, exercise bikes, and a device specially developed to allow astronauts to do resistive or strength training called an Interim Resistive Exercise Device. Medications, too, help with some problems: Midodrine, a pill that temporarily increases blood pressure, counters disorientation for returning astronauts, and several types of sedatives to counter sleep deprivation are available. In addition, the body in low gravity needs much less salt and more potassium due to fluid imbalances and bone-loss issues. Salt is a problematic astronaut contraband, especially because food seems to taste bad or worse in space due to a variety of factors. Astronauts grind up salt tablets before reentry for return to gravity. There is a possibility that some blurry vision problems are due to too much salt intake, but this is under investigation and there may be other causes still to be determined.

The field of medicine in space has developed significantly since its advent in the 1950s with great benefits to not only astronauts but to society in general. Advances in several areas of medical technology were a result of technologies developed by NASA such as MRI and CAT scans, which stemmed from NASA lunar imaging techniques, and the heart pacemaker, which has components that were developed for NASA **satellite** communication. Other spinoffs include foam cushioning, foldable walkers, and voice-controlled wheelchairs.

See also INVENTIONS.

MERCURY (PLANET). Mercury is the smallest and closest planet to the sun in the solar system. It is a hot, rocky planet that consists of approximately 70 percent metallic and 30 percent silicate material. Temperatures on Mercury can range from −173°C up to 427°C. It has no known satellites and maintains an orbital period of 88 days, the fastest of the eight planets.

The first spacecraft to visit Mercury was NASA's ***Mariner** 10* (1974–1975). *Mariner 10* provided the first close-up images of Mercury's surface, which immediately showed its heavily cratered nature, and revealed many other types of geological features. Due to the length of *Mariner 10*'s orbital period, the same face of the planet was lit at each of *Mariner 10*'s close approaches, resulting in only 45 percent of the planet's surface being mapped. A second **probe**, **MESSENGER**, was launched in 2004 and entered into an elliptical orbit around the planet in 2011. This mission successfully mapped most of the hemisphere not captured by *Mariner 10*.

The BepiColumbo **satellite**, a joint mission by the **European Space Agency** and the **Japan Aerospace Exploration Agency**, is set to launch in July 2016 toward Mercury, and will arrive in 2024 on a one-year mission, with a possible one-year extension. BepiColumbo will be the third spacecraft to visit Mercury.

MERCURY (PROGRAM). Mercury was the first **human spaceflight program** of the United States. It ran from 1959 through 1963 with the objective of putting a person in orbit around the Earth before the **Soviet Union**, as part of the early **space race**. It involved seven **astronauts** flying a total of six solo trips. On May 5, 1961, **Alan Shepard** became the first American in space in a suborbital flight after the Soviet Union had put **Yuri Gagarin** into orbit one month earlier. **John Glenn** became the first American to reach orbit on February 20, 1962. He was the third person to do so, after Soviet **Gherman Titov** made a daylong flight in August 1961.

The Mercury Project was popular worldwide and each of the manned flights was followed by millions on radio and TV around the world. Apart from the manned missions, Mercury had a total of 20 unmanned launches as a part of the development of the project. This also involved test animals, most famously the chimpanzees Ham and Enos. Mercury laid the groundwork for Project **Gemini** and the follow-on **Apollo moon**-landing program, which was announced a few weeks after the first manned flight. The astronauts were collectively known as the "**Mercury Seven**."

The total cost of the program was $384 million (or approximately $3 billion today). The capsule itself held only one astronaut, at most, and the astronaut had to remain in his seat for the duration of the flight.

See also ANIMALS IN SPACE.

MERCURY SEVEN. The group of seven **Mercury** astronauts selected by NASA on April 9, 1959. They are also referred to as the Original Seven or Astronaut Group 1. The seven men piloted the manned spaceflights of the Mercury program from May 1961 to May 1963. These seven original American **astronauts** were **Alan Shepard**, **Gus Grissom**, **John Glenn**, Scott Carpenter, **Wally Schirra**, Gordon Cooper, and Deke Slayton. Members of the group flew on all classes of NASA manned orbital spacecraft of the 20th century—Mercury, **Gemini**, **Apollo**, and the **space shuttle**. Of the seven, only John Glenn, who was the oldest, is still living; Grissom died in 1967, in the ***Apollo 1*** fire. The others all survived past retirement from service. All of their capsules carried the number 7: *Freedom 7*, *Friendship 7*, *Liberty Bell 7*, and *Fail 7* in honor of their original group.

MESSENGER SPACECRAFT. MERcury, Surface, Space ENvironment, GEochemistry, and Ranging (MESSENGER) is a robotic NASA spacecraft orbiting the planet **Mercury**, the first spacecraft ever to do so. The 1,069-pound (485 kg) spacecraft was launched on August 3, 2004, to study Mercury's chemical composition, geology, and magnetic field. MESSENGER became the second mission after 1975's ***Mariner 10*** to reach Mercury when it made several flybys from 2008 to 2009 and eventually entered orbit around Mercury on March 18, 2011. Among the many scientific instruments that the spacecraft carried was a camera with which it was able to return the first photo from Mercury orbit on March 29, 2011. It operated under its second **extended mission** until summer 2015, when it exhausted its fuel supply and was deliberately crashed onto the surface on April 30, 2015. See appendix B for a list of its instruments.

MICHOUD ASSEMBLY FACILITY (MAF). MAF is a site owned by NASA in New Orleans, Louisiana, and part of NASA's **Marshall Space Flight Center**. MAF is one of the largest manufacturing plants in the world with 43 environmentally controlled acres under one roof. From September 1961 to the end of the **Apollo** program in December 1972, the site was utilized by Chrysler Corp. to build the first stages of the **Saturn** I and Saturn IB, later joined by **Boeing** to build the first stage of the Saturn Vs. From 1973 to 2010, it was used for the construction of the **space shuttle**'s external fuel tanks. The **Vertical Assembly Center**, the world's largest spacecraft welding tool, is located at MAF. Approximately 2,500 individuals work at one of the facilities at Michoud, down from a peak of 7,500 during the 1990s.

MID-ATLANTIC REGIONAL SPACEPORT (MARS). Located at the southern end of **Wallops Flight Facility**, MARS is a commercial spaceport facility used by private contractors such as **Orbital ATK** for space launches. Previously known as the Virginia Space Flight Center, it began current launch operations from its two launch pads in December 2006, with the launch of an **Orbital Sciences Minotaur** rocket carrying cargo for the **U.S. Air Force** and for NASA. A failed launch of an Orbital ATK's **Antares** rocket on October 28, 2014, significantly damaged the spaceport; repairs were completed in September 2015, and the next launch will probably be in 2016.

MINIATURE RADIO FREQUENCY (MINI-RF). The Miniature Radio-Frequency instrument (Mini-RF) is a synthetic aperture radar (SAR) instrument on board the **Lunar Reconnaissance Orbiter**, currently in orbit around the **moon**. Because previous SAR instruments had been bulky, power hungry, and costly, the Mini-RF instrument was built to be cheaper, lighter weight, and to use fewer resources. The Mini-RF was included in the LRO because it can distinguish water-ice from other forms of water and can identify geologic deposits from its remote, orbiting position. In January 2011, after completion of Mini-RF's primary mission objectives, NASA announced that the transmitter had suffered a critical failure, although it sporadically continues to provide data. See appendix B for some of its instruments.

MINITRACK. The first U.S. **satellite** tracking network made operational in 1957. It was implemented as a string of stations along the 75th meridian and was used to track the flights of early space effort satellites such as ***Sputnik***, ***Vanguard***, and ***Explorer***. In 1961, due to tracking issues as satellites grew more sophisticated, Minitrack was replaced by the **Spacecraft Tracking and Data Acquisition Network**.

MINOTAUR. Built by **Orbital Sciences**, the Minotaur family of solid fuel rockets is based on converted Minuteman (Minotaur 1 and 2) and Peacekeeper (Minotaur 3, 4, and 5) intercontinental ballistic missiles. NASA's **Lunar Atmospheric and Dust Environment Explorer** was launched from a Minotaur 5 in September 2013 from **Wallops Island**.

MIR SPACE STATION. A space station that operated in low Earth orbit from 1986 to 2001, owned by the **Soviet Union** and later by **Russia**, Mir was the first modular space station and the first continuously inhabited long-term research station in orbit. It was assembled in orbit from 1986 to 1996 and was occupied for a total of 12-1/2 years out of its 15-year lifespan. Until March 21, 2001, it was the largest **satellite** in orbit until its orbit decayed, after which it was succeeded by the **International Space Station**. The station served as a microgravity research laboratory and had the capacity to support a crew of three. Experiments in biology, astronomy, physics, and spacecraft systems were conducted with the aim of providing research and technology for the permanent occupation of space.

MIR-2 STATION. A Soviet-era manned space station project that began in February 1976, the Mir-2 station was centered on a core module, the DOS-8, and was designed to be a replacement for the **Mir Space Station** that would begin operation in the 1990s. With the collapse of the **Soviet Union** and the economic chaos that ensued, the station was never completed. However, some of the modules built for Mir-2 have since been incorporated into the **International Space Station**.

MISSION CONTROL CENTER FOR THE 21ST CENTURY (MCC-21). *See* CHRISTOPHER C. KRAFT, JR. MISSION CONTROL CENTER.

MISSION: SCIENCE. Mission: Science is provided by NASA's **Science Mission Directorate** to engage students in the science, technology, engineering, and mathematical fields and to showcase educational resources created by NASA and its partners. The program features a website for elementary students through high school age with engaging multimedia, educational information, and various games and activities geared toward teaching scientific concepts (see http://missionscience.nasa.gov/).

MISSION TO PLANET EARTH (MTPE). MTPE was NASA's study of the Earth as a single unit of the competing systems, pressures, and changes, both natural and human induced, that affect our environment. By studying our globe from space, without regard to boundaries and divisions, NASA obtained comprehensive data about the interactions between and among our land, atmosphere, ice, oceans, and biota. It was incorporated into the Earth Systems Science Directorate, which also includes such programs as **Fire and Smoke**, **Earth Observing** system, and **Orbiting Carbon Observatory**.

See also SCIENCE MISSION DIRECTORATE (SMD).

MOON. (a) A solid, distinct, and natural object that orbits a more massive object.

(b) The Earth's only natural **satellite** and the most heavily explored object in the solar system. It is in a synchronous orbit around the Earth, always showing the same face, with a semi-major axis of 384,399 km. Although it is a quarter of the diameter of Earth, it is not the largest moon in the solar system; however, it is the largest moon relative to the object around which it orbits. The moon makes one complete orbit around the Earth every 27.3 days, but since the Earth is moving at the same time around the sun, it takes 29.5 days for the moon to go through one complete phase.

Although humankind has been observing the moon for millennia, actual exploration did not commence until the **space race** during the Cold War. Spacecraft from the **Soviet Union**'s Luna program were the first to accomplish a number of goals, including the first manufactured object to escape Earth's gravity and pass close to the moon (*Luna 1*), the first to impact the lunar surface (*Luna 2*), and the first object to take photographs of the far side of the moon (*Luna 3*), all of which occurred in 1959. Additionally, in 1966 the Soviet Union launched the first spacecraft to successfully make a lunar soft landing, *Luna 9*, followed by the first unmanned vehicle to orbit the moon, *Luna 10*. Throughout the 1970s, the Soviet Union also launched several missions that included lunar sample return missions as well as operating two robotic rovers on the surface of the moon in 1970 and 1973 as part of the Lunokhod program.

During the initial years of the space race, NASA essentially played a game of catch-up with the Soviet efforts of space exploration. The **Surveyor** program landed its first spacecraft on the lunar surface four months after the Soviets in 1966. NASA's manned **Apollo** program was developed alongside other space exploration efforts focusing on launching satellites and **probes**; however, after a series of unmanned and manned tests of the Apollo spacecraft in Earth orbit and prompted by a potential Soviet lunar flight, NASA launched *Apollo 8* in 1968 and made the first crewed mission to lunar orbit. The *Apollo 11* landing of the first humans on the moon in 1969 is seen as the height of the space race. Currently, the last human to stand on the moon was Eugene Cernan in December 1972 as part of the *Apollo 17* mission. The Apollo missions 11 through 17, with the exception of *Apollo 13* (which aborted its planned lunar landing), returned a total of 842 lbs. (382 kg) of lunar rock and soil.

Recent exploration of the moon by NASA has focused on examining the overall topography and composition of the lunar surface. In 1994, the United States sent the **Clementine** spacecraft to lunar orbit, which obtained the first near-global topographic map of the moon. Following this mission, in 1998 the Lunar Prospector mission indicated the presence of excess hydrogen at the lunar poles, likely caused by the presence of water-ice in the upper few meters of the **regolith**. In 2009, the **Lunar Reconnaissance Orbiter** and the **Lunar Crater Observation and Sensing Satellite** impactor were launched to obtain high-resolution imagery and study the effects of impact craters on the lunar surface, respectively. In recent years, NASA has launched the **Gravity Recovery and Interior Laboratory** spacecraft in 2012 to study the interior of the moon as well as the **Lunar Atmosphere and Dust Environment Explorer** probe to study the lunar exosphere in 2013.

Research been conducted by several other organizations and countries from around the world, including Japan, the European Union, **China**, and **India**. China, in particular, has pursued an ambitious lunar exploration program starting in 2007 with its first lunar orbiter, *Chang'e 1*. More recently, China launched the *Chang'e 3* spacecraft in December 2013, which made a soft lunar landing and deployed the country's first lunar rover, *Yutu*. Several countries have stated that they have plans to send manned missions to the moon in the future, with the **Japanese Aerospace Exploration Agency** proposing a 2020 manned lunar landing date.

MOON MINERALOGY MAPPER (M^3). M^3 is one of two instruments that NASA contributed to **India**'s first mission to the **moon**, *Chandrayaan-1*, launched October 22, 2008. M^3 was an imaging spectrometer that provided the first high-resolution spatial and spectral map of the entire lunar surface. *Chandrayaan-1* ended on August 29, 2009, due to several technical issues after operating for 312 days as opposed to the intended two years, but the mission achieved 95 percent of its planned objectives, including mapping over 95 percent of the lunar surface with the M^3 instrument. Refer to appendix B for its instruments.

MOON ROCK. "Moon rocks" refer to any rock that is found on the Earth's **moon** or to lunar material collected during the course of human exploration of the moon. Currently, there are three sources of moon rocks on Earth: (1) rocks that were ejected naturally from the lunar surface by cratering events and subsequently fell to Earth as lunar meteorites, (2) those collected by **Apollo** missions, and (3) samples returned by the **Soviet Union**'s Luna missions. Three new minerals were found in the returned moon rocks: armalcolite (named for the **astronauts** on the *Apollo 11* mission—**Armstrong**, **Al-**

drin, and **Collins**), tranquilityite (named for the Sea of Tranquility where it was found), and pyroxferroite ("fire stranger," thought to be from volcanic residue).

In 2011, an internal audit revealed that NASA had loaned out many moon rocks that had subsequently been lost to tracking and are considered lost. Because actual moon rocks are rare, they have been subject to theft and vandalism. Others had legitimately been gifted to politicians and state figures in honor of their accomplishments or given to scientists and laboratories for research purposes.

MORNING CONSTELLATION. A set of five **satellites** (**Landsat**-5, Landsat-7, **Terra**, **Earth Observing-1** (EO-1), and the Satelite de Aplicanciones Cientificas-C (SAC-C) flying in tandem around the Earth. The launching of satellites into tandem orbits, or **constellations**, began with the launch of the U.S. Geological Survey's Landsat-7 and NASA's Earth Observing System (EOS) Terra satellites in 1999. Landsat-7 and Terra were joined the following year by the Earth Observing-1 (EO-1) satellite from NASA and the SAC-C satellite from the Argentine Commission on Space Activities, resulting in a full-fledged orbiting constellation. Landsat-5, launched in 1984, also flies in the constellation at 438 miles (705 km) above Earth's surface and is phased eight days from Landsat-7. These five satellites cross the equator within minutes of each other at the mean local time (MLT) range of 10 to 10:30 a.m., hence the name "Morning Constellation."

See also A-TRAIN CONSTELLATION.

N

NAMING OF SPACECRAFT. NASA originally allowed each of its spacecrafts' crews the opportunity to name their spacecraft to go along with their mission number, so, for example, **Mercury**-Redstone 4 was named by **Gus Grissom** and his crew the *Liberty Bell 7*. However, when it came time for Grissom to name his spacecraft for the ***Gemini*** *3* flight, which would launch March 23, 1965, he named it *Molly Brown* in reference to the popular *Unsinkable Molly Brown* Broadway musical as well as to his earlier near-escape when the landing module on his first flight flooded and sank. NASA rejected the name, but when Grissom's next selection was *Titanic*, they quickly agreed to *Molly Brown* but never used it in public. After Grissom's safe return, NASA did not give individual names to its spacecraft until the **Apollo** program when they needed a way to distinguish between the **command module** and the **lunar module**, but they retained the right to veto suggestions.

NANORACKS. NanoRacks is a private company founded in 2008 by Jeffrey Manber and Michael Johnson that provides standard rack-mounted laboratory facilities and microgravity space access to commercial customers at the **International Space Station**.

NASA AUTHORIZATION ACTS. A series of laws passed every few years since the establishment of NASA in 1958 that authorize appropriations for the organization's budget. Each is passed by Congress and signed into law by the sitting president of the United States of America.

The most recent NASA Authorization Act of 2015 confirms NASA's goal for a human spaceflight program to Mars and requires the creation of a road map to achieve such a mission. According to the **House Committee on Science, Space, and Technology**, its goals are:

- Realizing the research potential of the **International Space Station** (ISS) with an Office of Science & Technology Policy-led strategic plan for all science agencies to conduct research on the Station. NASA will study the cost and feasibility of continuing its operational lifespan beyond 2020.
- Continued commitment to develop the **Space Launch System** and **Orion Crew Vehicle** and reiteration of Congressional direction that Orion serve as a backup system to support the ISS if necessary.
- Assist in building at least one **Commercial Crew** system (with NASA funds) to carry American astronauts on American rockets safely, reliably, and affordably to and from the ISS so that we are no longer reliant on **Russia** for crew access. . . .
- Maintains funding to support a launch date goal of the **James Webb Space Telescope** by 2018.
- Continues survey for potentially-hazardous Earth-crossing objects [especially with regard to **near-Earth objects**].
- Continues exciting search for planets around other stars and life on other worlds.

In addition, the House Committee directs NASA to keep Congress informed on its **Europa mission**, the **Mars 2020** rover, the maintenance of its **aeronautics research** program, the state of its infrastructure, the continuance of its **education and public outreach** program, and oversight against waste, fraud, abuse, and mismanagement.

NASA AWARDS AND MEDALS. NASA awards a variety of achievement and recognition medals. Ranked from the most prestigious, these are: **Distinguished Service Medal**, Distinguished Public Service Medal, Outstanding Leadership Medal, Outstanding Public Leadership Medal, **Exceptional Service Medal**, Exceptional Public Service Medal, Exceptional Engineering Achievement Medal, Exceptional Scientific Achievement Medal, Exceptional Technology Achievement Medal, Exceptional Achievement Medal, Exceptional Public Achievement Medal, Early Career Achievement Medal, Equal Employment Opportunity Medal, Exceptional Administrative Achievement Medal, Group Achievement Award, **Space Flight Medal**, Exceptional Bravery Medal, and Silver Achievement Medal. They are presented at an annual awards ceremony by NASA's highest officials.

NASA ENGINEERING DIRECTORATES. These sections of NASA at each **field center** provide the full spectrum of mission engineering support, including concept exploration, design, development and manufacturing, testing, verification, certification, flight/mission sustainment, and disposal. The

Engineering Directorate at **Glenn Research Center** is typical and is comprised of four divisions: Mechanical and Fluid Systems Division, Manufacturing Division, Avionics and Electrical System Division, and the Systems Engineering and Analysis Division.

NASA EXTREME ENVIRONMENT MISSION OPERATIONS (NEEMO). NEEMO is a NASA program for studying human survival in an underwater laboratory, the Aquarius, located 62 feet below the surface of the Atlantic Ocean near Key Largo, Florida, in preparation for future space exploration. NASA has used the program since 2001 for a series of missions, usually lasting 10 to 14 days, with research conducted by **astronauts** and other NASA employees. The crew members are called aquanauts instead of "divers," and they perform **extravehicular activities** in the underwater environment.

NEEMO 20, concluded in July 2015, was commanded by **European Space Agency** (ESA) astronaut Luca Parmitano, a veteran of **International Space Station** Expeditions 36 and 37. Aquanauts during the most recent mission evaluated tools and techniques proposed for use on future spacewalks and experienced a variety of gravity levels to help adjust to life on **Mars** or on asteroids. In addition, they experimented with ESA hardware that allowed them to read data from a tablet, smartphone, and head-mounted device without taking their hands or eyes away from the task at hand.

NASA HEADQUARTERS. Also known as Two Independence Square, NASA Headquarters houses NASA leadership who provide the oversight and direction to the space agency, under the direction of the NASA administrator. To effectively carry out NASA's mission, NASA Headquarters is organized into four Mission Directorates: **Aeronautics**, **Exploration Systems**, **Science**, and **Space Operations**. Currently, it is located at 300 E Street SW in Washington, D.C.

Among many other administrative functions, the headquarters building also includes a newsroom where public relations and media are concentrated, an information center covering much of NASA's history, a library that includes Freedom of Information Act access, and an exchange store that stocks items for both NASA employees and for tourists, making a small profit.

NASA INNOVATIVE ADVANCED CONCEPTS (NIAC). NIAC is focused on identifying and promoting new trends and concepts to encourage the development and advancement of technology in the United States, particularly in universities and research institutes. Established in 1998 under the name of the NASA Institute for Advanced Concepts (and under the management of the Universities Space Research Association on behalf of NASA), the group funded proposals for visionary technological concepts such as the space elevator, a magnetic sail, and bio-nano machines. Due to **budget** constraints, the program was terminated in 2007. A review in 2009 suggested that the program (or one like it) be reestablished, so in 2011 the program, with its name changed to the NASA Innovative Advanced Concepts, began soliciting proposals for revolutionary projects that would contribute to NASA's strategic goals with regard to national space or aeronautics programs. The project must represent something entirely new (or a great leap in capability), be unexplored, and be technically sound. In Phase I, the awards are nine-month funding projects that explore new concepts and proposals; Phase II awards continue the best of Phase I for two years. Phase II, announced in August 2014, includes five broad areas: (1) the concept for a 10-meter suborbital large balloon reflector that might be used as a telescope inside a high-altitude balloon; (2) a spacecraft-rover hybrid concept for the exploration of small solar-system bodies (asteroids or **moons**); (3) a concept for deep mapping of small solar system bodies using subatomic particles to map the interior and small surface features; (4) the concept for a low-mass planar photonic imaging sensor, an innovative sensor and spectrometer design to replace traditional, bulkier telescopes; and (5) a granular media imager concept ("Orbiting Rainbows") that would use an orbiting cloud of dust-like matter to image distant astronomical objects at very high resolution. These projects were identified through a peer-review process and are more than 10 years away from use.

In 2015, Phase II awards were for Swarm Flyby Gravimetry, 3D photocatalytic air processor for dramatic reduction of life-support mass and complexity, PERISCOPE (perilapsis subsurface cave optical explorer), Titan submarine (exploring the depths of Kraken Mare), SCEPS in space (nonradioisotope power systems for sunless solar system exploration missions), Transformers for lunar extreme environments (ensuring long-term operations in regions of darkness and low temperatures), and HERTS (heliopause electrostatic rapid transit system).

The eight FY 2015 Phase II awards (up to $500,000 over two years) include potential methods for astronauts to travel in space by lowering their core temperatures, developing faster interstellar spaceflight, an alternative method to decelerate probes nearing a planetary surface, cryogenic fuel storage, and four other programs.

NASA-ISRO SYNTHETIC APERTURE RADAR (NISAR). NISAR, targeted to launch in 2019–2020, is a joint project between NASA and the **Indian Space Research Organization** to observe Earth, making global measurements of the causes and consequences of changes to the surface of our planet, such as ice sheet collapse. It will be the first **satellite** mission to use two different radar frequencies—L-band and S-band—resulting in greater accuracy (to the centimeter) in measurements.

NASA POSTDOCTORAL PROGRAM (NPP). The NASA Postdoctoral Program consists of two components: the NASA Postdoctoral Research Program and the NASA Postdoctoral Management Program. The NASA Postdoctoral Management Program is an adjunct to the NASA Postdoctoral Research Program and is designed to provide valuable research management experience for some of the research program Fellows. It provides PhDs the opportunity to participate and gain firsthand experience with many of NASA's programs.

NASA SHARED SERVICES CENTER (NSSC). *See* SHARED SERVICES CENTER (NSSC).

NASA TV. The official television service for NASA that is broadcast by **satellite** with a simulcast over the Internet. Formally created in the early 1980s to provide NASA managers and engineers with real-time video of missions, it now carries a variety of regularly scheduled educational and public relations programming 24 hours a day on its public channel, including live coverage of launches, spacewalks, and other mission events. NASA also maintains an Education Channel, a Media Channel, and an internal-eyes-only Space Operations Channel. It has received two Emmy awards. NASA TV is managed by the NASA newsroom at **NASA Headquarters**.

NATIONAL ADVISORY COMMITTEE FOR AERONAUTICS (NACA). NACA was a U.S. federal agency founded on March 3, 1915, to undertake, promote, and institutionalize **aeronautical research**. Its first testing and research facility, **Langley Memorial Aeronautical Laboratory**, was founded in 1920, followed 20 years later by **Ames Aeronautical Laboratory** and by the Cleveland laboratory (later **Lewis Flight Propulsion Laboratory**). The first supersonic flight, conducted by **Chuck Yeager** in the experimental X-1, was monitored by NACA. Looking ahead to the development of manned spaceflight, NACA was working on heat shields when, on October 1, 1958, the agency was dissolved, and its assets and personnel transferred to the newly created National Aeronautics and Space Administration. **Hugh Dryden**, NACA's director, became the first deputy administrator of NASA, with **T. Keith Glennan**, a member of the **Atomic Energy Commission**, as its administrator.

NATIONAL AERONAUTICS AND SPACE ACT OF 1958. The National Aeronautics and Space Act of 1958 is the U.S. federal statute that created the National Aeronautics and Space Administration. The act, which followed soon after the **Soviet Union**'s launch of ***Sputnik*** *1*, was signed by President **Dwight Eisenhower** on July 29, 1958. Before enactment, the responsibility for space exploration was seen as belonging to the military, similar to the **Soviet** model that had launched the first orbital **satellite**. However, the U.S. military was seen as incapable of performing both its military duties and the "space race" performance that the executive and legislative bodies required, and so a civilian agency was created.

The act has been amended over the years, most recently in 2012.

NATIONAL AERONAUTICS AND SPACE COUNCIL (NASC). The NASC was a space science advisory council established by the **National Aeronautics and Space Act of 1958**. It reported to and was chaired by the president of the United States, although President Dwight D. Eisenhower did not often call upon it. Other members included the secretaries of state and defense, the NASA administrator, the chairman of the **Atomic Energy Commission**, and other presidential appointees.

At the end of his last year in office, Eisenhower recommended that the NASC be abolished. Just before assuming office, President-elect **John F. Kennedy** announced that he wanted his vice president, **Lyndon B. Johnson**, to become chairman of the NASC, requiring an amendment of the National Aeronautics and Space Act. After Johnson became president, the NASC shrank in prominence as Johnson gradually chose to establish special interagency groups to address specific issues, or to bypass the NASC altogether.

President Richard Nixon did not eliminate the NASC upon taking office in 1969; however, he chose to circumvent it when he established a task group to evaluate the future of American civilian space policy. After Nixon's reelection in 1972, he suggested the elimination of the NASC as part of a reorganization of the Executive Office of the President. Finally, in 1973, the NASC was disbanded.

A modified version of the NASC, the **National Space Council**, was created in 1989 by President George H. W. Bush. It existed until 1993.

NATIONAL AERO-SPACE PLANE (NASP). Part of a U.S. project to create a single-stage-to-orbit spacecraft and passenger spaceliner, the Rockwell **X-30**, NASP was cancelled in the early 1990s before a prototype was finished, although a great deal of development work in advanced materials and aerospace design was completed. One of the goals of NASP was for it to be a passenger liner capable of short-duration international trips (e.g., two hours from Washington to Tokyo) while the X-30 was planned for a crew of two and geared toward testing. In 1990, **Rockwell International** started development on the craft to deal with the technical and budgetary obstacles. Development on the X-30, as it was then designated, continued until 1993, when it was terminated amid **budget** cuts and technical concerns.

NATIONAL OCEANIC AND ATMOSPHERIC ADMINISTRATION (NOAA). A scientific agency within the U.S. Department of Commerce formed on October 3, 1970, NOAA focuses on the conditions of the oceans and the atmosphere. NOAA operates several atmospheric and weather monitoring **satellites** that provide NASA with satellite imagery and **space weather** forecasting data. NOAA's **Geostationary Operational Environmental Satellites** program, which is managed by NASA's **Goddard Space Flight Center**, and **NOAA-19**, the last of NOAA's **Polar-Orbiting Operational Environmental Satellites**, both provide images of trends in atmospheric conditions. Additionally, NOAA owns the Aquarius underwater laboratory that is used during **NASA Extreme Environment Mission Operations** for space simulations.

NATIONAL RECONNAISSANCE OFFICE (NRO). The NRO is an intelligence agency of the U.S. government under the **Department of Defense** tasked with designing, building, and operating spy **satellites** while also providing satellite intelligence to several government agencies. The NRO was established on August 25, 1960, under the recommendation of President **Dwight Eisenhower** to act as the coordinating agency for the reconnaissance activities of the **U.S. Air Force** and the CIA. The NRO's first photo reconnaissance satellite program was the **Corona Satellite Project**, which would yield the first image taken from space.

In 2012, the NRO donated two used space telescopes—originally used as reconnaissance satellites—to NASA, each of which was a significant improvement over the aging **Hubble Space Telescope**. The spacecraft were provided at no cost to NASA, but NASA must maintain the operating costs. Disposition of the two telescopes is still uncertain, but it is likely that they will be used as part of the **Wide Field Infrared Survey Telescope** program.

NATIONAL SCIENCE FOUNDATION (NSF). A U.S. government agency established by the National Science Foundation Act of 1950, the NSF supports fundamental research and education in all the nonmedical fields of science and engineering. The NSF's astronomy program maintains a close working relationship with NASA in that the NSF provides virtually all the U.S. federal support for ground-based astronomy, while NASA's responsibility is the U.S. effort in space-based astronomy.

NATIONAL SCIENTIFIC BALLOON FACILITY. *See* COLUMBIA SCIENTIFIC BALLOON FACILITY.

NATIONAL SPACE COUNCIL (NSC). A council within the Executive Office of the President of the United States charged with re-examining the U.S. space program, it existed from 1989 to 1993 during the administration of George H. W. Bush. It was a modified version of the earlier **National Aeronautics and Space Council** that was disbanded in 1973 under President Nixon. The Space Council had four main missions: establish broad goals and objectives for the U.S. space program, establish strategies to implement these goals and objectives through an integrated nationwide set of activities, monitor the implementation of these strategies, and resolve specific program or policy issues arising from ambiguities or disagreements in implementing the strategies. In 1993, it was disbanded due to friction with NASA and its functions were absorbed by the National Science and Technology Council.

NAVAL RESEARCH LABORATORY (NRL). NRL is the corporate research laboratory for the **U.S. Navy** and the U.S. Marine Corps. It conducts a broad range of scientific research specifically focused on naval operations, including applied research, technology and advanced development, particularly research in ocean and atmospheric sciences and remote oceanic and atmospheric sensing. The NRL opened in 1923 and is one of the first U.S. government scientific research and development laboratories. It is currently operating under the Office of Naval Research. The laboratory's involvement in the U.S. space program includes development of the first **satellite** tracking system, **Minitrack**, overseeing Project **Vanguard**, and the navy's **Global Position System**. NRL consolidated with the Naval Oceanographic and Atmospheric Research Laboratory in the 1990s.

NEAR EARTH ASTEROID RENDEZVOUS-SHOEMAKER (NEAR SHOEMAKER). A robotic space **probe** designed by **Applied Physics Laboratory** for NASA to study the near-Earth asteroid Eros from close orbit over a period of one year, NEAR Shoemaker was the first launch of the **Discovery** program. It succeeded in approaching Eros and orbited it several times over more than 13 months before ultimately touching down on the asteroid on February 12, 2001. NEAR Shoemaker's primary goal was to return data on the compositional properties and magnetic field of Eros, but to scientists' surprise, NEAR Shoemaker continued to broadcast after soft-landing the asteroid until February 28, 2001. An attempt was made to communicate with the spacecraft in December 2002, but given the estimated −279°F temperature, the attempt did not succeed. Other objectives included studies of **regolith** properties, interactions with the **solar wind**, possible ongoing activity, and the asteroid's spin. It was named after planetary scientist Eugene Shoemaker. See appendix B for a list of its instruments.

NEAR EARTH NETWORK (NEN). NEN, located at **Goddard Space Flight Center**, is a series of orbital communications support for near-Earth orbiting customer platforms throughout the world via various NASA ground stations. Formerly known as Ground Network (GN), it consists of the Merritt Island Launch Annex (MILA) and Ponce de Leon Ground Station (PDL) in Florida, **Wallops Island** in Virginia, and McMurdo Station in Antarctica.

NEAR-EARTH OBJECTS (NEOs). NEOs are solar system objects whose orbits bring them within 1.3 **astronomical units** from Earth (or 1.3 times greater than the distance to the sun). Congress has given NASA a mandate to catalog all NEOs that are at least 1 kilometer wide, because such an impact would be disastrous and this was confirmed in the recent **NASA Authorization Act** of 2016. As of June 2015, there have been 872 near-Earth asteroids larger than 1 km discovered, of which 153 are potentially hazardous asteroids. **NEOWISE** (2011) estimated that 93 percent of large NEOs (greater than 1 km/0.6 mi) have been found. Cooperatively, the United States, the European Union, and other countries are participating in Spaceguard to scan for hazards. As of June 2015, almost 13,000 NEOs have been discovered.

Asteroids between 1 and 50 meters generally do not warrant attention, and JPL's database lists approximately 2,360 of these. Near-Earth asteroids smaller than 1 meter are often referred to as meteoroids.

See also POTENTIALLY HAZARDOUS OBJECT (PHO).

NEIL A. ARMSTRONG FLIGHT RESEARCH CENTER. *See* ARMSTRONG FLIGHT RESEARCH CENTER.

NEOWISE. The four-month mission extension to the **Wide-field Infrared Survey Explorer**, NEOWISE searched for small solar system bodies close to Earth's orbit. During this time, it discovered many new objects in the solar system, including 20 comets. The spacecraft was placed into hibernation in 2011 and was reactivated in 2013 for three years to assist NASA with identifying the population of potentially hazardous **near-Earth objects**. See appendix B for a list of its instruments.

NEW FRONTIERS PROGRAM. New Frontiers is a series of space exploration missions conducted by NASA with the purpose of studying several planets, including **Venus**, **Jupiter**, and the dwarf planet **Pluto**. New Frontiers was built on the innovative approach used by the **Discovery** and **Ex-**

plorer programs of principal investigator-led missions. There are currently two New Frontiers missions in progress: **New Horizons**, which launched on January 19, 2006, and **Juno**, which launched on August 5, 2011; a third New Frontiers mission, OSIRIS-REx, has been selected for launch in September 2016.

NEW HORIZONS. A NASA space **probe** launched to study the dwarf planet **Pluto**, its natural **satellites**, and one or two Kuiper Belt objects, depending on which are in position to be explored. Part of the **New Frontiers Program**, the mission was approved in 2001 following the **cancellation** of the **Pluto Fast Flyby** and **Pluto Kuiper Express**. The probe was launched on January 19, 2005, and visited Pluto on July 14, 2015. Data collection by New Horizons instrumentation was successful and is undergoing a process of download that will take approximately 16 months after the time of flyby to complete. Early results indicate a geologically interesting world; see entry for Pluto for preliminary results. After the flyby of Pluto by New Horizons, it is continuing farther into the Kuiper Belt to observe objects that are situated close to its flight path and within the constraints of its remaining hydrazine fuel. Currently, New Horizons is scheduled to visit 2014 MU_{69}, a 30–45-km sized object that orbits the sun at 43.4 AU. See appendix B for a list of New Horizons's instruments.

NEW MILLENNIUM PROGRAM (NMP). NMP was a series of **satellites** geared toward the testing of advanced technology in space flight, and included **Deep Space 1**, **Deep Space 2**, **Earth Observing-1**, **Space Technology 5**, and Space Technology 6. The program was terminated due to **budget** cuts in 2009. Some **cancelled** programs, such as Deep Space 4, which was scheduled to launch in 2003 and to orbit and land on comet Tempel 1, have been resurrected under different names.

See also DEEP IMPACT; DEEP SPACE NETWORK (DSN).

NIGHT SKY NETWORK. An educational effort sponsored by NASA to help educate the public through amateur astronomy clubs across the United States, Night Sky Network is managed by **JPL**. Members host assorted activities, maintain calendars of meteor showers and significant astronomical events, and encourage the public's participation.

NOAA. *See* NATIONAL OCEANIC AND ATMOSPHERIC ADMINISTRATION (NOAA).

NOAA-19. NOAA-19 is the last of the United States' **National Oceanic and Atmospheric Administration**'s **Polar-Orbiting Operational Environmental Satellites** (POES) series of **weather satellites** that launched on February 6, 2009. Designated NOAA-N′ or NOAA-N Prime prior to launch, its instruments provide data for weather and climate predictions. On September 6, 2006, the **satellite** was badly damaged at the **Lockheed Martin** Space Systems factory when it fell over after a team of workers failed to notice it was not secured properly, costing an additional $135 million in repairs. See appendix B for a list of its instruments.

See also FAILURES.

NOAA-N PRIME. *See* NOAA-19.

NORTH AMERICAN AVIATION. North American Aviation was a major American aerospace manufacturer founded in 1928 and responsible for a number of historic aircraft, including the **X-15** rocket plane, and the **XB-70**, as well as Apollo **command and service module**, the second stage of the **Saturn** V rocket, and the **space shuttle** orbiter. Through a series of mergers and sales, North American Aviation became part of **Rockwell International** and is now part of **Boeing**.

NORTHROP-GRUMMAN. Northrop-Grumman is an American global aerospace and defense technology company formed by the 1994 purchase of Grumman Aerospace Corporation by Northrop Corporation, an aircraft manufacturer. Both companies were well established in the airplane manufacturing industry prior to their merger. Grumman was the chief contractor on the Apollo **Lunar Module** that landed men on the **moon**. They received the contract on November 7, 1962, and built 13 lunar modules. As the **Apollo** program came to a close, Grumman was one of the leading contenders for the contract to design and build the **space shuttle**, but lost to **Rockwell International**. The company ended up involved in the shuttle program regardless, as a subcontractor to Rockwell, providing the wings and vertical stabilizer sections. Northrop-Grumman's recent involvement in the space industry includes the **James Webb Space Telescope** and the **XS-1** space plane.

NUCLEAR SPECTROSCOPIC TELESCOPE ARRAY (NuSTAR). NuSTAR is a space-based X-ray telescope that uses focused high-energy X-rays from astrophysical sources. It is the 11th mission of the NASA **Small Explorer satellite** program and the first space-based direct-imaging X-ray telescope at energies beyond those of the **Chandra X-ray Observatory** and **XMM-Newton**. It was launched on June 13, 2012, having previously been delayed from March 21, 2012, due to software issues with the launch vehicle.

Its primary scientific goals are to conduct a deep survey for black holes a billion times more massive than the sun, to investigate how particles are accelerated to very high energy in active galaxies, and to understand how the elements are created in the explosions of massive stars by imaging the remains. Currently, it is in a two-year **extended mission** to 2016. See appendix B for a list of its instruments.

O

OCEAN SURFACE TOPOGRAPHY MISSION (OSTM). An international Earth observation satellite mission on board the **Jason**-2 satellite that continues the sea surface height measurements begun in 1992 by the joint NASA/**CNES TOPEX/Poseidon** mission and followed by the NASA/CNES Jason-1 mission launched in 2001. Commonly abbreviated as OSTM/Jason-2, the mission uses high-precision ocean altimetry to measure the distance between the **satellite** and the ocean surface to within a few inches. These very accurate observations of variations in sea surface height provide information about worldwide sea level, speed and direction of ocean currents, and heat stored in the ocean. See appendix B for a list of its instruments.

OPERATION ICEBRIDGE. Operation IceBridge aims to monitor changes in polar ice from a fixed-wing aircraft from 2009 to 2016. It is a temporary replacement for the **Ice, Cloud, and Land Elevation Satellite**. The project uses several science instruments to get a three-dimensional view of Arctic and Antarctic ice so that it can monitor changes in polar ice and gather data for predictive models of ice and sea-level rise. IceBridge flights began in October 2009 using a DC-8 jet airliner. See appendix B for a list of its instruments.

OPPORTUNITY. Also known as MER-B (Mars Exploration Rover–B) or MER-1, *Opportunity* is a robotic **rover** on **Mars** that has been active since 2004. Launched on July 7, 2003, *Opportunity* landed on Mars's Meridiani Planum on January 25, 2004, three weeks after its twin ***Spirit*** (MER-A), also part of NASA's **Mars Exploration Rover Mission**, touched down on the other side of the planet. *Opportunity* has continued to gather scientific observations and report back to Earth for over 40 times its intended lifespan. In 2005, *Opportunity* was meandering over the surface when it became mired in the sand for five weeks, and in 2007 barely survived a severe regional dust storm in the Victoria Crater when its solar panels became covered in dust, and its power dipped to critical levels. In 2011, *Opportunity* found a thin vein of gypsum on the rim of the Endeavour Crater, leading scientists to believe that it was deposited by liquid water billions of years ago. Nearby Matijevic Hill contains clay materials, another indication of a long-ago watery environment. On January 24, 2014, NASA reported that the ***Curiosity*** and *Opportunity* rovers would begin searching for evidence of early life. On July 28, 2014, NASA made an announcement that *Opportunity* had set a new record of having driven the greatest "off-world" distance of over 25 miles (40 km), surpassing the previous record held by the **Soviet Union**'s *Lunokhod 2* rover, which had traveled 39 km (24 mi) on the **moon**, and by March 2015 it had traveled a further 1.2 miles, completing the distance of a marathon. In December 2014, *Opportunity* suffered age-related amnesia in some of its memory, so it is currently operating in RAM-only mode. Asteroid 39382 Opportunity was named in the rover's honor.

ORBITAL ATK. In February 2015, **Orbital Sciences Corporation** merged with parts of Alliant Techsystems and became Orbital ATK. Its Flight Systems Group (which includes the **Antares** launch vehicles) is headquartered in Chandler, Arizona, and its Space Systems Group, which produces the **Cygnus** spacecraft and delivers cargo to the **International Space Station**, operates out of Dulles, Virginia.

ORBITAL DEBRIS. *See* SPACE DEBRIS.

ORBITAL SCIENCES CORPORATION. Commonly referred to as Orbital, it is a U.S. based company specializing in the design, manufacture, and launch of small- and medium-class space and rocket systems for commercial, military, and civil government customers. Orbital's primary products are **satellites** and launch vehicles, including the **Orbiting Carbon Observatory-2**, which monitors carbon dioxide concentration in the atmosphere. Orbital also provides satellite subsystems and space-related technical services to government agencies and laboratories.

In late October 2014, an Orbital Antares rocket exploded on liftoff from **Wallops Island**, carrying a payload of food, supplies, and research equipment to the **International Space Station**, setting back their delivery schedule by some months. The vehicle and its cargo was valued at roughly $200 million.

Founded by three Harvard University friends in 1982, Orbital began its work with NASA in 1985 with a contract for up to four transfer orbit stage vehicles, the first of which successfully launched the **Mars Observer** into space in 1997. **Dawn**, Orbital's first interplanetary spacecraft, launched in 2007 on its way to **Mars** and **Jupiter**. Orbital completed its first cargo delivery mission to the **International Space Station** in 2013, part of NASA's attempt to **contract** out those services to private industry.

ORBITING CARBON OBSERVATORY-2 (OCO-2). The NASA environmental science satellite, launched on July 2, 2014, is a replacement for the Orbiting Carbon Observatory (OCO) that was lost in a launch failure in 2009 due to a failed payload shroud deployment on a Taurus XL. OCO-2 was built by **Orbital Sciences Corporation** and launched by **United Launch Alliance** and is the first spacecraft dedicated to studying atmospheric carbon dioxide concentrations and distributions in the atmosphere, leading to a greater understanding of climate change.

ORIGAMI. Origami is an ancient Japanese method of folding paper into different shapes and designs. Recently, NASA has been testing different origami techniques to be used for space-based solar panels. Techniques include the Miura fold, developed by Japanese astrophysicist Koryo Miura, which utilizes an array of checkered parallelograms, and a more circular design inspired by the Miura fold and designed by Brian Trease, a mechanical engineer at NASA's **Jet Propulsion Laboratory**.

ORIGINS PROGRAM. A decades-long study by NASA addressing the origins of the universe, various astronomical bodies, and life, the Origins program was started in the 1990s and consists of the following missions:
Airborne:

- **Stratospheric Observatory for Infrared Astronomy**

Ground Based:

- **W. M. Keck Observatory** telescopes
- **Large Binocular Telescope** Interferometer

Space Based:

- **Hubble Space Telescope**
- **Far Ultraviolet Spectroscopic Explorer**
- **Spitzer Space Telescope**
- **Kepler Space Observatory**
- **James Webb Space Telescope**
- **Single Aperture Far Infrared Observatory**

ORIGINS SPECTRAL INTERPRETATION RESOURCE IDENTIFICATION SECURITY REGOLITH EXPLORER (OSIRIS-REx). OSIRIS-REx, set to launch in September 2016, is a planned asteroid study and sample return mission, the third mission of the **New Frontiers Program**, after **Juno** and **New Horizons**. The mission will study and return a sample of asteroid 101955 Bennu, a B-type carbonaceous asteroid and potentially hazardous **near-Earth object**, to Earth for detailed analyses in 2023. Material returned is expected to allow scientists to learn more about the formation and evolution of the solar system, including the initial stages of planet formation, and the source of organic compounds that led to the formation of life.

OSIRIS-REx 2 will perform a similar function for the two **moons** of **Mars**, if it is funded. See appendix B for a list of its instruments.

ORION LAUNCH ABORT SYSTEM (LAS). In the wake of the ***Challenger* tragedy**, NASA's efforts at providing a safe crew module in future flights have resulted in **Orion**'s Launch Abort System, managed by **Langley Research Center**. Its purpose is to protect the crew during launch by pulling the crew module away from the failing rocket within milliseconds, and position the module for a safe landing. Its fairing surface protects the capsule from heat, wind, and acoustics, and Exploration Flight Test-1 (EFT-1) successfully tested the hardware during a flight on December 5, 2014. During EFT-1, the spacecraft traveled to an altitude of 3,600 miles and returned at an estimated speed of 20,000 mph, thereby subjecting the heat shield to approximately 4,000°F before landing in the Pacific Ocean. The **Space Launch System**, which is a heavy-lift launch vehicle, will launch Orion on Exploration Mission-1 around the **moon**, estimated for late 2017 or later.

ORION MULTI-PURPOSE CREW VEHICLE (MPCV). The Orion Multi-Purpose Crew vehicle is a beyond-**low Earth orbit** manned spacecraft that is being built for crewed missions to the **moon**, asteroids, and **Mars**. **Lockheed Martin** was contracted for the design and manufacture for NASA and, using the same basic plans, Airbus Defence and Space will build it for the **European Space Agency**. It is planned to be launched by the **Space Launch System** with the first crewed mission expected to take place after 2020. Each Orion spacecraft is projected to carry a crew of up to six astronauts. The MPCV's debut unmanned test flight, known as Exploration Flight Test 1, successfully launched aboard a **Delta** IV Heavy rocket on December 5, 2014. Orion reached an altitude of 3,600 mi (5,800 km) and speeds of up to 20,000 mph (8,900 m/s) during its 4.5-hour flight that tested its heat shield, parachutes, jettisoning components, and on-board computers.

Much of the MPCV's design and testing had already been completed as part of the **cancelled Constellation** program and known as the Crew Exploration Vehicle. Based on the **Apollo** program's **command and service module**, the design also includes a deep space habitat in which **astronauts** could travel and live for missions of 60 to 500 days. Such times are necessary for the proposed missions, which require six months or more in transit.

OUTER SPACE TREATY. Formally known as the Treaty on Principles Governing the Activities of States in the Exploration and Use of Outer Space, including the Moon and Other Celestial Bodies, it forms the basis of international space law. The treaty between the United States, the United Kingdom, and the **Soviet Union** was opened for signature on January 27, 1967, and officially enacted on October 10, 1967. Currently, 103 countries are parties to the treaty, while another 26 have signed the treaty but not completed ratification.

The Outer Space Treaty bans the use or placement of nuclear weapons or any other weapons of mass destruction in orbit around Earth, from installing such weapons on the **moon** or any other celestial body, or to otherwise station them in outer space. It limits the use of the moon and other celestial bodies to peaceful purposes and prohibits the use of such bodies for testing weapons of any kind or for establishing any type of military base of operations. The treaty forbids any government from claiming a celestial body such as the moon or a planet, claiming that they belong to all of mankind.

One of the issues that the treaty does not specifically cover is that of the ownership of any resources extracted from the celestial body. For example, if a private company were to establish a mining site on an asteroid and extract a valuable mineral, to whom would those mineral rights belong? Theoretically, a 79-foot-wide M-type asteroid (metallic class) might hold 33,000 tons of metals. If that included platinum, it could be worth millions. Asteroid 1986 DA is an M-class asteroid made up of iron, nickel, gold, and platinum; its worth could be measured in the many trillions of dollars. Several entrepreneurial companies are gearing up for just these possibilities, such as Deep Space Industries and Planetary Resources, two U.S. firms that would benefit from the American Space Technology for Exploring Resource Opportunities in Deep Space (ASTEROIDS) Act currently in the House of Representatives.

OZONE MAPPING AND PROFILER SUITE (OMPS). OMPS is a suite of **instruments** intended to examine the ozone layer, contributing to the more than 30 years of total ozone and ozone profile records. Managed by **Goddard Space Flight Center**, OMPS takes daily measurements of ozone, continuing the earlier work of the Nimbus 4 satellite in the 1970s through the **Total Ozone Mapping Spectrometer** instruments on board various NASA, **NOAA**, and international **satellites**. OMPS can also measure dust clouds from the Sahara and smoke from fires. The current OMPS is flying onboard the **Suomi NPP** spacecraft.

P

PARASOL. *See* POLARIZATION AND ANISOTROPY OF REFLECTANCES FOR ATMOSPHERIC SCIENCES COUPLED WITH OBSERVATIONS FROM A LIDAR (PARASOL).

PERSONAL EGRESS AIR PACK (PEAP). PEAPs were devices on board the **space shuttles** that provided crew members with approximately six minutes of breathable air in the case of an accident while the vehicle was still located on the ground. PEAPs did not provide pressurized air, meaning that they were only intended to be used should the air inside the shuttle cabin become unbreathable by way of harmful gases. PEAPs have been replaced by the **Advanced Crew Escape Suits**, which contain pressurized air.

PHOENIX. A robotic spacecraft on **Mars** under the **Mars Scout Program** with the purpose of searching for environments suitable for microbial life and to research the history of water there, the *Phoenix* lander launched on August 4, 2007, and landed on Mars on May 25, 2008. *Phoenix* was NASA's first successful landing in a Martian polar region and its sixth successful landing out of seven attempts. The mission lasted for approximately three months and made a last brief communication with Earth on November 2, 2008, after which available solar power dropped with the Martian winter. Engineers were unsuccessful with recontacting the spacecraft, and the mission was officially ended on November 10, 2008.

The program was seen as a success because it accomplished all scheduled science experiments and observations. The most noteworthy of these observations included the discovery of subsurface water-ice sublimation, moderately alkaline soil composition, and perchlorates, which have the potential to be used as rocket fuel and a source of oxygen for future explorers. Perchlorates were also discovered by the ***Curiosity* rover**, suggesting that Mars has a potential global distribution of the compound. See appendix B for a list of its instruments.

***PIONEER 10* AND *11*.** These two American space **probes** completed the first missions to the planets **Jupiter** and **Saturn**, respectively. *Pioneer 10* was launched on March 3, 1972, and made its closest approach to Jupiter on December 4, 1973. Thereafter, *Pioneer 10* became the first spacecraft to achieve escape velocity from the solar system. Radio communications were lost with *Pioneer 10* on January 23, 2003, due to the loss of electric power for its radio transmitter, with the probe at a distance of 12 billion kilometers (80 AU) from Earth, and more than 40 years since its launch. *Pioneer 11* was launched on April 6, 1973, and made its closest approach to Jupiter on December 2, 1974. Several years later, on September 1, 1979, *Pioneer 11* made its closest approach to Saturn and tested the area for ring particles for the **Voyager** missions. Because of power constraints and the vast distance to the probe, last contact with the spacecraft was on September 30, 1995. At the request of **Carl Sagan**, *Pioneer 10* and Pioneer *11* carry gold-anodized aluminum plaques in case either spacecraft is ever found by intelligent extraterrestrial life-forms. The plaques feature the nude figures of a human male and female along with several symbols that are designed to provide information about the origin of the spacecraft. See appendix B for a list of some of their instruments.

PIONEER PROGRAM. A series of United States unmanned space missions lasting from 1958 to 1978 that were designed for planetary exploration, the initial Pioneer missions from 1958 to 1960 focused on proof of concepts, such as achieving Earth's escape velocity and reaching the moon for various studies. From 1965 to 1969, Pioneer missions 6 through 9 were placed into orbit around the sun to act as an interplanetary **space weather** network and currently remain in orbit. The most notable spacecraft of the Pioneer mission were ***Pioneer 10*** and ***Pioneer 11***, which explored the outer gas planets and left the solar system. *Pioneer 10* and *Pioneer 11* each carries a golden plaque depicting a man and a woman and information about the origin and the creators of the probes, should any extraterrestrials find them someday. Additionally, the Pioneer Program included the **Pioneer Venus Project** with the goal of studying **Venus**'s atmosphere and topography using the **Pioneer Venus Orbiter** and the **Pioneer Venus Multiprobe**.

PIONEER VENUS MULTIPROBE. Also known as *Pioneer Venus 2* or *Pioneer 13*, it was a spacecraft launched on August 18, 1978, to explore Venus as part of NASA's **Pioneer program**. The spacecraft carried one large and three small atmospheric **probes**, designed to collect data on atmospheric composition as well as temperature and pressure as they descended into the atmosphere of **Venus**. The probes did not carry photographic instruments and were not designed to survive landing—the smaller probes were not equipped with parachutes, and the larger probe's parachute was expected to detach as it neared the ground. All four probes arrived at Venus on December 9, 1978, and continued transmitting data until impact; one probe continued to transmit data from the surface for over an hour.

PIONEER VENUS ORBITER. Also known as *Pioneer Venus 1* or *Pioneer 12*, it was a mission to **Venus** conducted as part of the **Pioneer program**. Launched on May 20, 1978, the spacecraft was inserted into an elliptical orbit around Venus on December 4, 1978. The spacecraft's scientific mission was to study Venus's atmosphere, topography, and magnetic field. From its orbit of Venus, the Pioneer Venus Orbiter was able to observe Halley's Comet when it was unobservable from Earth due to its proximity to the sun in February 1986. The spacecraft continued to return data until October 8, 1992, whereupon the orbiter disintegrated upon entering the atmosphere of Venus on October 22, 1992.

PIONEER VENUS PROJECT. A NASA mission included within the **Pioneer program** to study the planet **Venus**, the Pioneer Venus Project consisted of the **Pioneer Venus Multiprobe** and the **Pioneer Venus Orbiter**, which were designed to study the atmosphere and topography of the planet. Both spacecraft were launched in 1978 and successfully carried out their missions, with the Venus Orbiter returning data until 1992. See appendix B for a list of its instruments.

PLANCK. Launched in May 2009, Planck is a space observatory operated by the **European Space Agency** designed to observe the cosmic microwave background (CMB) at microwave and infrared frequencies. The mission supplements and expands upon observations made by NASA's **Wilkinson Microwave Anisotropy Probe**, which previously measured the CMB with lower sensitivity. The Planck observatory also provides information applicable to other cosmological and astrophysical issues, such as testing theories of the early universe and the origin of the structure of the cosmos. At the end of its mission, Planck was put into a heliocentric orbit and decommissioned in October 2013 to prevent it from jeopardizing any future missions. The project, initially called COBRAS/SAMBA, is named in honor of the German physicist Max Planck (1858–1947), who won the Nobel Prize in physics in 1918. See appendix B for a list of Planck's instruments.

PLANET CLASSIFICATION. As of 2006, the International Astronomical Union (IAU) officially adopted the definition of classifying an object as a planet if it orbits a star or stellar remnant and meets the three following criteria:

- It is massive enough to be rounded by its own gravity
- It is not massive enough to cause thermonuclear fusion
- It has cleared its neighboring region of planetesimals

Within the solar system there are several classifications for astronomical objects, including the sun, the rocky planets (**Mercury**, **Venus**, Earth, and **Mars**), the gas giant planets (**Jupiter** and **Saturn**), the ice giant planets (**Uranus** and **Neptune**), dwarf planets (Ceres, **Pluto**, Haumea, Makemake, and Eris), moons, asteroids, and comets.

The term "planet" stems from the ancient Greek ἀστὴρ πλανήτης (astēr planētēs), meaning "wandering star." The definition of what constitutes a planet has changed significantly since they were first observed by the Babylonians. Up until the Scientific Revolution, planets were simply objects that moved across the sky as opposed to "stationary" stars. After Copernicus's heliocentric model gained sway in the 18th century, the definition of planet changed to mean an object that orbited the sun. The use of more sophisticated telescopes starting in the 19th century led to the gradual development of refining the definition of what it means to be a planet. After the discovery of Eris, an object 27 percent larger and almost 27 AU more distant than Pluto, the IAU redefined the term planet in 2006 so as to more clearly distinguish larger objects in asteroid belts from those that were not.

Current efforts to locate planets around other stars, or exoplanets, have proven very successful. The **Kepler** spacecraft, whose mission was to locate exoplanets, had detected 974 exoplanets as of June 2014, including 3,601 unconfirmed candidates. Several new classifications for planets have been created to describe recent discoveries including "hot Jupiters" and "super Earths."

PLUM BROOK STATION (PBS). A field station near Sandusky, Ohio, and part of **Glenn Research Center**, PBS is located about 50 miles (80 km) from the main campus. It specializes in large-scale tests that would be hazardous within the confines of the Glenn Research Center. The Spacecraft Propulsion Facility at Plum Brook Station is the world's only facility capable of testing full-scale upper-stage launch vehicles and rocket engines under simulated high-altitude settings. The PBS Space Power Facility houses the world's largest space environment vacuum chamber as well as a reverberant chamber for acoustic testing and an electrodynamic shaker, both of which are capable of testing the **Orion** spacecraft.

PLUTO. Discovered in 1930, Pluto was originally classified as the ninth planet from the sun. Its orbits around the sun are highly inclined and eccentric, averaging a distance of 39 AU. It is relatively small, with a diameter of about 2,372 km or about 68 percent that of the Earth's **moon**. It has five known natural satellites: Charon, the largest, followed by Nix, Hydra, Kerberos, and Styx. Its orbit is 248 Earth years. The further discoveries of similarly sized objects in the distant Kuiper Belt that have eccentric orbits like Pluto led to a re-examination of Pluto's planetary status. In 2006, the International Astronomical Union formally reclassified Pluto as a dwarf planet.

Further research about Pluto is limited due to its small size and immense distance from Earth observation. Several **satellite** missions to study Pluto have been planned in recent years, such as the **Pluto Fast Flyby** and the **Pluto Kuiper Express**, both of which were ultimately **cancelled** and replaced by **New Horizons**, launched in January 2005 and which became the first spacecraft to visit Pluto in July 2015.

New Horizons found a geologically interesting world, with smooth plains evidencing a geologically young age of less 100 million years and mountain ranges likely composed of water-ice up to 11,000 feet (3,500 meters). Icy plains on the western edge of the Tombaugh region in the Sputnik Planum indicate exotic ice flows composed of methane, nitrogen, and carbon monoxide ices. A haze layer or series of haze layers was detected up to 100 miles above the surface with the overall atmosphere of Pluto detectable to a height of 1,000 miles (1,600 km) and is composed of nitrogen (N_2), methane (CH_4) and carbon monoxide (CO).

The reclassification of Pluto was greeted with dismay by many, occasioned by petitions to the IAU, boycotts, resolutions, and even songs ("I'm Your Moon" by Jonathan Coulton, "Bring Back Pluto" by Aesop Rock).

PLUTO FAST FLYBY. A space mission meant to perform a flyby of the dwarf planet **Pluto**, this spacecraft was meant to be launched by 2000 and reach Pluto by 2010, so it could examine Pluto's atmosphere before it froze as Pluto moves away from the sun. The mission was later **cancelled** due to a lack of funding, but it was replaced by the **Pluto Kuiper Express**. The Pluto Kuiper Express would also later be cancelled for the same reasons. A mission to Pluto was eventually realized as **New Horizons**, which launched in January 2005 and which successfully visited Pluto on July 14, 2015.

PLUTO KUIPER EXPRESS. The Pluto Kuiper Express is a space mission designed to fly by the **Pluto** system and at least one other large object in the Kuiper Belt beyond Pluto's orbit. Originally called the **Pluto Fast Flyby**, it was scheduled to reach Pluto by 2013. The spacecraft was to have been a simple hexagonal prism-shaped structure powered by radioisotope thermal generators similar to those used on the **Galileo** and **Cassini** missions. The mission was **cancelled** for **budgetary** reasons, but later replaced by the similar **New Horizons** mission, which launched in January 2005.

POLAR. Launched into orbit on February 24, 1996, Polar is one of two NASA missions of the **Global Geospace Science** (GGS) initiative, which in turn is part of the **International Solar-Terrestrial Physics** program. It is designed to study the polar magnetosphere and aurora. It continued to collect data and its mission was **extended** several times until the program was terminated in April 2008. The spacecraft remains in orbit, though it is now inactive. Polar is the sister ship to GGS **WIND**. Refer to appendix B for a list of its instruments.

POLAR-ORBITING OPERATIONAL ENVIRONMENTAL SATELLITES (POES). A **constellation** of **weather satellites** in polar orbits around the Earth, POES is a joint effort between the **National Oceanic and Atmospheric Administration** and the European Organization for the Exploitation of Meteorological Satellites. Spacecraft for the project are provided by NASA and the **European Space Agency**. Data from the POES **satellites** are used for a broad range of environmental monitoring applications, such as weather analysis and forecasting, climate research, global sea surface temperature measurements, forest fire detection, and volcanic eruption monitoring.

POLARIZATION AND ANISOTROPY OF REFLECTANCES FOR ATMOSPHERIC SCIENCES COUPLED WITH OBSERVATIONS FROM A LIDAR (PARASOL). A French-built Earth-observing research **satellite**, PARASOL studied the radiative and microphysical properties of clouds and aerosols through an instrument called POLDER. PARASOL was launched from the French spaceport in Kourou, French Guiana, on December 18, 2004. It flew in formation in the **A-Train constellation** until December 2, 2009, when it was moved out of the **constellation**.

POTENTIALLY HAZARDOUS OBJECT (PHO). A PHO is a near-Earth asteroid or comet with an orbit such that it has the potential to make close approaches to the Earth (within 0.5 AU) and is of a size large enough—330–490 feet in diameter (100–150 m)—to cause significant regional damage in the event of impact. NASA has identified approximately 1,500 such objects.

See also NEAR-EARTH OBJECTS (NEOs).

PRESIDENT'S SCIENCE ADVISORY COMMITTEE (PSAC). Established as the Science Advisory Committee and part of the Office of Defense Mobilization, it was upgraded by President Eisenhower to the President's Science Advisory Committee (PSAC) and moved to the White House on November 21, 1957. The committee had no operating responsibilities and was originally established as a direct response to the launches of the **Soviet Union**'s **satellites**, ***Sputnik** 1* and *Sputnik 2*. Its purpose was to provide advisory opinions and analysis on science and technology matters to the entire federal government and specifically to the president.

Shortly after his reelection in 1972, President Nixon eliminated the President's Science Advisory Committee, forcing the White House Office of Science and Technology and the U.S. Congress to rely on federal agencies for guidance in scientific policy. A similar entity, the U.S. President's Council of Advisors on Science and Technology (PCAST), was established by President George H. W. Bush in 1990 that enabled the president to receive advice directly from the private and academic sectors on technology, scientific research priorities, and mathematics and science education.

PROBE. A probe is a robotic spacecraft that explores some part of outer space, whether it flies by, orbits, or lands on another celestial body or simply travels through space, returning information to Earth. It is distinguished from a **satellite**, which generally orbits a larger celestial body, although some probes are also considered satellites when they perform both functions, and some that are referred to colloquially as satellites are in fact probes, such as ***Voyager** 1* and *2*.

PROJECT ROVER. *See* ROVER.

PROJECT VANGUARD. *See* VANGUARD.

Q

QUICK SCATTEROMETER (QUIKSCAT). Launched on June 19, 1999, on a two-year mission, QuikSCAT carried a SeaWinds instrument to measure near-surface wind speed and direction over the oceans. QuikSCAT was intended to fill the gap when the **satellite** on which the previous NASA scatterometer was flying lost power in June 1997. In a single day, QuikSCAT made approximately 400,000 measurements and covered 90 percent of Earth's surface from an altitude of about 500 miles in a sun-synchronous orbit. It remained fully operational until November 2009 when SeaWinds's antenna stopped rotating, leaving it to cover a much narrower swath of the Earth's surface although still providing valuable measurements and overlap. The balance of data was then obtained through the **Indian Space Research Organization**'s Oceansat-2 scatterometer, until the ISS-**RapidScat** instrument launched on September 20, 2014, and now provides near-global coverage from the **International Space Station**. See appendix B for a list of QuickScat's instruments.

R

RADIATION BELT STORM PROBES. *See* VAN ALLEN PROBES.

RANGE SAFETY OFFICER. Public safety is a major concern of NASA, even to the extent of the destruction of spaceflight missions that would cause an immediate threat to lives on the ground. For that reason, every major vehicle that lifts off from NASA properties or on NASA missions—up to and including those with a human crew—has carried an explosive destruct system that could be triggered by the range safety officer. In addition, the solid rocket **boosters** and external tanks are also fit with explosive devices that could be triggered during uncontrolled flight toward urban areas. For example, after the breakup of the **space shuttle *Challenger*** on January 28, 1986, the solid rocket boosters appeared to be heading toward land or shipping areas and the self-destruct was ordered so that it would break up over uninhabited, empty ocean. The **Orion** spacecraft will have a rocket-powered escape pod available to crew members in an attempt to avoid having the range safety officer having to terminate a flight with humans on board.

RANGER. Ranger was a lunar **probe** mission to gather data about the **moon** that began in 1959, named by **Jet Propulsion Laboratory**. Ranger was one of NASA's earliest projects. The spacecraft were designed to take photos of the moon (each had six cameras) and to transmit the pictures directly back to Earth before crashing onto the surface. *Rangers 1* through *6* (launched between August 1961 and January 1964) were failures, but missions *Ranger 7* through *9* (launched between July 1964 and March 1965) were successful. Postmortem studies show it is likely that gold-plated flaking on the diodes was problematic in space, causing at least some of the failures. Relayed images were 1000 times better than those that were made by telescopes on Earth. See appendix B for a list of instruments on some of the missions.

RAPIDSCAT. An **International Space Station** project, RapidScat (also known as ISS-RapidScat) is a wind-watching scatterometer launched September 20, 2014, by **SpaceX** to be the first payload completely robotically assembled in space since the space station itself was constructed. It will be attached to the space station by a robotic arm using many of the test parts from the earlier **QuickScat** mission. RapidScat was a last-minute addition to the payload delivery schedule and was designed and built by **Jet Propulsion Laboratory** in two years. It will be used to support near-global weather forecasting and is part of the **Science Mission Directorate**.

REDSTONE. Redstone was originally a surface-to-surface missile with a range of 200–300 miles developed by the **U.S. Army** and modeled after the V-2 rocket. Named for the Redstone Arsenal where it was developed, the missile was adapted by NASA as an expendable launch vehicle for suborbital flights during Project **Mercury** (often called Mercury-Redstone).The later **Jupiter**-C and **Juno** 1 were modified from the Redstone, as well as the **Saturn** I and IB, used during the **Apollo** program.

REGOLITH. Merriam-Webster's defines regolith as "unconsolidated residual or transported material that overlies the solid rock on the Earth, moon, or a planet." It is also found on asteroids. Essentially, it is the loose dust, broken rocks, and similar material found on the surface of any celestial body, formed over thousands and even millions of years. "Soil" is often used as a description, and while technically not correct (soil is assumed to have organic content), the terms "Martian soil," "lunar soil," and similar descriptions can often be found. It is possible that large quantities of water-ice can be found within the regolith on **Mars**, particularly near the equator and at higher latitudes. Regolith on the **moon** or Mars, however, cannot be used as a complete growing medium as it lacks the necessary organic matter. In order to study regolith and its possibilities, simulates have been developed. **Johnson Space Center** Number One (JSC-1) was developed in 1994 by NASA to be like lunar soil, and Fuji Japanese Simulant 3, developed in Japan, is close to ***Apollo 11*** samples. Several Mars regolith simulants have also been developed, generally very fine basalt sands. The fine particles of particularly Martian regolith may end up posing health risks due to inhalation or eye contact.

RESEARCH AND DEVELOPMENT (R&D) IN SPACE. Not only does NASA conduct research *on* space, it conducts research—medical, biological, chemical, materials science, and so on—*in* space. NASA's research goals for the microgravity environment of the **International Space Station** (which is designated by the U.S. government as a federal laboratory) are focused in four areas: human health and exploration, technology testing for enabling future exploration, basic life and physical sciences, and Earth and space science. Specific projects have included numerous studies on the human body (cardiovascular, musculature, respiration, immune systems, and radiation), on biology and biotechnology (cellular biotechnology, genetic responses), microbiology (vaccines, drug resistance), plant biology, crystal growth, binary colloidal alloy tests, spacecraft materials and systems, and assorted contracted studies and evaluations. **Education and public outreach** is a large part of their research, and almost every crew participates in some sort of **CubeSat** or similar project designed by or for students to encourage science, technology, engineering, and mathematical exploration. The United States also partners with Canada, Europe, Japan, and **Russia** on other projects. Social media and live video allow partners on Earth as well as interested parties to participate in the discussion of the projects.

One example of a study on the ISS that has benefits on Earth is the development of a new targeted method of chemotherapy drug delivery (clinical breast cancer trials are now in development at M. D. Anderson Cancer Center in Houston). In 2002, scientists were wondering if microencapsulation of chemotherapy drugs would better deliver the treatment directly to a tumor, and the **astronauts** found that developing the microballoons without gravity made this possible. For the next five years here on Earth, studies continued (with mice) and the technologies were licensed to a private company for further development; FDA drug approval is next.

In another study, astronauts found that mice with osteoporosis treated with osteoprotegerin decreased bone resorption compared to treated mice. Although studies were underway on Earth, the studies could be done faster by allowing scientists to actually slow the process down to better understand the drug and how it functions.

NASA maintains a complete listing of all of the space research done on the ISS at www.nasa.gov/mission_pages/station/research/experiments_category/index.html#.VCMQNPldVDQ.

RESEARCH AND ENGINEERING DIRECTORATES. Research and engineering directorates can be found at each of NASA's major **field centers**, and each has a specific focus. **Glenn Research Center**'s Engineering Directorate focuses on mechanical and fluid systems, manufacturing, avionics and electrical systems, and systems engineering and analysis. **Johnson Space Center**'s Engineering Directorate, in line with its mission control function, focuses on the design, development, and test support for spaceflight programs, as well as crew support. **Ames Research Center** provides multidisciplinary engineering design and analysis, and **Langley Research Center** focuses on aviation and all of its related areas as well as spaceflight programs. LaRC Engineering developed CERES, **CALIPSO**, and is currently finalizing development of **Stratospheric Aerosol and Gas Experiment** 3 (and the previous SAGE instruments). Langley also led the entry descent and landing for the **Mars Science Laboratory** and the **Mars Reconnaissance Orbiter**. **Armstrong Flight Research Center**'s expertise concentrates on the software, hardware, analysis, modeling, planning, and execution of flight, while **Goddard Space Flight Center**'s Applied Engineering and Technology Directorate develops and operates unmanned scientific spacecraft. **Jet Propulsion Laboratory** is the lead U.S. center for robotic exploration of the solar system and has major programs in space-based Earth sciences. **Kennedy Space Center** concentrates on human spaceflight launches, but also specializes in meteorology, corrosion research, space life sciences, and electromagnetic physics. **Marshall Space Flight Center** develops and operates the engines, vehicles, space systems, and science payloads for a wide variety of missions, and **Stennis Space Center** is a rocket testing facility. The centers all work together to provide the highest level of safety and technology available, and a single project often travels from one field center to another at different stages in its development.

REUVEN RAMATY HIGH-ENERGY SOLAR SPECTROSCOPIC IMAGER (RHESSI). A **Small Explorer** mission, RHESSI was launched on February 5, 2002, to explore particle acceleration and energy release in **solar flares**. The name Reuven Ramaty was prefixed to the original name to honor one of NASA's pioneers of high-energy solar physics. RHESSI is helping scientists understand how the electrons (emitting mostly X-rays) and protons and heavier ions (emitting mostly gamma rays) drive **space weather** by combining the images of those emissions with high-resolution spectroscopy. Understanding what drives solar processes is one of the major goals of space physics and astrophysics. RHESSI imaged solar gamma rays from a solar flare for the first time, and its original two-year mission has been extended several times. In June 2014, its detectors were turned off and it underwent a regular anneal procedure to recover lost energy resolution and sensitivity due to radiation damage, and was brought back to normal operation in August 2014, with a slight reduction in image quality for soft X-rays. See appendix B for a list of its instruments.

RIDE, SALLY KRISTEN (May 26, 1951–July 23, 2012). Ride was the first American woman in space when she flew on board the **space shuttle *Challenger*** on June 18, 1983; she was preceded in space by two Soviet women, **Valentina Tereshkova** (1963) and Svetlana Savitskaya (1982). Born in Los Angeles, Ride received her bachelor's degree in English and physics from Stanford University, followed by a master's degree and doctorate in physics; she specialized in the interaction of X-rays with the interstellar medium. She joined NASA in 1978 and was the first person to use a robotic arm to retrieve a **satellite** in space, a skill she excelled at given her tennis-playing skills. She later served on the investigative boards of both the ***Challenger* tragedy** and the ***Columbia* disaster**, the only person to serve on both, and one of the few public figures to support **Roger Boisjoly**. After her NASA career, she worked in education and as director of the California Space Institute, while also working on **education and public outreach** programs for NASA, including the **Gravity Recovery and Interior Laboratory** mission. She wrote seven books on space for children, and was survived by her partner of 27 years.

ROCKETS, MULTISTAGE. Modern multistage rockets were developed independently by three scientists—**Konstantin Tsiolkovsky** in Russia, **Robert Goddard** in the United States, and Hermann Oberth in Germany (the fathers of modern rocketry)—although evidence shows earlier successful but simpler attempts in **China** in the 14th century, Korea in the 16th century, and Poland in the 17th century. Multistage means that one rocket (engine and fuel) is attached to another (whether on top of or next to the other), one or more of which is jettisoned later as it becomes empty or is used up. There are two-stage (**Delta** IV, **Saturn** IB), three-stage (Saturn V, **Ariane** 2), and even four-stage (Proton, Ariane 1) rockets. Multistage rockets can also have **boosters**, which increase thrust and payload capacity, that can be recovered and reused (such as the **space shuttle**'s solid rocket boosters) or expendable. Most of the weight on liftoff consists of the propellant; once each stage depletes its fuel, it becomes a drag on the rocket; jettisoning that stage lightens the spacecraft and thus lessens the total fuel needed to reach the velocity or altitude. Each stage can also be modified for the atmospheric conditions it will meet (higher pressure near Earth, near vacuum in space). Multistage rockets can be designed with the stages in parallel, serial, or a mix of the two (**Titan** III, for example). With serial staging, one stage finishes and the next begins. With parallel stages, one or more engines are operating simultaneously. Each stage can have its own type of propulsion—for example, the Titan III has a liquid-powered, two-stage engine with two solid rocket strap-ons during launch, that are quickly discarded after launch. Each additional stage increases some degree of risk for the spacecraft and its crew (if any). With multistage rockets also comes the problem of what to do with the jettisoned engines and housing. Some are left to fall to Earth, hopefully over the ocean. The upper stages of the rocket are often jettisoned in space and are a significant portion of **space debris**.

Crew placement depends on the rocket and spacecraft. The space shuttle is attached to the side of the rocket, with the crew seated with their backs parallel with the surface of Earth. Others such as **Apollo** and **Soyuz** are stacked with one or more rockets beneath them.

ROCKETS, SINGLE-STAGE. Single-stage rockets (or single-stage-to-orbit, SSTO) vehicles are complete spacecraft (fuel, housing, instruments, engine, etc.), without any parts to be jettisoned; unfortunately, they (to date) are not powerful enough to reach orbit. SSTOs would cut costs significantly for **low Earth orbit** satellites; with current materials (lighter, stronger) and better construction, it is theoretically possible for an SSTO to reach orbit—but without a payload. The high-profile failure of the Delta Clipper (DC-XA), an attempt in the 1990s, precluded further efforts at the time. Part of the problem is the amount and weight of fuel that would need to be used, and another is that the same engine would have to work both under atmospheric pressure on Earth and in the near vacuum of space. Research is underway, and the British government in collaboration with the **European Space Agency** is working on Skylon, expected to be in the testing phase by 2019. SSTOs are generally reusable. Because of lower gravity, SSTOs are already possible from the **moon** (for example, **Apollo**'s **lunar module**). NASA's Centennial Challenges program has announced a suborbital Lunar Lander Challenge that might well include SSTOs, and other private efforts are underway as the method still has potential.

NASA does have some single-stage sounding rockets, which are 30-minute or so airborne missions to collect scientific data and conduct engineering tests. The rocket motor separates from the payload and falls back to Earth. The payload continues on its 5–20-minute mission, collecting data that is returned via telemetry, before it too returns to Earth. **Wallops Flight Facility** and its Sounding Rocket Program is the primary host of about 25 launches each year.

ROCKWELL INTERNATIONAL. Rockwell, a major American manufacturing conglomerate in the latter half of the 20th century, was directly involved in the aircraft and space industries, among many others. Rockwell manufactured the **Apollo command and service module**, which landed the first astronauts on the **moon**, as well all six of the **space shuttles**. In the 1980s, it designed the **X-30** as part of the **National Aero-Space Plane** project to be used as a passenger spaceliner; however, the project was ultimately cancelled in the early 1990s. The company split into two companies in 2001: Rockwell Automation and Rockwell Collins.

ROENTGEN SATELLITE (ROSAT). A **German Aerospace Center satellite**, NASA and the United Kingdom contributed instruments to the telescope, which launched on June 1, 1990, and operated for over eight years before it reentered Earth's atmosphere on October 23, 2011, over the Bay of Bengal. Originally an 18-month mission, it ceased operations in February 1999. ROSAT contributed significantly to the all-sky survey catalog during its years in space. Although details have not been released and there is no proof of causation, its termination was at least coincident with a cyber intrusion at **Goddard Space Flight Center** from **Russia**, and much speculation remains about the hacking activities. ROSAT was named in honor of Wilhelm Conrad Röntgen, the German physicist who was awarded the Nobel Prize in physics in 1901 for his work on X-rays. See appendix B for a list of ROSAT's instruments.

ROGERS COMMISSION. President Ronald Reagan created a commission chaired by former secretary of state William Rogers (hence the name) to investigate the cause of the **space shuttle *Challenger* tragedy**. After weeks of hearings, the five-volume report was released on June 9, 1986, six months after the accident. Its main findings urged NASA—and particularly its **Marshall Space Flight Center**—to modify safety features and to change its organizational communication and handling of space shuttle deployments, and faulted Morton Thiokol for poor engineering in the booster motors and assembly. The commission, with help from physicist **Richard Feynman**, identified the immediate cause of the accident as a failure of the rubber O-rings that sealed the joint between two parts of the solid rocket **booster**, causing a leak that then expanded within seconds and exploded the external fuel tank. The potential O-ring failure was a known problem, as confirmed by **Roger Boisjoly**, a mechanical engineer and fluid dynamicist for Morton Thiokol. The commission found that NASA's attempt to force launch 24 shuttle missions a year was beyond its capability with regard to **budget**, personnel, and spare parts. Because of its enhanced schedule, the **U.S. Air Force** had begun buying **Titan launch vehicle**s instead of relying on access to launch facilities for its national security satellites the year before, another sign of NASA's preoccupation with the space shuttles.

In its report, the commission found nine broad areas of recommended improvement:

- Design of the solid rocket motor and independent oversight of its design
- Review of the shuttle management structure, with input from **astronauts** and the establishment of a safety panel
- Analysis and review of all potential hazards on critical aspects of flight
- Establishment of an Office of Safety, Reliability, and Quality Assurance
- Improved communication structure, particularly at Marshall Space Flight Center
- Improved landing safety at all abort landing sites and at **Kennedy Space Center**
- Implementation of launch abort and crew escape systems
- Review of NASA's reliance on the shuttle and its flight rate

NASA published its list of Actions to Implement the Recommendations of the Presidential Commission on the Space Shuttle *Challenger* Accident on July 14, 1986. By June 1987, NASA had completed (or nearly so) and published a record of those implementations, and shuttle flights resumed in 1988.

The 14 commission members included **Neil Armstrong**, **Sally Ride**, and **Chuck Yeager**, among other notable individuals.

ROSCOSMOS. *See* RUSSIAN FEDERAL SPACE AGENCY (RKA or FKA).

ROSETTA. Rosetta is a **European Space Agency** (ESA) mission to investigate comets. NASA contributed three instruments, one (**ALICE**) to map the comet's nucleus for dust and ice, and two (MIRO and IES) to check for water coming off the nucleus. NASA also contributed part of the electronics for a fourth instrument (ROSINA). Rosetta's primary mission, launched March 2, 2004, is to catch and track at close range comet 67P/Churyumov-Gerasimenko as it makes its way through our solar system and as it gets closer to the sun. The spacecraft reached the comet on August 6, 2014, and will continue orbiting the comet for at least 17 months. It sent a lander, Philae, to the comet's surface (site J) on November 12, 2014, but Philae landed in shadow. Philae awoke after sunlight reached the comet in June 2015 and transmitted data through late July 2015, when too much dust accumulated on its surface and its solar panels could not be recharged. Secondary missions included observation of **Deep Impact** in 2005, of **Mars** during the ESA's Mars Gravity Assist maneuver in 2007, and of two asteroids in 2008 and 2010. See appendix B for a list of the NASA-contributed instruments.

ROVER. (a) Project Rover was a program to develop a nuclear thermal rocket that NASA inherited from the **Atomic Energy Commission**. It had three phases: Kiwi (named after the flightless bird) from 1955 to 1964; Phoebus from 1964 to 1969, and Pewee, from 1969 until 1972, when the project was **canceled.** Most of the work was conducted at Los Alamos and testing was completed at the Nevada Test Site. The active part of the program was followed by the Nuclear Engine for Rocket Vehicle Application (NERVA) program that in 1969, following **Apollo**'s success, encouraged **Wernher von Braun** to propose sending 12 men to **Mars** on two rockets, each with three NERVA engines, in November 1981.

(b) Rovers can also be a (usually) six-wheeled vehicle landed on another celestial body (the **moon**, Mars) for exploratory and scientific purposes. **Mars Exploration Rovers *Spirit*** and ***Opportunity*** have spent 10 years on

Mars, and ***Curiosity*** has been there since August 2012; rovers from *Apollo 15*, *16*, and *17* remain on the **moon**, as well as two from the **Soviet Union**, and one from **China**.

RUSSIA. Although relations between NASA and Russia are generally civil, there have been instances where cooperation was less than ideal, such as in the wake of the Ukrainian unrest in spring 2014 when NASA suspended cooperation and travel between the two agencies, unless they were specifically excepted (such as work on the **International Space Station** [ISS] as well as travel to and from the space station in **Soyuz** spacecraft). The unease was one of many factors—perhaps the primary one—that led to the announcement some months later by NASA that **Boeing** and **SpaceX** would be returning spaceflight to American soil in 2017 as part of its **Commercial Crew Program**.

Russia had earlier indicated it might end its involvement with the ISS in 2020, an issue the United States had also been considering; the ISS at that point will be more than 20 years old, and the **Soviet Union**'s experience with **Mir** indicated that 30 years was a maximum lifetime for a space station, both in terms of use (the damages that comes from everyday activities) and more modern equipment (newer, incompatible instruments and modules). The problem with what to do with a disabled ISS is yet to be solved, as each country is responsible for its own modules and equipment.

RUSSIAN FEDERAL SPACE AGENCY (RKA or FKA). Also known as Roscosmos, RKA is located in Moscow, with their **Cosmonauts** Training (**astronaut training**) Center in nearby Star City. Launch facilities (which NASA is currently using) are located in Kazakhstan. Founded in 1992, having been preceded by an equivalent organization of the **Soviet Union**, RKA is one of the major partners of the **International Space Station** (ISS) and in 2008 agreed to sell to NASA spots on its **Soyuz** spacecraft for transit to the ISS at $21 million per person, per trip. The RKA also provides **space tourism** flights to the ISS at an equivalent cost.

In September 2014, RKA announced that it will begin a full-scale exploration of the **moon**, expected to begin in the late 2020s. Before that, it has a major **Mars** mission, ExoMars, in which it is collaborating with the **European Space Agency**, to launch in 2016–2018, and a mission to land a probe on **Europa** (one of **Jupiter**'s moons) as part of the ESA's Jupiter Icy Moons Explorer.

See also RUSSIA.

S

SAGAN, CARL EDWARD (November 9, 1934–December 20, 1996). Born in New York City, Sagan did more for the popularization of astronomy and space science than almost anyone else. He earned four degrees from the University of Chicago in the arts, physics, astronomy, and astrophysics. Through his work with the **Jet Propulsion Laboratory** as a visiting scientist, he participated in the design and management of ***Mariners*** *2* and *9*, ***Vikings*** *1* and *2*, ***Voyagers*** *1* and *2*, ***Galileo***, and many others. He wrote more than 600 scientific articles and more than 24 books, winning a Pulitzer Prize for his *The Dragons of Eden*. *Cosmos*, the television show he hosted beginning in 1980 and for which he received two Emmy Awards (and recently resurrected by **Neil deGrasse Tyson**), was seen by more than 500 million people in 60 countries. He was a founding member of the Planetary Society.

Working as an advisor with NASA beginning with its inception, Sagan briefed the **Apollo astronauts** before their trip to the **moon** and was the designer of the gold plaque attached to ***Pioneers*** *10* and *11* as well as the gold disk on *Voyagers 1* and *2* that could potentially be understood by extraterrestrial beings, should the spacecraft encounter any. Sagan was greatly involved in **education and public outreach** for astronomy and the sciences.

SAMPLE RETURN MISSIONS. Samples have been collected from five celestial bodies other than Earth: the **moon**, comet Wild 2, asteroid 25143 Itokawa, and two meteorites (one from **Mars**, and one from Vesta, an asteroid). **Moon rocks** were the first sample returned to Earth when ***Apollo*** *11* brought back almost 50 lbs. of lunar material in 1969. In June 2010, **Japan Aerospace Exploration Agency**'s **probe** returned dust from an asteroid for the first time, and they are scheduled to launch a probe for such purposes again in 2015. Samples have also been returned of **solar wind**.

See also SOIL SAMPLES.

SATELLITE. A satellite is a natural or man-made object that orbits a much larger celestial body. Natural satellites include our **moon** as well as the moons of other planets. Man-made satellites include those that circle the

Earth for various reasons—weather, communication, data, research, and so forth—as well as those that orbit other celestial bodies. A **probe** sent to another celestial body (planet, asteroid, etc.) is also considered a satellite when as part of its mission it orbits the larger body, though probes are generally thought of as smaller and more disposable (i.e., sent on self-destructing exploratory plunges).

SATURN (PLANET). Known for its rings, Saturn is the sixth planet from the sun and home to more than 150 moons and moonlets that comprise the rings. Titan, one of its moons, is larger than the planet **Mercury**. Along with **Jupiter**, Saturn is known as a gas giant for its thick atmosphere of both hydrogen (metallic and liquid) and helium, and its dense core. It orbits an average of 9.4 **astronomical units** (869 million miles, 1.4 billion km) from the sun, and takes about 29 Earth years to complete one orbit. The origin of the rings is uncertain; their particles range from specks to bodies of water-ice, tholin, and carbon about 30 feet in diameter.

Pioneer *11* flew by Saturn in September 1979, but most of our data comes from the subsequent flybys of ***Voyager*** *1* and *2* in 1980/1981 and especially from the **Cassini–Huygens** spacecraft and probe since 2004. Cassini has been orbiting Saturn since then, while Huygens was a probe landing on Titan. Among other findings, we now know that Saturn's hurricanes are 20 times larger than those found on Earth, lightning is 1,000 times more powerful, and there are geysers on its moon Enceladus.

SATURN (ROCKETS). Saturn launch vehicles have been the only ones to send humans beyond **low Earth orbit** to date (i.e., to the moon). Somewhat equivalent rockets are the N1 rocket family from **Russia** and the **Falcon** Heavy from **SpaceX**. Although initially developed as a military **satellite** launcher before NASA's formation in the **National Air and Space Act** in 1957, they were modified from the **Juno** V for the **Apollo** program during the mid-1960s through the early 1970s. Three versions were flown: Saturn I (1961–1965) and Saturn IB (1966–1975) during early tests and manned flights as well as **Skylab** 2–4 and **Apollo–Soyuz Test Project**, and the more powerful Saturn V, which was used beginning with *Apollo 8* between late 1969 and 1972, and which included all the flights to the **moon**.

The Saturn family of launch vehicles were all multistage rockets. Ten Saturn I rockets were launched; the first four were designated Block 1, which included an S-1 first stage (manufactured by Chrysler) and dummy upper stages. Block 2, the remaining six flights, included an uprated S-I stage and an S-IV second stage (built by Douglas Aircraft). All Block 2 S-IV engines used liquid hydrogen and liquid oxygen as propellants. The following version, Saturn IB, featured an uprated S-1 first stage and a new S-IVB second stage, developed by McDonnell Douglas. These S-IVB second stages become the third stage for the Saturn V rockets, while the second stage was an S-II and the first stage an S-IC. Apollo's **command and service module** (CSM) sat at the very top, above the instrument unit and the **lunar module** (LM). An emergency escape system was above the command module.

Saturn V's first stage (five F-1 engines) ignited at liftoff and lasted for about 2.5 minutes, carrying the entire spacecraft to an altitude of 38 miles, before it separated and burned up. The second stage (five J-2 engines) burned for six minutes, taking the spacecraft to outer space, or an altitude of 115 miles, before it too was jettisoned. The final and third stage was one J-2 engine, which took what was left of the spacecraft to orbital velocity (17,500 mph) in about 3 minutes. It was then shut down until the spacecraft reached the moon, when it was reignited to propel the CSM/LM to a translunar trajectory before it was also discarded. The CSM was the only part designed to return to Earth in one piece.

SCHIRRA, WALLY (March 12, 1923–May 3, 2007). One of the original **Mercury Seven** astronauts, Schirra was the first person to go into space three times, as well as the only **astronaut** to have taken part in the **Mercury**, **Gemini**, and **Apollo** programs. Born in Hackensack, New Jersey, he later received a BS degree in aeronautical engineering from the U.S. Naval Academy in 1945 and served in the last months of World War II as well as the Korean War, where he worked as a pilot on loan to the newly formed **U.S. Air Force** in 1951. One of 110 military test pilot candidates for NASA's Project Mercury, he specialized in life support systems after he was selected. During the Gemini program's test rendezvous, he piloted his *Gemini 6* spacecraft as close as 1 foot to *Gemini 7*, an amazing feat at the time. Later, during the Apollo program, Schirra and his crew became backup to the ill-fated ***Apollo 1*** mission in January 1967, thus taking part in the next Apollo flight (renumbered as *Apollo 7*) on October 11, 1968. He took the redesign of the **command and service module** seriously and hounded NASA management mercilessly until they complied with his engineering requirements, going so far—from the safe remove of space—to let Houston know that he was first and foremost an engineer and pilot, and only after that a scientist and TV crew. He received an Emmy award for the first public broadcast from inside a crewed spacecraft. After NASA, he had a variety of business and entrepreneurial ventures, a part-time television career, and co-authored several books dealing with the **U.S. Navy** and space. He was the recipient of many awards, medals, and honorary degrees, and a Navy ship, the USNS *Wally Schirra*, was named for him in March 2009.

SCIENCE ADVISORY COMMITTEE. *See* PRESIDENT'S SCIENCE ADVISORY COMMITTEE (PSAC).

SCIENCE MISSION DIRECTORATE (SMD). Also known as Science Directorate, the SMD is a division within NASA that coordinates many of the "big picture" areas of scientific discovery and exploration—the Earth, solar system, the sun, and the universe. SMD's focus is particularly aimed at remote sensing, **satellite** technology, and robot exploration, especially regarding future exploration of space. SMD is itself comprised of four divisions: (1) Earth Sciences: atmosphere, climate, and oceans; (2) Planetary Science: scientific knowledge of the solar system, its origins, potential for life, hazards to human life, and foundations for exploration; (3) Heliophysics: the study of the sun, its atmosphere, magnetic fields, cycles, effects on Earth, and exploration; and (4) Astrophysics: the universe and its structure, evolution, history, origin, destiny, and the search for Earth-like planets.

SELECT COMMITTEE ON ASTRONAUTICS AND SPACE EXPLORATION. In the aftermath of ***Sputnik 1***'s historic flight in October 1957, the U.S. Congress established the Select Committee on Astronautics and Space Exploration early the following year. This was the organization that created the **National Aeronautics and Space Act** that then chartered NASA as well as the permanent **House Committee on Science and Astronautics** (now known as the Committee on Science, Space, and Technology).

SENATE COMMITTEE ON COMMERCE, SCIENCE, AND TRANSPORTATION. Through its Subcommittee on Science and Space, this committee has responsibility for the oversight of NASA, as well as the **National Science Foundation**, the National Institute of Standards and Technology, and the Office of Science and Technology Policy. One of its more recent activities (spring 2014) was the "From Here to **Mars**" hearing featuring experts on NASA's exploration strategy, on international collaboration in space, and on commercial space efforts. The hearing featured William Gerstenmaier, associate administrator of NASA's **Human Exploration and Operations**; Leroy Chiao, a former NASA **astronaut**; Jeffrey Manber, managing director of **NanoRacks**; and Susan Eisenhower of the Eisenhower Institute, a Washington, D.C., center for leadership and public policy.

SERVIR. SERVIR (Spanish for "to serve") is a NASA mission in partnership with the U.S. Agency for International Development focused on providing climate **satellite** imagery, geospatial data, visualization, and decision support tools to developing nations to improve their planning and response to natural disasters, climate change, air quality, and weather prediction. SERVIR operates from regional hubs to provide access to data and form partnerships with governments, disaster responders, and other regional groups.

SHARED SERVICES CENTER (NSSC). NSSC (or NASA Shared Services Center) was established at **Stennis Space Center** in Mississippi in 2006 to consolidate NASA business services. NSSC provides central business functions for NASA in terms of human resources, information technology, finances, procurement, and other business support services. NSSC received a Best New Captive Shared Services Organization Excellence Award in 2009, a Leadership in Energy and Environmental Design silver certification in 2010, and a Government Customer Support Excellence Award in 2011, in addition to other nominations and awards. It employs approximately 125 people.

SHEPARD, ALAN (November 18, 1923–July 21, 1998). Shepard was the first American to fly in space and the only **astronaut** of the original **Mercury Seven** to walk on the **moon**. In addition to Project **Mercury**, he also commanded the **Apollo** 14 mission and is the oldest person (at age 47) to have walked on the moon. Born in Derry, New Hampshire, Shepard graduated from the U.S. Naval Academy with a BS in engineering in 1944 and served during World War II in the Pacific theater. He learned to fly in the **U.S. Navy** and eventually became an instructor in the navy's Test Pilot School. After being chosen as an astronaut, he was scheduled to fly in October 1960 and would have become the first person in space, but the launch was delayed several times. On April 12, 1961, **Yuri Gagarin** beat him into space, and Shepard followed three weeks later in the ***Freedom*** *7*, traveling some 300 miles in space during a 15-minute flight. He was originally scheduled to be part of the **Gemini** mission, but health problems (inner ear pressure) for five years kept him grounded, so he served as chief of the astronaut office, responsible for NASA's coordination among the astronauts. He returned to flight status in 1969, and commanded the *Apollo 14* mission in early 1971, for the third landing on the moon—during which he also managed to hit two golf balls on the moon. After his successful Apollo mission, Shepard created several business ventures and foundations after his retirement from the navy and NASA (after 15 years, longer than any other Mercury Seven astronaut) in 1974. He was the recipient of many medals, awards, and honorary doctorates, and in 2011 the U.S. Postal Service issued a first-class stamp in his honor, the first time an astronaut had been so honored.

SHIP–AIRCRAFT BIO-OPTICAL RESEARCH (SABOR). SABOR was a NASA experiment over the summer of 2014 that brought together two different science platforms—airborne remote sensing and ocean ships—to study the carbon cycle, phytoplankton, and organic and inorganic material in the ocean. The SABOR experiment used measurements from reflected light, polarization, and aerosol laser radar (LIDAR) from **satellites**, aircraft, and surface ships to gather this data. Beyond the study of phytoplankton and ocean particles, this combination of platforms validates instruments, models, and measurements and will be useful for planning future instrumentation and satellite technology. SABOR was managed by the **Earth Science Project Office** in Moffett Field, California.

SHUTTLE RADAR TOPOGRAPHY MISSION (SRTM). SRTM was a NASA **space shuttle** mission sponsored by NASA and the National Geospatial-Intelligence Agency in February 2000 on STS-99 aboard the shuttle ***Endeavour*** to create a digital elevation model of the Earth's surface from 60° north to 56° south. This mission used interferometric synthetic aperture radar by deploying two radar antennas from the *Endeavour*'s payload bay and achieved one arc second (30-meter resolution) in the raw data. Public data was released with three arc second (90-meter resolution) with the exception of the United States and Australia, where 30-meter resolution data was released. These digital elevation models are commonly used in geographic information systems and are freely available. See appendix B for a list of its instruments.

SHUTTLE-MIR. Initiated in 1993, Shuttle-Mir was a joint program between NASA and **Russia** that would allow American **astronauts** to learn from the more-experienced Russians in space in their **Mir** space station and for Russian cosmonauts to fly in the **space shuttle**. During the five-year first phase of the mission, however, tensions rose after a fire aboard Mir caused by the malfunction of the solid-fuel oxygen generator revealed that backup protection was limited and faulty (broken respirator masks, unavailable fire extinguishers, and blocked access to the emergency vehicle). A near-collision with a resupply spacecraft and a later collision with solar arrays that caused Mir to lose pressure led to concerns about safety, and a lack of communication between Russian and U.S. administrators during those incidents led to a lack of trust. In addition, some felt that the $325 million paid to Russia for U.S. involvement was the only thing keeping Mir afloat, given the status of the Russian economy at the time. Still, the beginnings of cooperation for what would eventually become the **International Space Station** emerged during the seven long-term residencies aboard Mir.

SINGLE APERTURE FAR-INFRARED OBSERVATORY (SAFIR). SAFIR is a proposed cryogenic space observatory for detecting far-infrared light as part of NASA's **Origins Program**. The plan calls for a single large mirror, 5–10 meters in diameter, cooled to −268°C. SAFIR will study the earliest phases of forming galaxies, stars, and planetary systems at wavelengths where these objects are brightest. Most of this section of the electromagnetic spectrum is not observable from the ground due to absorption by moisture in Earth's atmosphere. Its five-year mission time does not yet have a launch date, although it had been planned for no earlier than 2015.

SKYLAB. Skylab was NASA's first space station and was deployed from 1973 until 1979. The Skylab program made considerable reuse of **Apollo**-era equipment. The space station itself was built from a modified S-IVB, the third stage of a **Saturn** V, which weighed approximately 199,750 lbs. (90,606 kgs) and provided an approximate habitable space of 12,700 cubic feet (354 cubic meters). The station contained an orbital workshop that housed crew quarters and a large experiment center, an airlock module, a multiple docking adaptor, and the Apollo telescope mount, which contained solar telescopes and primary attitude-control gyros. Skylab was launched on a two-stage Saturn V rocket and crew members were shuttled to Skylab using Apollo-era **command and service modules** launched on Saturn IBs. Three crew rotations with three crew members per mission occurred: Skylab 2 with 28 days on station, Skylab 3 at 59 days, and Skylab 4 at 84 days. The first crewed Skylab mission, Skylab 2 or SL-2, initially focused on repairing critical solar panel and solar shade damage sustained by the station during launch. Subsequent science experiments by the SL-2 crew and the crews that followed accomplished many space station firsts, especially in the area of solar studies and long-term human space habitation. The Skylab program incorporated human factors into the interior design such as the food service area, smooth and easy-to-clean paneling, a shower and toilet facilities (which were an upgrade from previous designs), and the large viewing portal. At times, the many station repair and maintenance activities involved difficult **extravehicular activities**. A solar observatory attached to Skylab, the Apollo telescope mount, hosted hydrogen-alpha telescopes, various ultraviolet spectrographs and sensors, a visible light coronagraph, and X-ray telescopes that were first to confirm the existence of coronal holes. In addition, crews collected Earth observation data, performed space materials science experiments, zero-gravity studies, spacecraft environment studies, and conducted 19 specially selected high school student experiments. It was intended that Skylab remain functional into the **space shuttle** era but higher-than-anticipated solar activity heated and expanded Earth's atmosphere, increasing drag on the station and eventually forcing it to reenter the atmosphere on July 11, 1979, over southwestern Australia. Refer to appendix B for a list of instruments.

SMALL EXPLORER MISSIONS. *See* EXPLORERS PROGRAM.

SMALL SATELLITES. Small Satellites (nanosatellites) generally weigh between 1 and 200 kilograms (2 and 440 pounds) and are used by NASA to cheaply test new and emerging technology often using off-the-shelf technology. Small **satellites** typically ride as secondary payload and have been launched from the **International Space Station**.

See also CUBESAT LAUNCH INITIATIVE (CSLI).

SMALL SPACECRAFT TECHNOLOGY PROGRAM (SSTP). SSTP is a program within the **Space Technology Mission Directorate** with an objective of developing new small spacecraft technologies, capabilities, and to use small spacecraft as a test bed. This technology program collaborates with universities and private companies to develop small spacecraft experiments and tests in a variety of areas such as propulsion, guidance, communications, and scientific study. One example of an SSTP project is the Edison Demonstration of SmallSat Networks, a swarm of eight **CubeSats**, each about the size of a tissue box, that would travel as a group measuring space radiation for a period of about 60 days from **low Earth orbit**. SSTP selections generally involved university competitions.

NASA defined small spacecraft in varying terms, but generally they are:

- Minisatellite, 100 kg or more
- Microsatellite, 10–100 kg
- Nanosatellite, 1–10 kg
- Picosatellite, 0.01–1 kg
- Femtosatellite, <0.01 kg

SOHO. *See* SOLAR AND HELIOSPHERIC OBSERVATORY (SOHO).

SOIL SAMPLES. Analysis of samples from nonterrestrial origins contribute to core NASA missions such as origin of the solar system, presence of the chemical building blocks of life, and the possible existence of extraterrestrial life forms. NASA first returned soil samples from another solar system body on ***Apollo 11*** in 1969 from Earth's **moon**. Subsequent Apollo missions returned more lunar material (including **moon rocks**), which ended with *Apollo 17* in 1972. NASA has returned comet particles in 2006 from the **Stardust** sample return capsule, and **solar wind** particles from the **Genesis** spacecraft in 2004. Soil samples have also been analyzed on site by past and current NASA **rovers** and spacecraft using on-board instruments such as mass and laser spectrometers and gas chromatographs. Beyond sample return and on-location study, other sources of extraterrestrial samples for study come from meteorites.

See also CURIOSITY; OPPORTUNITY.

SOJOURNER. *Sojourner* is the **Mars Pathfinder** robotic **rover** that landed on **Mars** on July 4, 1997, and explored its surrounding area for around three months. It has front and rear cameras and hardware to conduct several scientific experiments, including X-ray spectrometer observations. The rover was active for 83 sols (Martian days) and traveled a distance of just over 330 feet (100 m) by the time communication was lost on September 27, 1997.

SOLAR AND HELIOSPHERIC OBSERVATORY (SOHO). A spacecraft built in collaboration between the **European Space Agency** and NASA and launched in 1995, SOHO resides between the Earth and the sun in **Lagrange point 1** (L1), a point in space 1.5 million kilometers from Earth that allows the spacecraft to remain in a nearly stationary location. SOHO was designed to study the sun's helioseismology, **solar wind**, chromosphere, and corona. SOHO provides a three-day warning for Earth-facing solar disturbances and transmits approximately 200–300 kbits of data through NASA's **Deep Space Network**. SOHO lost its gyroscopes in 1998 and the subsequent recovery stabilized the spacecraft using a software update and its on-board reaction wheels, making it the first spacecraft to be stabilized without gyroscopes. SOHO instrumentation enables the discovery of comets by scientists and amateur astronomers, with a comet discovered on average every 2.5 days. See appendix B for a list of SOHO's instruments.

SOLAR ANOMALOUS AND MAGNETOSPHERIC PARTICLE EXPLORER (SAMPEX). SAMPEX was an Earth near-polar orbiting **satellite** designed to study particles from supernova cosmic rays, anomalous cosmic rays, solar energetic particles, and magnetospheric electrons from the **solar wind**. Built in a collaboration between NASA and the Max Planck Institute for Extraterrestrial Physics of Germany, its primary mission operated from launch in July 1992 to mission end in June 2004. The German-built heavy ion large telescope continued to collect data until November 2012, shortly before the satellite reentered Earth's atmosphere on November 13, 2012. SAMPEX advanced understanding of cosmic rays and trapped components in Earth's radiation belts, solar energetic particles, and related solar events, the behavior of electrons in Earth's radiation belts, and other detailed findings around particle composition of the middle atmosphere. See appendix B for a list of SAMPEX's instruments.

SOLAR DYNAMICS OBSERVATORY (SDO). The SDO is a **geosynchronous satellite** and part of the **Living with a Star** program, launched on February 11, 2010, to study the sun in multiple wavelengths and especially in regards to the impact of solar variability on Earth. The combination of instruments allows for detailed study of solar cycles, magnetic changes, solar irradiance, Doppler shifts, and surface ripples, and energetic events such as **solar flares**. The SDO generates approximately 1.5 terabytes of data a day. It does not store data onboard but downloads it directly to a semi-automated ground station at the **White Sands Complex**. On October 1, 2015, SDO began its first extended mission of two years.

See also SOLAR DYNAMICS OBSERVATORY GROUND STATION (SDOGS).

SOLAR DYNAMICS OBSERVATORY GROUND STATION (SDOGS). SDOGS is a NASA-run semi-automated ground station consisting of two 18.3-meter Ka-band radio antennas located within the **White Sands Complex**. SDOGS is operated remotely from NASA's **Goddard Space Flight Center** and was designed with failover capabilities, real-time data capture and transmission, high data volume, and 24-7 operations. The success of the SDOGS design led to considerable reuse of its hardware and software in the **Lunar Reconnaissance Observatory** ground station program and represented a new level of NASA ground station capability. SDOGS is also a backup to the Lunar Reconnaissance Ka-band ground station.

SOLAR FLARE. A solar flare is an intense burst of energy associated with sunspots and appears to us as a flash of light, taking as little as eight minutes to reach the Earth (unlike **coronal mass ejections**, which can take from one to three days). It is the release of magnetic energy across many wavelengths and, like optical light, that energy travels toward Earth. Solar flares can last for minutes or hours, depending on their intensity. Once that magnetic energy reaches Earth's atmosphere, it can affect long-range radio communications, radar, and similar equipment, causing disruptions and even temporary blackouts, and it strongly influences **space weather**. Although scientists cannot yet predict when solar flares will occur, they have found some indications of potential flares; several missions (such as **Solar Dynamics Observatory**,

Genesis, and **Advanced Composition Explorer**) are specifically studying solar flares, coronal mass ejections, and the **solar winds** that carry the magnetic particles.

SOLAR PROBE PLUS. NASA's Solar Probe Plus will study the sun's outer atmosphere, **solar wind** and creation, inventory, and movement of solar particles by flying through the sun's corona as close as 5.9 million kilometers from the solar surface. It will accomplish this close approach by retracting its solar arrays and protecting the spacecraft with a 4.5-inch composite shield and a whole-craft Thermal Protection System. Solar Probe Plus will directly analyze particles and take 3D images of the sun's corona, take direct measurements of magnetic and electrical fields, and transfer its data using X/Ka-band high-speed downlinks. It is currently under development by John Hopkins University's **Applied Physics Laboratory** with an original goal of launching no later than 2015, but now anticipated for 2018, and with a planned mission of just under seven years. See appendix B for a list of its proposed instruments.

SOLAR RADIATION AND CLIMATE EXPERIMENT (SORCE). SORCE is a NASA **satellite** that measures solar radiation received by the Earth in the ultraviolet spectrum, the full spectrum of visible and near infrared wavelengths, X-rays, and a measure of total solar irradiance. The measurements of solar radiation by SORCE contribute to Earth–sun studies and provide analysis and prediction of climate change, natural solar variability, ozone, and UV-B radiation. SORCE downloads its data through NASA's **Wallops Flight Facility** in Virginia and a station in Santiago, Chile, and is operated by Laboratory for Atmospheric and Space Physics at the University of Colorado at Boulder. See appendix B for a list of SORCE's instruments.

SOLAR STUDIES. NASA studies the sun and solar interaction with the Earth, solar system, and galaxy to understand fundamental processes that affect life on Earth, **astronaut** and **satellite** safety, and key physical interactions of the sun and solar system that model processes throughout the galaxy. Areas of solar study include solar magnetic fields, changes in energy and matter within solar processes, the Earth-sun interaction, **solar wind**, energetic particles, changes in solar output and the bubble of the heliosphere overall that defines the boundaries of the solar plasma bubble and its interaction with the interstellar medium.

See also LIVING WITH A STAR (LWS).

SOLAR TERRESTRIAL PROBES (STP) PROGRAM. The STP program is a part of the **Science Mission Directorate**'s heliophysics division; science priorities for the program focus on filling knowledge gaps in the fundamental connections between the sun and solar system, the sun and Earth, and conditions of solar system space. Objectives in the STP program include understanding the impact of solar variability on planets and human society, technology, and predicting conditions in space in order to maximize productivity of humans, **probes**, and robotic exploration. The program gathers data from multiple platforms, sensors, **satellites**, and ground-based observations.

Current STP missions include the **Magnetospheric Multiscale**, the **Solar Terrestrial Relations Observatory**, **Hinode (Solar-B)**, and **Thermosphere, Ionosphere, Mesosphere Energetics and Dynamics** missions.

SOLAR TERRESTRIAL RELATIONS OBSERVATIONS (STEREO). The third of NASA's **Solar Terrestrial Probes** program, STEREO is a set of two identical spacecraft launched on October 25, 2006, one ahead of the Earth on its orbit around the sun (STEREO-Ahead) and the other trailing after Earth (STEREO-Behind), to study the sun and its **coronal mass ejections (CMEs)**. Because of their two positions, the observatories provide a three-dimensional view of the geomagnetic storms from the sun by mapping the CMEs and studying effects on Earth. Those effects can damage or destroy **satellites**, are extremely hazardous to **astronauts**, and have caused electrical power outages. The two work with two other observatories—the **Solar Dynamics Observatory** and the joint **European Space Agency**/NASA **Solar and Heliospheric Observatory**. The combination allows them to model CMEs, particularly of the July 2012 eruption as it exploded from the sun at 1800–2200 miles per second. STEREO is operating in an **extended mission** capacity, following its original three-year mission.

Contact was reestablished in July 2015 with STEREO-A as it emerged from the far side of the sun after it had been put in "safe mode." However, contact with STEREO-B was lost in October 2014 and has not been reestablished. See appendix B for a list of their instruments.

SOLAR WIND. Temperatures of the sun's corona become so hot that gravity alone cannot hang on to the plasma, so it streams off from the sun from **solar flares** and **coronal mass ejections** at speeds from about 300,000 to 1,000,000 miles per hour, the driving force behind **space weather**. This streaming mass of particles carries with it magnetic clouds that, as they pass by Earth, wreak havoc with our magnetosphere, affecting the flight paths of spacecraft by causing geomagnetic storms and affecting power grids and communication on Earth.

Ulysses, retired on June 30, 2009, measured solar wind speed, magnetic field strength, and composition. The **Advanced Composition Explorer**, located at Earth-sun **Lagrange point** 1 (L1), now monitors solar wind and reports data on a real-time basis, and from the **European Space Agency**'s **Cluster II** mission, scientists have found that it is potentially easier for the solar wind to infiltrate the magnetosphere than they had earlier believed. **Solar Terrestrial Relations Observations**' STEREO-A released new images in October 2015 that contributed to understanding solar "windsocks." Because the sun is such an important part of life as we know it, solar studies and studies of the winds that bring the sun's properties to us are vital to understanding our planet's continuing existence.

SOUNDS OF EARTH. Sounds of Earth is the name of the **Voyager** phonographic Golden Records that contain sounds and images selected to portray the diversity of life and culture on Earth. They are intended for any intelligent extraterrestrial life form, or for future humans, who may find them. The committee that chose the sounds—ranging from wind, rain, and surf to tractors and Morse code—was chaired by **Carl Sagan**.

SOVIET UNION. Although the United States was already preparing **satellites** for space exploration and observation, the Soviet Union's ***Sputnik 1*** provided the impetus for increased attention to an expansion upward into space, beginning the **space race** between the two countries. Both countries had long histories of scientific research and breakthroughs. The **International Geophysical Year** (1957–1958) highlighted accomplishments from 67 countries, but it was clear that the Soviet Union and the United States—already in a cold war—were poised to make significant discoveries.

The Soviet Union's push into space had a number of firsts after *Sputnik 1*: the first **animal in space** (on *Sputnik 2*), the first human in space (**Yuri Gagarin**, who completed the first orbit around Earth), the first woman in space (**Valentina Tereshkova**), the first **extravehicular activity**, the first impact on the **moon** (September 14, 1959), the first image of the far side of the moon (October 7, 1959), the first robotic soft landing on the moon (February 3, 1966), the first space **rover** (*Lunokhod 1*, November 1970), the first **space station** (Salyut 1, April 19, 1971; it reentered Earth's atmosphere six months later), and the first interplanetary **probe** (Luna 9, January 31, 1966), among many others. The Soviet Union maintained a vital and brave program until its breakup in 1992.

See also RUSSIA; RUSSIAN FEDERAL SPACE AGENCY (RKA or FKA).

SOYUZ. Soyuz ("Union" in Russian) is both a launch vehicle and a spacecraft developed in the **Soviet Union** in the 1960s. Currently, it is the only way to reach the **International Space Station** (ISS), and NASA has contracted for $424 million in astronaut delivery services through early 2017. Each Soyuz spacecraft now has capacity for a crew of three and consists of command, orbital, and service modules. There have been several upgrades of the Soyuz over the years, and one Soyuz spacecraft is kept docked at the ISS at all times in case of emergency. The **Russian Federal Space Agency** is currently working on the Prospective Piloted Transport System, which is expected to replace the Soyuz spacecraft and is similar to the **Orion** spacecraft of the United States.

The Soyuz multistage rocket is one of the world's most reliable launch vehicles. Soyuz 1 launched on April 23, 1967, but the parachute failed on reentry, killing its **cosmonaut**. Since then, there has only been one accident, in 1971, when the crew was also killed on reentry. There have been more than 125 successful launches using the Soyuz rockets, making it extremely reliable, and all but one have used RP-1 and liquid oxygen as a propellant, unlike most multistage rockets launching from the United States.

SPACE CAPSULE. A space capsule often refers to that portion of a spacecraft that can host a crew for a set period of time, including life support systems, water, and food as well as communication systems, protection from the environment, and some method to restrain **astronauts** during buffeting while on reentry or in spaceflight. Capsules cannot be flown on their own for an extended period of time. Almost all crewed rocket flights to **low Earth orbit** (with the exception of the **space shuttle**) and beyond have carried their human passengers in a capsule. Some capsules do not require human crew, but they are rare. At the current time, NASA has no unmanned capsules, but reentry capsules do exist on **SpaceX**'s **Dragon**; they were used on the former **Apollo command and service module** and some **Mercury** and **Gemini** spacecraft.

SPACE COMMUNICATIONS AND NAVIGATION (SCaN). SCaN is the program office for all of NASA's space communications activities. It manages and directs the ground-based facilities for the **Deep Space Network**, the **Near Earth Network**, and the **Space Network**. Together, the three networks support more than 100 NASA and non-NASA programs. Current research efforts in optical communications include the Lunar Laser Communications Demonstration and the Laser Communications Relay Demonstration. It has a strong **education and public outreach** program, focusing on grants and internships, as well as education in the science, technology, engineering, and mathematics fields.

SPACE DEBRIS. Of increasing concern is the amount of orbital debris floating around our planet. Latest estimates show approximately 1,000,000 pieces, although most of these are less than one cubic half-inch in volume and less than 150 miles above the Earth's surface. The greatest concentration of debris is at an altitude of about 500 miles. (For reference purposes, the **International Space Station** [ISS] operates about 250 miles above the surface.) The debris comes from expended rocket stages, aged **satellites**, old test weapons, fragments from disintegration, and so forth. The two greatest sources of space debris are the intentional destruction of the Fengyun-1C **weather satellite** in 2007 by **China** and the accidental collision of two communications satellites belonging to the United States and **Russia** in 2009 (the Cosmos-Iridium collision), which sent thousands of pieces of debris into orbit. Spacecraft such as the ISS have room to maneuver around debris greater than four inches that might pose a risk. The higher the altitude, the longer a piece of debris remains in orbit. At 400 miles, a piece might remain for decades, but at 250 miles, it will usually reenter Earth's atmosphere within a few years. At some point, the number of new pieces of space debris will outnumber the pieces that have eroded or deorbited, and collisions would escalate. Donald Kessler, an American astrophysicist and former NASA employee for whom the **Kessler syndrome** is named, initially estimated in 1978 that collisions would be imminent by the year 2000.

With the launch of ***Sputnik 1*** in 1957, the North American Aerospace Defense Command kept the Space Object Catalog, a list of all known launches as well as any objects that also might reach **low Earth orbit**, such as upper-stage rockets. Radar, LIDAR, and radio waves are used to track space debris, but these work best for larger pieces. The U.S. Strategic Command continues a catalog of known orbital objects. Tracking, however, is not enough. To reduce future space debris, the **United Nations Committee on the Peaceful Uses of Outer Space** published voluntary guidelines in 2007, and the International Standards Organization is planning a new standard to deal with mitigation. Russia, China, Japan, France, and the **European Space Agency** have all set mitigation guidelines. The ISS, for example, will have to eventually be dismantled and destroyed or returned to Earth, but this represents both technological and **budgetary** problems. Each country will be responsible for its own portion of the ISS at that time. The European Space Agency is planning a mission, called e.deorbit, for 2021 that will remove debris heavier than 8,800 lbs. (4,000 kg), but the method of capture—net, harpoon, robot arm, clamping—is yet to be determined. The **Japan Aerospace Exploration Agency** is planning a satellite to track debris from space, and a Swiss company is developing CleanSpace One, a spacecraft that would collect space debris and then return itself and the debris to Earth's atmosphere, where it would burn up.

Space debris was a major plot point in 2013's successful film *Gravity*.

SPACE ENVIRONMENT TESTBEDS (SET). SET is part of the **Living with a Star** program, which is itself part of the Heliophysics Division, and is concerned with how the sun–Earth interactions affect humans, both on Earth and in space. SET is studying how to improve hardware performance in the radiation environment of space and how to counteract the problems of **space weather**. The SET payload is scheduled for launch in September 2016.

SPACE EXPLORATION INITIATIVE (SEI). President George H. W. Bush announced the SEI on July 20, 1989—the 20th anniversary of ***Apollo*** *11*'s **moon** landing. His goals set forth an ambitious plan to return humans to the moon and to explore **Mars**. The initial study estimated its long-term cost at approximately $500 billion over 20–30 years. Because the cost was too prohibitive, NASA was then asked to focus on space and Earth sciences, forgoing for a time any manned missions beyond the **space shuttle** and the **International Space Station**. President Bill Clinton allowed the program to expire before anything was completed.

See also BUDGET/FINANCING.

SPACE EXPLORATION TECHNOLOGIES (SpaceX). Founded in 2002 by Elon Musk, known for his entrepreneurial involvement with PayPal and Tesla Motors, SpaceX's "ultimate goal [is] enabling people to live on other planets." To that end, they have created a family of launch vehicles (**Falcon**, **Dragon**) to send cargo and, later, humans to the **International Space Station** (ISS) and beyond. SpaceX is best known for its launch of the first liquid-fueled rocket, the Falcon 1, by a private corporation, to reach orbit, which it did on September 28, 2008. NASA **contracts** with SpaceX for some of its deliveries to the ISS as part of its **Commercial Orbital Transportation Services** program and in September 2014 was contracted along with Boeing to fly humans to the ISS as part of NASA's **Commercial Crew Program**. (Sierra Nevada Corporation filed suit against NASA's choice later that month.)

SpaceX's first major accident occurred on June 28, 2015, when a Falcon 9 rocket with a Dragon cargo ship carrying more than 4,000 pounds disintegrated shortly after liftoff as it was bound for the International Space Station. Later investigation showed that the failure of a single metal strut holding down a helium bottle allowed the helium to break free, which led to the explosion. This was the first major accident after 18 successful launches.

SpaceX also maintains contracts with other private companies as well as the U.S. military to launch **satellites** into **low Earth orbit**. Eventually, Musk hopes to send a **Mars** landing mission as the forerunner of a permanent human presence on Mars. Headquartered in Hawthorne, California, in the heart of aerospace technology, SpaceX also maintains offices and facilities in McGregor, Texas, and at Spaceport America in New Mexico.

SPACE FLIGHT MEDAL. NASA's Space Flight Medal is awarded to any civilian or military **astronaut**, mission specialist, payload specialist, or other individual to recognize participation in **human spaceflight missions**.

SPACE LAUNCH COMPLEX 6. *See* VANDENBERG AIR FORCE BASE.

SPACE LAUNCH INITIATIVE (SLI). SLI was a NASA/**Department of Defense research and development** program initiated in 2000 managed by **Marshall Space Flight Center**. The program studied the feasibility of a reusable launch vehicle (**X-33**, X-34) with second-generation technologies to replace the **space shuttle** program and to simultaneously decrease the risks of space transportation. The program was **cancelled** when NASA began to focus on the **Constellation** program, which was itself cancelled in 2010.

SPACE LAUNCH SYSTEM (SLS). Set for initial launch in November 2018, the SLS is a powerful deep-space multistage **rocket** program that is intended to propel manned spacecraft (via the **Orion Multipurpose Crew Vehicle**) to **Mars** by the 2030s. It is NASA's first attempt at developing a heavy-lift launch vehicle for human exploration in 40 years (since **Saturn** V carried **astronauts** to the **moon**) and will have an unprecedented lift capacity of 143 tons (130 metric tons). NASA estimates that the total cost of developing the first of the three planned launch vehicles in the SLS family is an estimated $18 billion, although the Government Accountability Office stated that that estimate may be $400 million short in a report issued in July 2014. NASA's estimate does not include the Orion Multipurpose Crew Vehicle (another $6 billion) and upgrades to launch pads and other facilities (another $2 billion).

SLS is a replacement for the **Constellation** program that was **cancelled** in 2010 by President Barack Obama. The first of the rockets will have a 77-ton capacity, will provide more than 10 percent thrust than the Saturn V, and have more than three times the payload of the **space shuttle**. It is expected that the SLS will provide backup for commercial and private flights (**SpaceX**, **Boeing**, **Orbital ATK**) as part of its **Commercial Crew Program** and additional missions to the **International Space Station**.

Exploration Mission 1 (scheduled for 2018) will comprise an uncrewed Orion spacecraft with the SLS, launch to a stable orbit beyond the moon, and return to Earth. Exploration Mission 2 will include a crew of up to four astronauts and travel farther than the moon. NASA also plans on using the SLS to send astronauts to a relocated asteroid in the **Asteroid Redirect Mission** (scheduled for 2025), which will be a testing ground for flights to Mars (scheduled for the 2030s). Like the space shuttle engines, the RS-25 engines will be fueled by liquid hydrogen and liquid oxygen, and will use solid rocket **boosters** for added thrust.

Marshall Space Flight Center is working with **Johnson Space Center** (Orion program) and **Kennedy Space Center** (Ground Systems Development and Operations) on the SLS and managed by the Exploration Systems Development Division within the **Human Exploration and Operations Mission Directorate**. Other field centers have been involved in wind-tunnel analysis, engine testing, and payload fairing research, such as the **Michoud Assembly Facility**'s **Vertical Assembly Center**.

Criticism of the SLS centers on cost. Not only is the cost of each launch higher than that borne by the space shuttle, but the cost of the program will focus NASA's **budget** on SLS and not on other worthwhile and continuing projects. In addition, SpaceX has stated that it could build a similar vehicle (140- to 150-ton payload) for significantly less, and hence this program might lead to duplication of efforts.

Note: NASA's SLS is no relation to the Space Launcher System, a **U.S. Air Force** project in the 1960s.

SPACE LAW. Space law is that body of regulations that governs national and international activities in outer space. Beginning with the launch of ***Sputnik 1*** in 1957, some kind of agreement between nations was necessary to ensure international accord and the United Nations created the **UN Committee on the Peaceful Uses of Outer Space** (COPUOS) in 1959. The UN's 1963 Declaration of Legal Principles Governing the Activities of States in the Exploration and Uses of Outer Space declares that space exploration will be conducted with good intentions and will be available to all nations that act in accordance with international law. That same year, they created the Treaty Banning Nuclear Weapon Tests in the Atmosphere, in Outer Space, and Under Water, which is sometimes referred to as the Partial Test Ban Treaty. Four additional legal principles adopted by the UN in subsequent years have fine-tuned the general declarations: The Principles Governing the Use by States of Artificial Earth Satellites for International Direct Television Broadcasting (1982), The Principles Relating to Remote Sensing of the Earth from Outer Space (1986), The Principles Relevant to the Use of Nuclear Power Sources in Outer Space (1992), and the Declaration on International Cooperation in the Exploration and Use of Outer Space for the Benefit and in the Interest of All States, Taking into Particular Account the Needs of Developing Countries (1996). Two years later, the nations involved with the **International Space Station** (ISS) entered into their own Space Station Agreement, which provides that NASA is the lead agency in its activities but that each nation retains jurisdiction over its individual modules; this will become particularly important when it comes time to dismantle the ISS.

Although space is vast, the optimal area for **geostationary satellites** (a ring above the equator, approximately 22,000 miles high) is not, and this has led to competition among different countries. Disputes are mediated by the International Telecommunications Union, an international body located in Geneva, Switzerland. Those countries along the equator have also declared ownership of the space above them, most recently in 1976, though those claims were largely ignored.

Future issues relevant to space law involve the resources in outer space. Even though no one claims ownership of asteroids, to whom do the asteroid's resources belong if a private corporation or even a publicly funded nation develops a method of mining such potentially valuable minerals or metals? The field of space law is still in its infancy and will be complicated by the variety of national and international laws that may hold precedence.

See also OUTER SPACE TREATY.

SPACE NETWORK. Established in the 1980s to replace NASA's network of ground tracking stations, the Space Network consisted of a **constellation** of Earth-orbiting satellites (**Tracking and Data Relay Satellites**), **satellites** in **low Earth orbit**, and the ground systems and facilities that acted as relay systems between the satellites. Because this system is more than 25 years old, it has become more difficult to operate and maintain, and replacement parts are not always available. This has led the **Space Communications and Navigation** office to begin the Space Network Ground Segment sustainment project, which is a projected upgrade of the technology to maintain the current high level of service.

SPACE OPERATIONS. Space Operations is a line in NASA's budget that refers to the **space shuttle**, the **International Space Station**, and all the space and flight support (**Launch Services Program**, **Space Communications and Navigation**) needed for those programs. It has been on a downward trend since the **cancellation** of the space shuttle and the anticipated retirement of the International Space Station.

SPACE PLACE. Created in 1998, the Space Place (spaceplace.nasa.gov) is a joint website run by NASA's **Science Mission Directorate**, **JPL**, the California Institute of Technology, and the International Technology and Engineering Education Association as part of their **education and public outreach** programs. It is geared for elementary-age children and young adults to encourage interest in space and all of its related areas. Its related Facebook, YouTube, and Twitter feeds support that interest and provide information in English and Spanish on more than 40 missions.

SPACE RACE. A period of competition between the **Soviet Union** and the United States for supremacy in spaceflight capability, which lasted from 1955 to 1972. The space race produced innovative efforts to launch artificial **satellites**, unmanned probes of the **Moon**, **Venus**, and **Mars**, as well as **human spaceflight**. The period of competition officially began on August 2, 1955, when the Soviet Union declared that they would also launch a satellite in response to the U.S. announcement four days earlier of intent to launch artificial satellites "in the near future." The Soviet Union beat the United States in the initial stages of the space race with the launch of ***Sputnik*** *1* in 1955. The space race reached its peak in 1969 with the U.S. landing of the first humans on the moon with ***Apollo*** *11*, and finally ended in a truce between the two countries with the April 1972 agreement on a cooperative **Apollo-Soyuz Test Project**, resulting in the July 1975 rendezvous in Earth orbit of a U.S. astronaut crew with a Soviet **cosmonaut** crew.

SPACE SHUTTLE. Colloquially, "space shuttle" refers to both (a) the orbiter with its boosters and fuel tank and (b) the program that coordinated the vehicles. The space shuttle orbiters (of which there were five operational and one test vehicle: ***Atlantis***, ***Challenger***, ***Columbia***, ***Discovery***, ***Endeavour***, and ***Enterprise***) were reusable spacecraft used to ferry crews from Earth to a **low Earth orbit** destination, generally the **International Space Station**. The shuttles themselves ferried up to eight **astronauts**—a commander, a pilot, and mission specialists—although 11 people could be flown in the shuttle in the event of an emergency, and a high-capacity payload. The shuttles appeared nearly identical, though there were differences among them, mainly in internal upgrades. Several NASA offices worked together for a complete shuttle mission: **Kennedy Space Center** was responsible for the launch and landing; **Johnson Space Center** was the mission control office; **Marshall Space Flight Center** coordinated the three main engines, external fuel tank (liquid oxygen and hydrogen fuel), and solid rocket **boosters**; **Goddard Space Flight Center** monitored the tracking; and **Stennis Space Center** tested the main engines.

Each space shuttle had three main assemblies: the orbital vehicle, the external tank, and the two solid rocket boosters. Only the orbital vehicle and the solid rocket boosters were reusable: The orbital vehicle could glide to a horizontal landing and the boosters would parachute into the ocean for later refurbishing. Other pieces—**Spacelab**, modules, and so forth—could be added on as needed. The orbiter itself was 120 feet tall with a wingspan of 78 feet, and included a pressurized crew compartment, a cargo bay about the size of a school bus (60×15 feet), and three main engines. The crew compartment itself had three levels, with the flight deck (pilot and commander) at the top; galley, sanitation, storage, and sleep stations in the middle, with access to the cargo bay and to space via an airlock hatch; and utility (water and air tanks). Payload capacity averaged 50,000 lbs. in the cargo bay, and the heaviest was the **Chandra X-ray Observatory** at 56,000 lbs. All of the shuttles were built by **Rockwell International**.

The external fuel tank was more than 150 feet tall and weighed 1,670,000 lbs. on liftoff, carrying enough fuel to burn off 1,000 gallons of fuel each second in each engine. Originally, the fuel tank was painted white to protect the insulation, but later testing showed that the paint was not necessary and the tanks were left the dark orange color of their spray-on foam.

The two solid rocket boosters combined to provide thrust of 5.8 million lbs., getting the shuttle and fuel tank to an altitude of 28 miles and a speed of 3,000 mph in only eight minutes, after which the orbiter's three main engines kicked in to provide thrust. The boosters separated at that point, to be recovered and reused. When orbital maneuvering was later necessary, 44 small rockets in the orbiter's nose cone could provide more precise maneuvering, such as when the shuttle was involved in retrieving and repairing **satellites** or observatories while in orbit.

Before launch and after recovery, shuttles were transported via shuttle carriers, **Boeing** 747s converted to allow the shuttle to be attached to the top of the aircraft. Touchdown speeds of the shuttles were approximately 215 mph, and parachutes were jettisoned once the speed decreased to approximately 70 mph to slow the shuttle, and then, after the orbiter cooled, loaded onto the shuttle aircraft carrier. As later events proved, reentry into Earth's atmosphere was dangerous because of the heat generated during descent. The orbiter's surface was covered in 24,000 thermal protection tiles, each installed by hand. Each LI-900 silica tile was lightweight yet dissipated heat so quickly that, after reaching a temperature of 2,300°F (1,260°C) on descent, it could be touched on the edges with bare hands moments after arrival. Landings at Kennedy Space Center were most convenient, but because of weather conditions, landings at Edwards Air Force Base and White Sands Space Harbor also occurred.

More than 355 individual astronauts (civilian, military, and all other personnel) flew on the space shuttles, representing 16 different countries. Of these, 49 were women and 306 were men. Story Musgrave, a NASA astronaut, was the only one to have flown on all five orbiters.

After the ***Challenger* tragedy**, the ***Columbia* disaster**, and the cancellation of the program, the remaining shuttles were distributed to public institutions and museums around the country. About 7,000 thermal protection tiles were distributed to educational institutions on a first-come/first-served basis.

After 2011, crews were transported to the International Space Station by the **Russian Soyuz** spacecraft. In September 2014, NASA announced that crews would once again depart from the United States via private contractors, and named Boeing and **SpaceX** as eligible for the **Commercial Crew Program**. Sierra Nevada Corporation sued NASA to be included in the list, subsequently lost its case, but was given an extended deadline until March 31, 2016, to finish its work under the Commercial Crew Integrated Capability (CCiCap) agreement.

The space shuttle program, officially called the Space Transportation System, ran from 1981 through 2011, 15 years longer than it was originally designed to exist. Each shuttle had been designed for 100 launches or 10 years, although none of them achieved that many launches. Combined, the five operational shuttles carried out 135 missions. Total cost for the program ran to about $170 billion over the 30 years.

Preflight development started in 1972 as a "space truck" that would be used to build a space station in the 1980s, but delays on the International Space Station extended the period during which the space truck would be needed. NASA planned for the retirement of the space shuttle to be replaced by **Commercial Resupply Services** and the Commercial Crew Programs when private enterprise showed more ability to come in under the excessive costs that included development and design.

Early space shuttle missions included the Spacelab deliveries as well as construction of the International Space Station. Later missions included delivery of and repair to the **Hubble Space Telescope** and satellites to low Earth orbit, and the initial stages of interplanetary missions such as **Magellan** and **Galileo**.

Criticism of the space shuttle program centered mainly on the cost and safety issues. Average shuttle missions by the end of the program were approximately $450 million, whereas equivalent launches from Russia were estimated at $110 million. The program, however, had been estimated at its inception to cost a total of $54 billion (instead of the $170 billion it ended up costing).

SPACE STATION ALPHA. A temporary name for the **International Space Station**. Its use was discontinued when the Russians preferred a different name, feeling that Alpha implied "first," and suggested "Mir-2" or "Beta," since the **Soviet Union** legitimately claimed the first space station.

SPACE STATION *FREEDOM*. Efforts to design and construct a permanently manned Earth-orbiting space station by NASA took place throughout the 1980s. The **Soviet Union** had begun plans on their **Mir** space station in 1976 and successfully operated it from 1986 to 2001. In response to this, the

United States proposed plans for their own space station named *Freedom*, which was approved by President Ronald Reagan in 1984. Several other countries were asked to participate in Space Station *Freedom*'s construction, including the **European Space Agency**'s (ESA) Columbus laboratory as well as Japan's Japanese Experiment Module, or **Kibō**. Although several cutbacks resulted in the Space Station Freedom never being constructed, the international partnership between NASA, the ESA, Japan, and later Canada and **Russia** would later evolve into the **International Space Station** program.

SPACE SYSTEMS *ENGINEERING*. Space systems engineering is the application of traditional engineering systems to a space environment, with a focus on the design and management of complex systems over their life cycle. With spacecraft, the life cycle is not complete until the spacecraft has been returned to Earth, destroyed, or sent beyond the reach of communication efforts. In the wake of the ***Challenger* tragedy** and the ***Columbia* disaster**, NASA set in motion a system of stages of design, development, and management that involves all possible points of view and checks.

Every proposed mission goes through a series of mission phases of evaluation, and the process is typically followed in this order:

- Pre-Phase A, Conceptual Study
- Phase A, Preliminary Analysis
- Phase B, Definition
- Phase C/D, Design and Development
- Phase E, Operations Phase

The time it takes to proceed through these steps can vary, depending on the complexity of the project, from many months to more than a decade.

The system has proven extremely safe, but that safety comes at a cost of a bureaucratic complexity that can move with glacial deliberateness even when lives and equipment are not at risk. In the event of imminent danger to astronauts, the record is less certain. Add in **budget** concerns and political necessities and you have a system that—although safe, risk averse, and responsible—can be less innovative, courageous, and pioneering than they might otherwise be. Even so, NASA remains at the forefront of national inventive opportunities, supporting private and commercial concerns as it was designed to do.

Recently, NASA has been working toward "faster, smarter, cheaper" while maintaining a culture of safety. It remains to be seen if this will be effective in **human spaceflight missions**.

SPACE TECHNOLOGY 5 (ST5). Part of the **New Millennium Program**, ST5 launched on March 22, 2006, as a **constellation** of three **satellites** to test the design, development, integration, and operation of their miniaturized elements. Each microsatellite (microsat) weighed less than 55 lbs. (25 kg) and was a full-service satellite used to study **space weather**. The microsats were deployed on a single rocket and, as planned, continued operation for 90 days, after which the satellites were decommissioned. See appendix B for a list of their instruments.

See also SMALL SATELLITES.

SPACE TECHNOLOGY MISSION DIRECTORATE (STMD). The STMD is NASA's division charged with developing the innovative technologies needed for NASA's current and future missions. Highly competitive programs—challenges, grants, demo missions, and opportunities—encourage the rapid development of high-tech innovations. One of these is the Centennial Challenge, which began in 2005 and offers prizes for revolutionary solutions to problems NASA finds of interest. For example, in 2014 NASA held a Sample Return Robot Challenge; 17 teams competed in the final event for $1.5 million in prizes, but none of them made it to the second stage. In late 2014, NASA also announced the first 3D printer flown to space, where it now operates in microgravity, which has the potential to eliminate some portion of resupply missions. It is supported in part (along with the **International Space Station** Technology Department and the Advanced Exploration Systems at NASA's headquarters in Washington, D.C.) by the STMD, also located at **NASA Headquarters**.

SPACE TOURISM. Traveling in space for recreational purposes is a burgeoning field. The first space tourist was Dennis Tito, who flew on the **Russian** Soyuz for eight days in 2001. Charles Simonyi became the first repeat space tourist eight years later. Since 2001, seven space tourists have flown on the **Soyuz**, paying a total cost of $20 to $40 million. However, the field is about to get more crowded. A number of companies have proposed—and some have already sold tickets for travel on—spaceflights to suborbital and outer space. These include **Boeing**, Excalibur Almaz (Isle of Man), Reaction Engines Limited, Space Adventures, **SpaceX**, Virgin Galactic, and XCOR Aerospace, and others will undoubtedly spring up. Virgin Galactic has signed 200 potential space tourists as of June 2015, and the United Kingdom has invested $90 million into development of the Skylon space plane by Reaction Engines Limited. In May 2015, SpaceX received **U.S. Air Force** approval to launch military satellites, which further paves this company's route to space. The United States issued proposed rules for space tourism in 2005, but it is unclear how these will affect commercial tourists in spaceflight, international tourists, or departures of U.S. citizens from foreign spaceports. There are now university classes in space tourism, space tourism market studies, and even an announced reality show, *Space Race*.

Concerns about space tourism generally revolve around environmental effects, which could accelerate global warming, cost, and danger. None of these are yet specific enough to cause widespread concern. Health risks are a major concern; medical facilities are obviously not as complete as those on Earth, and some issues can be aggravated in space, such as kidney stones.

Concerns about risk were resurrected in late October 2014 when SpaceShip Two, belonging to Virgin Galactic, exploded, killing one of its pilots and severely injuring the other. Investigation is underway, and a second SpaceShip Two, the VSS *Voyager*, is under construction.

SPACE TRANSPORTATION SYSTEM (STS). *See* SPACE SHUTTLE.

SPACE WEATHER. Although we tend to think of interplanetary and interstellar space as empty, billions of tons of particles are moved through space by the continuous stream of plasma we call the **solar wind**. Those particles—released from **coronal mass ejections**—create problems for magnetic and communications equipment. The solar wind and the particles it moves can affect our planet, **satellites**, spacecraft, the atmosphere, the magnetosphere, and so forth on a massive scale; we call it space weather when referring to such large-scale systems.

The primary effects on and near Earth produced by these magnetic storms are:

- Aurora borealis, the northern lights, and aurora australis, the southern lights. When the charged particles and solar wind enter our atmosphere, drawn by the Earth's magnetosphere, the particles show up in the high altitudes of space, and we can most readily see them against the night sky.
- Communication and **GPS** disruptions. Radio waves travel through our ionosphere and are bent (think of how a pencil or a straw appears in a glass of water). These can be distorted enough to make GPS signals unreliable for minutes up to a few days. In a similar way, shortwave radio waves are bent by the ionosphere, making communications difficult.
- Radiation hazards to orbiting **astronauts** and spacecraft. In October 2003, a geomagnetic storm surged through space. Of the 70 spacecraft failures during that year, 46 of them occurred in October. Radiation penetrates the outer shell of spacecraft, and without additional protection, astronauts are also subjected to radiation of varying degrees. In

August 1972, a large solar energetic particle event occurred that would have led to serious problems if astronauts had been in space. Radiation can also cause erroneous signals in the spacecraft's equipment.

- Current surges in power lines. A coronal mass ejection on March 9, 1969, led to the collapse of the Hydro-Quebec electrical power system four days later when an overloaded transformer failed.
- Orbital degradation. Friction between the spacecraft's surface and Earth's atmosphere leads to resistance, and a spacecraft's orbit will eventually decay. Recent spacecraft have included a small power system to increase altitude to compensate for this decay.
- Corrosion in oil pipeline steel and pipeline drilling. Locating reservoirs of oil is generally done by magnetic surveys, which can often be misdirected by magnetic storms. Corrections are often necessary.

Both NASA and **NOAA** in the United States maintain space weather prediction offices. Space weather is monitored from ground stations that study the surface of the sun (sunspots) and the radio noise emitted by the sun, as well as magnetometers and magnetic observatories, and from numerous spacecraft.

See also ADVANCED COMPOSITION EXPLORER (ACE); GEOSTATIONARY OPERATIONAL ENVIRONMENTAL SATELLITES (GOES); SOLAR AND HELIOSPHERIC OBSERVATORY (SOHO); VAN ALLEN PROBES.

SPACECRAFT TRACKING AND DATA ACQUISITION NETWORK (STADAN). A U.S. **satellite** tracking network which replaced the **Minitrack** system in 1961, STADAN added new complete tracking, telemetry, and command sites around the world and created a network of 20 stations throughout the 1960s. Like Minitrack, STADAN was capable of tracking and receiving signals from orbiting satellites, but it was also able to send commands up to satellites because they were able to be commanded from Earth. In May 1971, STADAN was consolidated with the **Manned Space Flight Network** (MSFN) to form the **Spaceflight Tracking and Data Network** (STDN).

SPACEFLIGHT TRACKING AND DATA NETWORK (STDN). A U.S. **satellite** tracking system that was created from the merge of the **Spacecraft Tracking and Data Acquisition Network** (STADAN) and the **Manned Space Flight Network** (MSFN) in May 1971. It provided long-duration, highly available space-to-ground communications and reduced the total number of stations available to 18 while upgrading the remaining locations to accomplish specific tasks at higher data rates.

SPACEX. *See* SPACE EXPLORATION TECHNOLOGIES (SpaceX).

SPACELAB. Spacelab (not to be confused with **Skylab**) was a reusable space laboratory that was ferried on the various **space shuttle** orbiters in the cargo bay as part of a joint NASA/**European Space Agency** (then European Space Research Organization) project. It included both pressurized and unpressurized components and allowed **astronauts** to perform experiments while in **low Earth orbit**. The first Spacelab flew on *Columbia* as part of its mission in November 1981, and Spacelab was included on 21 other flights, the last in 2008, though later missions were rather limited as most of the work was then being performed in the **International Space Station**.

SPECIAL ASSISTANT FOR SCIENCE AND TECHNOLOGY TO THE PRESIDENT. The Office of Science and Technology Policy's (OSTP) function (2014) is to advise the president on science and technology as it relates to domestic and international concerns. The head of this office is currently known as the president's science advisor, although in prior years was also called the assistant to the president for science and technology. President **Dwight Eisenhower** established the special assistant position in 1957 in the wake of ***Sputnik 1***, and named Dr. **James R. Killian**, then president of the Massachusetts Institute of Technology, to the position. Killian's primary responsibility was to keep the president informed of the scientific and technological programs being conducted in the various governmental agencies and to make recommendations regarding those programs. He had a small staff that also served as the staff of the President's Science Advisory Committee. The position retained its name until President Nixon changed it to science advisor to the president in 1969. Over the years, the OSTP director is often also appointed the assistant to the president for science and technology. Although the name changes from one administration to another, the position remains essentially the same.

SPINOFFS. *See* INVENTIONS.

SPIRIT. Also known as MER-A (Mars Exploration Rover–A) or MER-2, *Spirit* is a robotic **rover** on **Mars** that was active from 2004 to 2010. It was one of two rovers of NASA's ongoing **Mars Exploration Rover Mission**. It landed on Mars on January 4, 2004, three weeks before its twin, ***Opportunity***, landed on the other side of the planet. Its name was chosen through a NASA-sponsored student essay competition won by Sofi Collis, a third-grade Russian-American student from Arizona. The rover has several scientific instruments on board, including several cameras, spectrometers for analyzing soil composition, and a rock abrasion tool for exposing fresh rock.

Spirit's discovery of basaltic rock in the Gusev Crater, where it landed, led scientists to believe that the area had once been covered with volcanic magma and water. In 2005, *Spirit* managed to photograph twin dust devils. Both *Spirit* and *Opportunity* recorded detailed temperature measurements, which will give researchers a better understanding of what processes are at work on the surface of the planet and to better design spacecraft for the Martian environment. In 2007, in one of *Spirit*'s most exciting discoveries, the rover found a bright white deposit of pure silica. Because such minerals are found on Earth only where very hot water and rocks meet, such as at Yellowstone National Park, it is thought that this was a site of an ancient thermal vent—showing that Mars once had both water and an energy source. The rover became stuck in late 2009 and continued to operate as a stationary science platform role until communication stopped on March 22, 2010. *Spirit* had operated more than 25 times its initial 90-day mission, and asteroid 37452 *Spirit* was named in its honor. See appendix B for a list of *Spirit*'s instruments.

SPITZER SPACE TELESCOPE. Launched on August 25, 2003, Spitzer follows a heliocentric orbit, trailing the Earth around the sun. Its 33-inch primary mirror operates in conjunction with an infrared array camera, an infrared spectrograph, and an imaging photometer. Spitzer completed its original 2.5-year mission, though it was expected to operate for five or more years, or until its liquid helium supply ran out, which happened on May 15, 2009. Since then, its infrared camera (which does not require liquid helium for cooling) has continued to operate in an **extended mission**, the Spitzer Warm Mission.

Spitzer is the fourth in NASA's Great Observatory series, following the **Hubble Space Telescope** (launched 1990), the **Compton Gamma Ray Observatory** (launched 1991), and the **Chandra X-ray Observatory** (launched 1999). Each of the four examined different wavelength regions, and Spitzer was assigned the infrared region that is for the most part unavailable to ground-based telescopes, particularly important for very distant objects where most visible light is redshifted to infrared. In addition, Spitzer is valuable for its abilities to "see through" dust and gases to the objects within, and its spectroscopy could determine the chemical composition of the object's surface.

NASA does not have a successor telescope to Spitzer, but the **James Webb Space Telescope** has the capacity to examine the near infrared, and for a while (until 2013) the **European Space Agency**'s **Herschel** Space Observatory examined the far infrared. The **Stratospheric Observatory for Infrared Astronomy** will examine near and mid infrared ranges. See appendix B for a list of its instruments.

SPUTNIK. Russian for "satellite," *Sputnik 1* launched on October 4, 1957, into **low Earth orbit** from the Baikonus Cosmodrome. Visible across the globe, with its radio signals easily detectable, the **satellite** surprised the United States and began the **space race** between the **Soviet Union** and the United States. Its batteries ran out 22 days later, and it burned up on reentry into Earth's atmosphere on January 4, 1968.

The United States was not far behind, but the blow to national pride was immense. *Sputnik 1* had launched as part of the **International Geophysical Year**, and the United States had expected to be first with its launch of Project **Vanguard**, scheduled for September 1957. However, the launch was delayed and *Sputnik 1* took to the air before the problems could be solved.

Sputnik 1 was a tiny satellite, less than two feet in diameter, with a polished metal surface. Four antennas broadcast its radio signals to amateur radio enthusiasts, who tracked its orbit; viewers—part of Project Moonwatch—tracked its progress by sight through binoculars and telescopes, when it was especially visible near dawn and dusk. Earlier attempts by the Soviet Union, like those in the United States, had failed, but *Sputnik*'s redesign and lighter weight, combined with the modified R-7 two-stage rocket, succeeded. It was powered by three batteries, two of which powered the radio transmitter and the third its temperature regulation system, and the satellite was expected to last for two weeks. Changes in temperature or pressure were indicated by changes in the radio signal's impulses.

Longer lasting than the satellite itself were the space race repercussions. People around the world reacted with both fear and amazement, depending on personal orientations and national policy. The United States created the **Advanced Research Projects Agency** and NASA, and pushed math and science education. **Lyndon Johnson**, then Senate majority leader, called for a congressional investigation of the impact of Sputnik and recommended that the United States increase its efforts in the space race. Nikita Khrushchev used *Sputnik* as a way to increase his standing in world politics and to promote socialism/communism.

Sputnik 2 followed on November 3, 1957, with the launch of Laika, a stray dog. Khrushchev declared that the Soviet Union would overtake the United States in industry and manufacturing, which helped him consolidate his power in the Soviet Union but increased American fears of a missile gap and of falling behind in the space race. *Sputnik 2* was more than six times heavier than *Sputnik 1*, and remained in orbit for 162 days until its orbit decayed. Laika survived for only a few hours, and probably died from heat exhaustion in the satellite.

Sputnik 3, a research satellite, launched on May 15, 1958, to study our upper atmosphere. It carried a dozen instruments that were to record the atmospheric pressure and composition of the atmosphere, and to measure magnetic and electrostatic fields, but its tape recorder failed before it could record evidence of, for example, the **Van Allen radiation belts**. It reentered Earth's atmosphere on April 6, 1960, longer lasting than any of its predecessors.

STARDUST/STARDUST-NEXT. Launched on February 7, 1999, Stardust was a space **probe** designed to collect dust samples from the Wild 2 comet (from a distance of 147 miles) and return the samples to Earth for study. Designed by **Lockheed Martin** and managed by **Jet Propulsion Laboratory**, the probe communicated via the **Deep Space Network** for more than 3.5 billion miles. When it neared its target, it opened a tray of aerogel to capture the interstellar and comet particles in its path. The **sample return** capsule (SRC) returned to Earth on January 15, 2006. Because there was a potential for contamination—either from the SRC to life on Earth, or from Earth to the SRC—the SRC was secretly transferred to a clean room at **Johnson Space Center**. NASA revealed that the samples had contained at least a million microscopic dust specks. Some of the contents indicated the presence of liquid water on Wild 2, and in March 2014 the presence of seven particles of interstellar dust was announced. With less than 44 lbs. of propellant after the SRC was jettisoned, the spacecraft entered hibernation mode.

However, on January 3, 2007, NASA approved a future reactivation. Stardust-NExT (Stardust-New Exploration of Tempel 1) reached the asteroid Tempel 1, earlier visited by **Deep Impact** in 2005. Deep Impact had sent an impactor into the asteroid and Stardust was perfectly positioned to relay images of the impact site and to measure and analyze the composition of the dust particles. On February 15, 2011, Stardust-NExT flew within 112 miles of Tempel 1 and relayed 72 images. Its remaining fuel was used up a month later. See appendix B for a list of its instruments.

STENNIS SPACE CENTER (SSC). Located in Hancock County, Mississippi, and originally known as the National Space Technology Laboratory, the facility was selected in 1961 as the site of a static test facility for launch vehicles used in the **Apollo** program. At that time, it was the largest construction project in the state, consisting of 125,000 acres of acoustical and environmental buffer zone and more than 13,000 acres dedicated to the testing facility, which was called the Mississippi Test Facility. In 1965, it became part of the **Marshall Space Flight Center**. It was renamed in 1995 after the death of Senator John C. Stennis. Stennis was designated NASA's lead center for implementing commercial remote sensing activities in 1997 and is also currently the lead center for geographic information systems and **small spacecraft technology**. The eye of Hurricane Katrina passed directly overhead on August 29, 2005, but the facility suffered only water and wind damage, mostly from roofs; the major problem was the affected personnel, with homes no longer habitable. The location provides convenient facilities for more than 30 tenants, both government agencies (**U.S. Navy**, **NOAA**, etc.) and private corporations (Rolls Royce, **United Launch Alliance**, and **Lockheed Martin**, among others). It is also the home of the INFINITY Science Center, which includes science and space exhibits, an Omega flight simulator, and a full-size **International Space Station** module.

STRATOSPHERIC AEROSOL AND GAS EXPERIMENT (SAGE). A series of Earth-observing instruments known as SAGE have been mounted on the **International Space Station** to make long-term measurements of water vapor, aerosols, ozone, and other gases as well as temperatures and pressures present in Earth's atmosphere. SAGE 1 (1979) and SAGE 2 (1984) assisted in the collection of data that led to the discovery of Earth's ozone hole and to the banning of chlorofluorocarbons. SAGE 3 (2001) was installed on the ISS but stopped working in 2006; the instrument was returned and stored until February 2011 when scientists at Langley Research Center coordinated the resurrection of the instrument, and it is scheduled to be sent into space in 2016.

STRATOSPHERIC OBSERVATORY FOR INFRARED ASTRONOMY (SOFIA). SOFIA is a science program and an airborne observatory that uses a 17-ton Cassegrain telescope with a 2.5-meter mirror mounted inside a **Boeing** 747 special performance (747 SP) aircraft. When operating, a port-side door exposes the telescope directly to the night sky. Designed primarily for infrared studies of planets, nebulae, and interstellar space, SOFIA operates at altitudes between 39,000 and 45,000 feet, above most of the Earth's moisture that interferes with infrared observations at lower altitudes. Vibrations and air turbulence are minimized with a combination of bearings in pressurized oil, a system of three gyroscopes, and minor aerodynamic shaping of the aircraft. SOFIA is a joint collaboration between NASA and Germany's space program (**DLR**) and maintains an **education and public outreach** program, SOFIA (E/PO), to involve educators, students, journalists, and scientists with direct and indirect access to SOFIA's research. See appendix B for a list of its instruments.

STUDENT SPACEFLIGHT EXPERIMENTS PROGRAM (SSEP). Through NASA's **education and public outreach** partnership with the National Center for Earth and Space Science, the Arthur C. Clarke Institute for Space Education, and **NanoRacks**, students compete for the opportunity to propose and design projects, first for the **space shuttle** and now for the **International Space Station**. Many of these projects are part of the **CubeSat Initiative**. The program is geared to increase students' awareness of science, technology, engineering, and mathematics fields. Recent projects have included "What Is the Effect of Microgravity on the Formation of Silly Putty and How Do the Characteristics of That Silly Putty Differ from the Silly Putty Made on Earth?" by the eighth grade of Lincoln Middle School, and "Will Microgravity Have a Significant Effect on Packed Synthetic HBOCs?" from 11th-grade Montachusett Regional Vocational Technical School. The SSEP came under fire in late 2014 for spending "billions" on elementary school projects, but further study showed that the total cost to taxpayers was closer to $50,000 per year when private funding was included in the calculation.

STUDIES OF EMISSIONS AND ATMOSPHERIC COMPOSITION, CLOUDS, AND CLIMATE COUPLING THROUGH REGIONAL SURVEYS (SEAC^{4}RS). SEAC^{4}RS is a NASA mission within its **Earth Science Project Office** with an objective of studying aerosol pollution, convection, cloud cover, and related atmospheric conditions to study and forecast pollution and climate impact on regional atmospheric conditions. This NASA mission primarily uses aircraft to survey the atmosphere with additional support from **satellite** data collection and other atmospheric sensors. SEAC^{4}RS was formerly the South East Asia Composition, Clouds and Climate Coupling Regional Survey but due to a lack of suitable regional hosting was renamed and became a U.S. mission in 2013. See appendix B for a list of its instruments.

SUBMILLIMETER WAVE ASTRONOMY SATELLITE (SWAS). Part of the **Small Explorer** program, SWAS launched on December 5, 1998, to begin a two-year mission to observe star formation. Interstellar clouds cool and collapse to form stars and planets, but the process is not completely understood. Over its six-year life, it found water in almost every dust cloud in space (high amounts of water in warm gases, low amounts in cold dense gases); measured the water vapor in interstellar clouds; discovered a group of comets evaporating around a red giant; and observed the **Deep Impact probe** as it collided with Comet P/Tempel 1, measuring about 730 pounds of water ejected every second during and immediately after the crash. The **satellite** is currently inactive in orbit. See appendix B for a list of its instruments.

SUOMI NATIONAL POLAR-ORBITING PARTNERSHIP (SUOMI NPP). Part of the **Earth Observing System**, Suomi NPP collects data on long-term climate change and short-term weather conditions in a joint project with the **Department of Defense** (through the DoD's Defense Meteorological Satellite Program), the **National Oceanic and Atmospheric Administration**, and NASA. Suomi NPP launched on October 28, 2011. Its five instruments transmit data on a real-time basis and the spacecraft operates on a sun-synchronous polar orbit 512 miles above Earth. As earlier generations of Earth-observing **satellites** age out, Suomi NPP is the first of a new group of satellites to take over the continuous observation records. NPP was renamed Suomi NPP in honor of Verner E. Suomi, considered the father of satellite meteorology. See appendix B for a list of its instruments.

SURFACE WATER AND OCEAN TOPOGRAPHY (SWOT). *See* CENTRE NATIONAL D'ÉTUDES SPATIALES (CNES).

SURVEYOR. A NASA program between 1966 and 1968 that followed the **Ranger** program, Surveyor consisted of seven robotic spacecraft sent to the **moon** prior to the ***Apollo 11*** landing in July 1969. The **satellites** tested soft landings on the moon as well as the ability of the spacecraft to make accurate midcourse corrections and end-course maneuverability, five of them successfully. They also took close-up photographs of the lunar surface, to be used to determine a safe landing spot for later Apollo landings, relayed via television through the **Deep Space Network**. All (in various after-flight configurations after descent) remain on the moon, with the exception of some parts of *Surveyor 3* that were returned on *Apollo 12*. The flights were the first that tested the radar systems needed to determine the lander's altitude and velocity, which were paramount prior to a manned landing. Each satellite slowed to about 40 feet per second by its main solid fuel retrorocket, about 50 miles above the surface of the moon, and was then jettisoned about four miles above the surface. It took the satellite 2.5 minutes to complete the journey using pressurized helium, and the last 10 feet was in low-gravity free fall. See appendix B for a list of typical instruments.

SUZAKU. *See* ASTRO-E2.

SWIFT. Launched on November 20, 2004, the Swift **satellite**—part of the Medium **Explorer** class and called the Swift Gamma-Ray Burst Mission—studied elusive gamma-ray bursts in an effort to determine their origin. Gamma-ray bursts, the most powerful explosions in the universe, occur from random locations, and scientists were considering the possibility that they were produced when a massive star dies and two neutron stars collide. Swift contained multi-wavelength telescopic instruments that were so fast they could relay the position of the burst to the ground within 20 seconds of its occurrence. On May 9, 2005, Swift detected a burst that lasted one-twentieth of a second, the first time that a gamma-ray burst's location had been identified. Swift has been on an **extended mission** since 2010. It has identified more than 800 gamma-ray bursts and new black holes, increasing our knowledge of such events tremendously. On April 23, 2014, Swift detected the strongest, hottest, and longest-lasting set of stellar flares ever seen, with an initial blast as much as 10,000 times more powerful than the largest **solar flare** ever recorded. It is estimated that the flare reached a temperature of 360 million°F (200 million°C). See appendix B for a list of its instruments.

T

TEACHER IN SPACE PROJECT (TISP). TISP was a six-year project (1984–1990) designed to encourage students in science, technology, engineering, and mathematics. The idea was that teachers would plan and instruct students from space, thus spurring their interest. However, the first teacher in space, **Christa McAuliffe**, died during the ***Challenger* tragedy**, and the program never recovered. The project was revived in 2005 by private industry, which has since trained several teachers as **astronaut** candidates, conducted a number of experiments in space, and delivered workshops for students and teachers. No teachers, however, have returned to space through that program. In the meantime, NASA has since created an Educator Astronaut project, which again seeks to inspire students and encourage teaching. This program, however, designates the teachers—who are full-time astronaut candidates instead of teachers on an extended project—as educator-astronauts who maintain normal mission duties in addition to their education-related projects. Three teachers flew on board **space shuttle *Discovery*** in 2009 and another in 2010. Barbara Morgan, McAuliffe's backup, went on to become a full-fledged astronaut on board the ***Endeavour*** in 2007.

TERESHKOVA, VALENTINA (March 6, 1937–). Tereshkova, a **Soviet** citizen, was the first woman to fly in space, on June 16, 1963. Her American counterpart, **Sally Ride**, would not fly in space until 20 years later, on June 18, 1983. Tereshkova logged 48 orbits of Earth during almost three days in space on board the ***Vostok*** *6*. She is the recipient of numerous awards, medals, recognitions, and honorary citizenships, and a lunar crater is named after her.

TERRA. Terra (Latin for "Earth") is the original **Earth Observing** System **satellite**, launched on December 18, 1999, to study the interactions among the atmosphere, land, snow and ice, ocean, and energy. Five specialized instruments are geared to give precise complementary observations, allowing them to be layered into a single image that connects all the Earth's systems. The instruments are used by other organizations for data specific to their needs: The Environmental Protection Agency and the **National Oceanic and Atmospheric Administration** use air quality information; the Federal Aviation Administration monitors volcanic ash eruptions; the U.S. Forest Services evaluates fire threats and potential for spread, for example. The April 2010 oil spill in the Gulf of Mexico was monitored by three of Terra's instruments to help determine the impact on surrounding wetlands and the spread of fires, as well as the boundaries of the surface oil spill. That same month, volcanic ash from Iceland created a disruption in air traffic, and Terra's instruments helped reroute air traffic and monitor ash content. In 2008, hackers gained unauthorized access to Terra's command and control systems but were content with simply monitoring its actions. Terra operates in a sun-synchronous orbit about 440 miles above the Earth. See appendix B for a list of its instruments.

THREE-DIMENSIONAL (3D) PRINTING. In September 2014, a 3D printer was delivered to the **International Space Station** (ISS) by a **SpaceX** resupply mission. Manufactured by a company called Made in Space, the machine was assembled in November by **astronauts**. In December, engineers and computer scientists created a CAD file on Earth, emailed the file to the ISS, and the ratcheting socket wrench was then manufactured in space. This technology has the potential to mean lower costs for experiments, faster design creation, and a method to replace broken parts or create new tools as needed for colonies and scientists far from Earth.

TIME HISTORY OF EVENTS AND MACROSCALE INTERACTIONS DURING SUBSTORMS (THEMIS). Originally a **constellation** of five **satellites** launched in 2007 for a two-year mission to study the Earth's magnetosphere, and in particular its substorms, THEMIS discovered the triggering events for such substorms by measuring magnetic reconnections. At the conclusion of its original mission, two **probes** were spun off to become **ARTEMIS** while the remaining three continue to study the space environment around Earth. In March 2014, THEMIS discovered a pool of dense particles that normally circles the Earth would move to concentrate the particles to block out the incoming solar material near the outer boundary of the magnetosphere during a **solar flare**; one of NASA's scientists described it as something like stacking up furniture against the door to keep an intruder from breaking into your room. How this affects **space weather** as well as global terrestrial weather is still being studied. In June 2015, THEMIS observed magnetosonic waves in a sharp rising tone for the first time, which is more complicated than the smooth tones it had earlier observed; the reasons for this are still unclear. The project is managed by University of California–Berkeley. See appendix B for a list of its instruments.

THERMOSPHERE, IONOSPHERE, MESOSPHERE ENERGETICS AND DYNAMICS (TIMED). TIMED was originally a two-year mission sponsored by NASA but designed by **Applied Physics Laboratory** to study the Earth's atmosphere. The mission has been extended four times after its launch on December 7, 2001, into **low Earth orbit.** TIMED's primary concern is an examination of the sun's solar cycle and how it affects Earth's upper atmosphere, which can affect **satellite** and radio communications. More than 20 instrument teams from universities as well as private and public organizations around the world study its data. See appendix B for a list of its instruments.

TITAN LAUNCH VEHICLES. The Titans were a set of related expendable rockets used to launch spacecraft between 1959 and 2005. Used for both civilian and military purposes, Titans launched more than 350 spacecraft—for human, robotic, **satellite**, **probe**, intelligence, and weather purposes. Titan I was used by the military in the early to mid-1960s. Titan II was both military (intercontinental ballistic missiles) and civilian (NASA) focused, and used for the **Gemini** program. Titan III (with seven versions) was a more powerful version of Titan II, and used for scientific spacecraft such as **Voyager** and **Viking**, as well as a number of heavy-lift launches. Finally, Titan IV was used for intelligence gathering as well as the **Cassini-Huygens** mission. They were manufactured by the Glenn L. Martin Company, a pioneering aircraft and aerospace manufacturing firm that merged with American-Marietta to become Martin Marietta, and later with Lockheed to form the current **Lockheed Martin** Corp. Due to the high cost (both in dollars and toxicity) of its propulsion fuel, Lockheed Martin began using its Atlas launch vehicles instead. Of Titan's more than 350 launches, fewer than 10 percent were failures, almost all of those in the early years or with test flights.

TITOV, GHERMAN (September 11, 1935–September 20, 2000). Titov, a citizen of the **Soviet Union**, was notable for the number of "firsts" in space: He was the first man to orbit Earth multiple times (17), the first to spend more than a day in space; the first to sleep during orbit; the first to have "space sickness" (trouble adapting to weightlessness); the first to manually pilot a spacecraft; the first to take a manual photograph from space; and the first to take photographs of Earth from space. Titov was also the second human to orbit Earth (after **Yuri Gagarin**), the fourth to reach outer space (after **Alan Shepard** and **Gus Grissom**), and the youngest person (25 years old) to fly in space. (That record might be broken when Virgin Galactic's flights to space take place, as there are a number of teenagers signed up on SpaceShip Two, though the date is uncertain.)

TOTAL OZONE MAPPING SPECTROMETER-EARTH PROBE (TOMS-EP). TOMS-EP was a mission coordinated by the **Earth Observing** System to study the Earth's atmospheric ozone and the sulfur dioxide released in volcanic eruptions, launched on July 2, 1996. The measurements were a continuation of NASA's Nimbus-7 **satellite** readings in 1978 and **Russia**'s Meteor-3 satellite in 1994. Comparison mapping of the "ozone hole" over the Antarctic (which forms September–November every year) as well as worldwide distribution of ozone helps scientists to determine how much of the decrease in ozone is due to human activities and how much is simply part of a short-term trend within a long-term cycle. To complete its mapping, TOMS measurements were taken every 8 seconds, and each measurement covered 30–125 miles on the surface, resulting in more than 200,000 daily measurements. However, because of calibration problems, data after 2002 has not been used for trend analysis. The TOMS program was managed by **Goddard Space Flight Center**. See appendix B for a list of its instruments.

See also OZONE MAPPING AND PROFILER SUITE (OMPS).

TOPEX/POSEIDON. TOPEX/Poseidon was a **satellite** launched on August 10, 1992, as a joint mission between NASA and **CNES** to map ocean surface topography. It studied the seasonal cycle of ocean currents and circulation with a high degree of accuracy (within 1.65 inches, or 4.2 cm), with particular emphasis on El Niño and the decadal oscillation in the Pacific Ocean, as well as studies in the Atlantic Ocean and Mediterranean Sea. TOPEX/Poseidon produced the most accurate global maps of tides used today, transmitted through the **Tracking and Data Relay Satellite System**. The satellite was decommissioned in January 2006 after more than 13 years, despite an original mission design of three to five years.

TRACKING AND DATA RELAY SATELLITE SYSTEM (TDRSS). A tracking system comprised of a set of communications **satellites** managed by **Goddard Space Flight Center** that NASA uses with its ground stations in the **Space Network** for communication in space. The first satellites were launched in the early 1970s, the last in 2014; later satellites have begun replacing the earlier ones. Like many other missions, the satellites are referred to by letter before launch and by number after deployment: TDRS-A, the first, became TDRS-1 after launching and deployment (TDRS-B was lost with the ***Challenger* tragedy**). TDRS 1–7 were part of the first-generation satellites, built by TRW; TDRS 8–10, the second generation, and third-generation TDRSs 11–12 and M–N, were all awarded to **Boeing**. TDRS-M, originally scheduled for late 2015, does not yet have a launch date. TDRS supports more than 30 NASA missions, including the **International Space Station**, **SpaceX**'s Falcon launch, **Landsat**, and many more. See appendix B for a list of their typical instruments.

TRANS-EARTH INJECTION. A trans-Earth injection is a spacecraft's maneuver to intersect the Earth's trajectory for either orbit or landing. The first trans-Earth injection was by ***Apollo** 8* on Christmas Day, 1968. Similarly, trans-lunar injections are maneuvers to intersect the **moon**'s orbit, and trans-Mars (such as that of the **Mars Atmosphere and Volatile Evolution** in 2014) with **Mars**, where the trajectory is called the Mars transfer orbit.

TRANSITION REGION AND CORONAL EXPLORER (TRACE). TRACE, a **Small Explorer** mission, was a space telescope launched on April 1, 1998, that investigated the three-dimensional magnetic fields and plasma structures of the sun by using ultraviolet wavelengths. Its last useful scientific image was relayed in 2010. During its mission, it examined periods of magnetic fluxes (which can morph into sunspots given enough energy) as well as more extended periods of limited solar activity, giving scientists and researchers a basis for comparison and data for modeling solar activity.

TRANSOCEANIC ABORT LANDING (TAL) SITES. In the event of an unscheduled contingency (engine failure, emergency mechanical problems, adverse weather, etc.) landing, the **space shuttle** needed alternative landing sites around the world. Depending on where and when the decision was made to abort the mission, whether it was a high- or low-inclination launch, and payload specifications, several sites were selected. Primary sites (called "augmented sites" because they were staffed with NASA, **Department of Defense**, and **contractor** personnel during launches and contingency landings) included Istres Air Base in France and Zaragoza Air Base and Moron Air Base in Spain, all of which were administered by separate international agreements. Backup sites included Banjul (in the Republic of the Gambia) until it was closed in 2002, and Ben Guerir Air Base in Morocco, replacing Casablanca in 1988 but which is itself now closed as well. Each TAL had specific shuttle-unique landing aids—equipment, runway, and personnel requirements—to be followed in the event of a contingency landing. None were ever required to be used.

TREATY ON PRINCIPLES GOVERNING THE ACTIVITIES OF STATES IN THE EXPLORATION AND USE OF OUTER SPACE, INCLUDING THE MOON AND OTHER CELESTIAL BODIES. *See* OUTER SPACE TREATY.

TROPICAL RAINFALL MEASURING MISSION (TRMM). TRMM is a joint mission between NASA and the **Japan Aerospace Exploration Agency** (JAXA) launched in December 1997 to monitor and study tropical rainfall, particularly over the oceans where measurements are scarce (between 35 degrees north and south latitudes). In July 2014, the **satellite** indicated it had consumed most of its fuel and has begun to lose altitude; it will reenter Earth's atmosphere sometime after February 2016. In the meantime, however, it is providing limited data, such as for Hurricane Odile in Baja California (Mexico) during September 2014. TRMM is part of NASA's **Science Mission Directorate** in a program called **Mission to Planet Earth**. Global atmospheric circulation depends largely on rainfall patterns, so an understanding of the water cycle will increase the accuracy of weather forecasts in both short- and long-term predictions. **Global Precipitation Measurement**, launched on February 27, 2014, also a joint mission between NASA and JAXA, is a successor mission.

TSIOLKOVSKY, KONSTANTIN (September 17, 1857–September 19, 1935). Tsiolkovsky, along with **Robert H. Goddard** and German scientist Hermann Oberth, was one of the founding members of rocket science and astronautics; he was also a source of inspiration for **Wernher von Braun**. Tsiolkovsky was particularly known for his pioneering work in jet propulsion, despite never having built a rocket; his most important work was The Exploration of Cosmic Space by Means of Reaction Devices (1903). He, Oberth, and Goddard all came to similar conclusions regarding multistage rocketry, though each worked independently. Tsiolkovsky also built the first Russian wind tunnel in 1897. A large crater on the far side of the **moon** is named after him.

TWO INDEPENDENCE SQUARE. *See* NASA HEADQUARTERS.

TYSON, NEIL deGRASSE (October 5, 1958–). A recipient of NASA's **Distinguished Service Medal**, Tyson is a popular media figure, astrophysicist, author, cosmologist, and scientist. He has served on the Commission on the Future of the United States Aerospace Industry and the President's Commission on Implementation of United States Space Exploration Policy, is the host of the radio talk show *StarTalk*, the star of the remade version of *Cosmos* popularized by **Carl Sagan**, and the recipient of numerous awards, honors, and honorary doctorates in both scientific and humanitarian fields. He was also the inspiration for the nonprofit group Penny4NASA, whose goal is to increase NASA's **budget** to 1 percent of the federal budget.

U

ULYSSES. A joint project of NASA and the **European Space Agency**, Ulysses was the first spacecraft to survey the space environment from the poles of the sun. It was launched from **space shuttle *Discovery*** on October 6, 1990. During its 18-year mission, Ulysses measured cosmic radiation, energetic particles created by solar storms, and **solar wind**. It measured dust particles in space, found helium atoms in space, and found that the sun's magnetic field is balanced across its latitudes. Its mission ended on June 30, 2009, when its transmitters began to fail as it traveled farther into space and the **Deep Space Network** lost its signal. Refer to appendix B for a list of its instruments.

UN COMMITTEE ON THE PEACEFUL USES OF OUTER SPACE (UNCOPUOS). UNCOPUOS (or simply COPUOS) was established by the UN General Assembly in 1959 to, among other matters, study legal problems regarding the exploration of outer space, to encourage scientific research in space, and to encourage the peaceful use of outer space. There are 76 member-states in the committee, and it has two subcommittees, the Scientific and Technical Subcommittee and the Legal Subcommittee. In addition, the UN Office for Outer Space Affairs, created in 1962, is part of the Department of Political Affairs; it is responsible for implementing decisions of the UNCOPUOS. Administrator **Charles Bolden** addressed the committee in 2013 and reiterated NASA's support for and encouragement of the committee's goals. As more nations become involved in space, the committee's role is likely to enlarge, particularly regarding matters relating to scarce resources.

In addition to the **Outer Space Treaty**, COPUOS is also involved with the Rescue Agreement ("The Agreement on the Rescue of Astronauts, the Return of Astronauts and the Return of Objects Launched into Outer Space"), the Liability Convention ("The Convention on International Liability for Damage Caused by Space Objects"), the Registration Convention ("The Convention on Registration of Objects Launched into Outer Space"), and the Moon Treaty ("The Agreement Concerning the Activities of States on the Moon and Other Celestial Bodies").

COPUOS's 58th session met from June 10–19, 2015, at the United Nations Office in Vienna, Austria; four individuals from NASA were listed among the participants.

See also ASTEROID MINING.

UNITED LAUNCH ALLIANCE (ULA). United Launch Alliance is a joint venture of **Lockheed Martin** and **Boeing** that was created in December 2006 to provide launch services for NASA and military partners. Using the well-established history of both **Delta** (Boeing) and Atlas (Lockheed Martin) rockets, and more recently its Vulcan rockets, the merger was initially challenged on an antitrust basis by **SpaceX** but was allowed. Since then, United Launch and SpaceX have competed on a number of contracts, finishing their 100th successful launch on October 2, 2015. ULA operates from **Cape Canaveral** and **Vandenberg Air Force Base**, depending on the proposed orbit. In September 2014, United Launch announced a contract from the **U.S. Air Force** for $938 million to provide military rocket launch services, and later the same month NASA announced both Boeing and SpaceX as candidate certificators for human spaceflight development.

United Space Alliance is an earlier cooperative effort between Lockheed Martin and Boeing for work and support to NASA for mission design, training, and other launch services; the contract was primarily for **space shuttle** operations.

UPPER ATMOSPHERE RADIATION SATELLITE (UARS). UARS was launched on September 12, 1991, from the **space shuttle *Discovery*** to provide data on the upper atmosphere's composition, dynamics, and energy input. It orbits Earth 363 miles above the surface in **low Earth orbit**. One of its goals was to study man-made disruptions in the upper atmosphere and to learn about long-term climate effects. UARS also provided important data on the amount of light that comes from the sun in ultraviolet and visible wavelengths. UARS ceased to function in 2005 and reentered Earth's atmosphere in September 2011 over the Pacific Ocean. Refer to appendix B for a list of its instruments.

URANUS. The seventh planet from the sun, Uranus was named after the ancient Greek god of the sky. Uranus is more like **Neptune** and is normally referred to as an "ice giant" to distinguish it from the larger gas giants **Jupiter** and **Saturn**. It has the coldest atmosphere in our solar system, near –371°F (–224°C, 49 K). Uranus is unique in that its axis is tilted sideways, as discovered from images from ***Voyager** 2*, and it is composed mostly of ice and rock. It takes Uranus 84 Earth years to complete its orbit around the sun. The planet also has the lowest escape velocity of the four gas and ice giants, and has been proposed—though not realistic yet—as a mining site for helium-3 for use as thermonuclear fuel. At this point, *Voyager 2* is the only spacecraft to have flown by Uranus; it was closest on January 24, 1986.

U.S. AIR FORCE. Because of the common focus on air and space, NASA has more joint ventures and projects with the U.S. Air Force than it does with other military departments. For example, one of the more recent joint studies is an effort to develop an affordable, next-generation rocket engine between **Marshall Space Flight Center** and the air force's Launch Systems Directorate in Los Angeles that is a candidate for the **Space Launch System**'s cryogenic propulsion stage. NASA often monitors air force equipment in space from one of its field centers. Like its agreement with the **U.S. Army**, NASA also has an aeronautics research partnership to ensure the exchange of research information and avoid duplication of effort. The agreement covers aircraft design, propulsion development, and **aviation safety**.

The civilian–military partnership, however, does have its drawbacks. The air force urgently needed to send classified Geosynchronous Space Situational Awareness Program **satellites** into **low Earth orbit**, which bumped the **Orion** capsule test flight from September to December 2014 since both were scheduled to use the same launch pad. Earlier, **space shuttle** services had been so oversubscribed that the air force contracted separately for **Titan launch vehicles**.

U.S. ARMY. NASA has several joint agency agreements with the U.S. Army. Most involve some aspect of **aeronautics research** partnerships with a goal of innovative tools to the benefit of both. Some of the recent projects include collaboration on developing and evaluating concepts for a variable/multispeed drive system; wind tunnel tests for rotor aircraft; and helicopter flight tests evaluating maneuverability. The partnership extends back to the founding of NASA when it absorbed elements of the **Army Ballistic Missile Agency** and its staff, including **Wernher von Braun**. NASA's civilian focus and the army's military emphasis allow the goals of each to overlap in the fields of new technology.

U.S. HOUSE COMMITTEE ON SCIENCE AND TECHNOLOGY. *See* HOUSE COMMITTEE ON SCIENCE AND ASTRONAUTICS.

U.S. NAVY. The U.S. Navy worked with NASA in the early days after NASA's founding on Project **Vanguard**. Since then, the navy cooperates primarily with NASA in the retrieval of spacecraft and is gearing up for additional support for the **Orion** project when **astronauts** are expected to splash down in the ocean. Tests in August 2014 were practice for the recovery of the Exploration Flight Test 1 scheduled for December 2014, which was Orion's first mission and which splashed down in the Pacific Ocean. In addition, the navy contracts with NASA for naval aircraft testing and performance measurements.

V

VAN ALLEN PROBES. The second mission of the **Living with a Star** program, the two Van Allen **probes** study Earth's radiation belts from near-Earth space, and in particular, to determine how the Van Allen radiation belts are affected by the sun. They were launched from **Cape Canaveral** on August 30, 2012, on a two-year mission to study the inner and outer Van Allen radiation belts, which present a radiation danger to **astronauts** as the magnetic fields expand and contract in reaction to solar activity. Two main donut-shaped clouds are layers of energetic charged particles that surround the Earth, held in place by Earth's magnetic field, and a third weaker belt comes and goes with solar activity. Data from the Van Allen probes is leading to design of spacecraft to better protect the astronauts and the equipment on board, and to a better understanding of **space weather**. The two probes have completed their initial two-year mission and continue to provide scientific data for researchers. The Van Allen probes were formerly called the Radiation Belt Storm Probes, but the name was changed in 2012 to honor the late James Van Allen, who discovered the belts in 1958. Refer to appendix B for a list of their instruments.

VAN ALLEN RADIATION BELTS. Reacting to Earth's magnetic fields, a layer of energetically charged particles is held in place around the planet. The belts—there are at least two and sometimes more—are located about 600–40,000 miles from the surface of Earth, in the inner region of our magnetosphere. Because of the high degree of radiation in the trapped particles that arrived on the **solar wind**, the Van Allen radiation belts create problems for humans in space as well as sensitive equipment. Electronic components are particularly vulnerable to the radiation, and some—such as the **Hubble Space Telescope**—will have their sensors turned off during periods of intense radiation. **Low Earth orbit** missions are generally unaffected by the Van Allen belts, but missions beyond that level are exposed to cosmic rays, radiation, and **solar flares**. This will be particularly important as NASA gears up for its **Asteroid Redirect Mission** and for any interplanetary voyages, such as those it plans to **Mars**.

Radiation belts surround other planets that have magnetic cores as well. The Van Allen radiation belts were first detected by the ***Explorer 1* satellite**, though their existence had been predicted. They are named after James Van Allen, an early space scientist.

VANDENBERG AIR FORCE BASE. Used primarily for earth science and military launches, Vandenberg's Space Launch Complex 6 (SLC-6) specializes in polar-orbiting **satellites** as its geography favors a southern direction for launches. Although initial plans called for the **space shuttle** to use its facilities, those plans never materialized. Instead, **Lockheed Martin** developed its Athena launch vehicle for use there and **Boeing** launched its **Delta** IV from SLC-6. **SpaceX** leases launch pad facilities from Vandenberg for its Falcon launch vehicle. Vandenberg, which is named after Air Force General Hoyt S. Vandenberg, is used by the Department of Defense as a space and missile testing base, although NASA maintains a presence for its **Earth Sciences Program**.

VANGUARD. Aware of the **Soviet Union**'s anticipated launch of ***Sputnik** 1* and as part of the **International Geophysical Year** (1957–1958), the United States was determined to launch a **satellite**. The **U.S. Army**, **U.S. Air Force**, and the **U.S. Navy** all submitted proposals, from which the U.S. Navy's Project Vanguard was selected and launch was anticipated for September 1957. Complications caused a delay, and on October 4, 1957, word came that *Sputnik 1* had launched. On November 3, *Sputnik 2* was launched. Eager to show that the country was not far behind, the navy launched TV-3, a test vehicle, on December 6 at **Cape Canaveral**, although mission control remained in Washington, D.C., at the **Naval Research Laboratory**'s headquarters. The Vanguard rocket lost thrust about four feet above the launch pad, fell back to the surface, and exploded. (The attempt was later referred to by the news media as "kaputnik," "flopnik," "stayputnik," and other such ironic descriptions.) The U.S. Army, anxious to show off its space acumen, launched ***Explorer 1*** on January 31, 1958, outdistancing their naval counterparts and, as they saw it, saving America's honor. TV-4 was then launched by the navy on March 17, 1958, and was renamed *Vanguard 1* after its successful launch. It remains the oldest satellite in orbit, as Vanguard's predecessors—*Sputnik 1*, *Sputnik 2*, and ***Explorer 1***—have decayed from orbit, and is expected to remain in orbit for about 240 years. It was the first solar-powered satellite and weighed just 3 lbs. (1.4 kg), being so small that the Soviets called it "the grapefruit satellite." In all, there were 11 launches of Project Vanguard between 1957 and 1959, of which only three achieved orbit. The Naval Research Laboratory staff formed the core of what would later become **Goddard Space Flight Center**.

Note: Project Vanguard is not to be confused with the Vanguard Six, a group of six trained **cosmonauts** part of the **Vostok** program.

VEHICLE ASSEMBLY BUILDING (VAB). Not to be confused with the **Vertical Assembly Center** in New Orleans that was completed in September 2014, the VAB is located at **Kennedy Space Center** in Florida. Its name was changed from Vertical Assembly Building in the early 1970s when the **space shuttle** program began. The VAB was used to assemble the large launch vehicles for the U.S. **human spaceflight program** between 1968 and 2011. Completed in 1966, the building stands 525 feet tall, covers eight acres, and accesses Launch Complex 39 from which the **Apollo** and space shuttle programs launched, as well as a multitude of commercial (**SpaceX**'s Falcon, for example) and **satellite** missions. In volume, it is equivalent to 3.75 Empire State Buildings. For the Apollo program, beginning several months before a launch, the component parts of the **Saturn** V launch vehicle and the Apollo spacecraft were delivered to the VAB where they were assembled on a mobile launcher that could then be moved to the launch pad (3.5–4.2 miles away) by a crawler transporter. The 456-foot high doors would open to allow the crawler-transporter to move the launch vehicle and command module to the launch pad at 1 mph. After modernization and modification of the building beginning in 2014, NASA anticipates that the VAB will be used to prepare commercial launch vehicles and in preparation for the **Space Launch System**.

VENUS. Venus is the planet next-closest to the sun from Earth, is our nearest planetary neighbor, and is normally the brightest object in our night sky after the **moon**. Although almost the same size as Earth, its atmosphere is much denser (90 atmospheres, or about what we would feel half a mile deep in the ocean), and its temperature is much hotter (863°F, 462°C)—hotter even than Mercury, which is closer to the sun but has a thinner atmosphere. Clouds composed of sulfuric acid cover the surface, completely obscuring our view. Because Venus rotates slowly, one day on Venus is 243 days on Earth; its year—its trip around the sun—is 225 Earth days, which means that its day is longer than its year (in Earth terms). Venus has the longest day in the solar system, and the planet rotates backward compared to the other planets (i.e., the sun rises in the west, sets in the east). Its surface is rocky and volcanic, and it has no moons.

The first spacecraft to travel to Venus was ***Mariner** 2* in 1962; *Venera 7*, a **Soviet** spacecraft, returned data from the surface of Venus in 1970, and *Venera 9* landed on Venus in 1975 and returned photos of its rocky surface—the first spacecraft to land on another planet. **Pioneer Venus** made the first high-quality map of the planet three years later. By 1994, **Magellan** had mapped most of the surface of Venus by radar, creating a gravity field map. Since then, MESSENGER performed two flybys of the planet, and the **European Space Agency**'s Venus Orbiter launched in 2005, but its operations ended on May 15, 2014, and it was allowed to drop into the atmosphere. Akatsuki, a **Japanese Aerospace Exploration Agency** spacecraft, was intended to orbit Venus in 2010, but it is thought a faulty valve malfunction created problems; the craft instead flew by the planet and is on course to fly by again in November 2015 with a possible orbital insertion. In total, more than 40 spacecraft have explored Venus, most from its orbit.

VENUS EXPRESS. Venus Express is a **European Space Agency** project to examine the atmosphere and surface of Venus. It was launched on November 9, 2005, and reached the planet five months later. Since then, it has helped scientists decide if Venus is geologically active (still unknown), has detected water molecules escaping into space, and provided infrared views of the planet's surface. The spacecraft is modeled after that of the **Mars Express**, modified to withstand the environment closer to the sun. The mission ended in December 2014 when data was no longer being transmitted by the spacecraft.

VERTICAL ASSEMBLY CENTER (VAC). Completed in September 2014, NASA's Vertical Assembly Center will be used to assemble elements of the **Space Launch System** to send humans to **Mars** as part of the **Orion** mission and to an asteroid. The VAC is located at the **Michoud Assembly Facility** in New Orleans, which is where the core stage (that part of the rocket that stores cryogenic liquid hydrogen and liquid oxygen to fuel the engines) is being built by **Boeing**. The VAC is the world largest (170 feet high, 78 feet wide) spacecraft welding tool. Problems were announced by Boeing in March 2015 regarding bonding process alignment, but no delay in the Space Launch System was anticipated at that time.

See also VEHICLE ASSEMBLY BUILDING (VAB).

VIKING. *Viking 1* and *2* were space **probes** launched to **Mars** in August and September 1975 and managed by **Langley Research Center** as extensions of the earlier Voyager Mars program. They reached Mars's orbit in June and August 1976 and orbited the planet for more than a month. Both had a lander to explore details on the surface as well as an orbiter to photograph the planet. The landers were deployed to the surface while the orbiters documented the landings and took photographs of the surface. One of the orbiters' first surprising discoveries was that there were huge valleys on the surface caused by flooding long ago, and that large areas in the southern hemisphere had dendritic patterns that indicated rainfall in the far distant past. The landers, meanwhile, conducted biological experiments that were hoped would indicate the presence or absence of life in Martian soil—but the results remain inconclusive. *Viking 2* gave out first: Its orbiter shut down after a fuel leak almost two years after it launched, but its lander lasted almost two years after that. *Viking 1* continued to orbit for four years, and its lander for six years. The program was terminated in May 1983. Then in 2010, experiments prompted by NASA's **Phoenix** Mars lander indicated that perhaps the soil examined earlier by the Viking landers did in fact contain carbon-based "building blocks." On the earlier trip, the Viking landers identified chloromethane and dichloromethane, which were felt to be contaminants from cleaning fluids. However, when perchlorate was added to desert soil from Chile containing organics, the soil revealed chloromethane and dichloromethane as well. This does not necessarily indicate life on Mars, albeit a very primitive life, as many meteorites contain organics. Refer to appendix B for a list of their instruments.

Note: There is also a Swedish **satellite** called Viking, which was launched on February 22, 1986, and operated until May 12, 1987, exploring plasma processes in Earth's magnetosphere.

VISIBLE INFRARED IMAGING RADIOMETER SUITE (VIIRS). Managed by **Goddard Space Flight Center**, VIIRS—a scanning radiometer—collects information about visible and infrared imagery and measures clouds, aerosols, ocean color, surface temperature, fires, and albedo (reflectivity). It is on board, for example, the Nimbus 7 spacecraft and is an important tool for climatologists.

See also FIRE AND SMOKE; WEATHER SATELLITES.

VISION FOR SPACE EXPLORATION. On January 14, 2004, less than a year after the ***Columbia*** **disaster**, President George H. W. Bush outlined his vision for NASA in a speech at **NASA Headquarters** in Washington, D.C. The fundamental goals of the program were fourfold: (1) to have a cost-effective and sustainable human and robotic program to explore the solar system and beyond; (2) to return humans to the **moon** by 2020, in preparation for human exploration of **Mars** and beyond; (3) to develop new technologies and infrastructures to support these explorations; and (4) to promote commercial participation and international partnerships to further U.S. scientific, security, and economic interests. A key part of the plan included the **Lunar Reconnaissance Orbiter**, which launched in June 2009 to look for water-ice on the moon. Flights to the moon would then be launched by the **Ares** rockets.

Unfortunately, the plan had several critics. **Buzz Aldrin** stated that returning to the moon was "like reaching for past glory," and the Aerospace Technology Working Group stated the Vision had major problems with financing, space policy, and politics. Others felt that NASA should concentrate on scientific research, **space tourism**, and business. Another issue was that the Vision called for the end of the **space shuttle** by 2010 and of U.S. participation in the **International Space Station** by 2017.

The Vision as a blueprint for future NASA direction was terminated by President Barack Obama in June 2010 and replaced with the **Space Launch System**.

VON BRAUN, WERNHER (March 23, 1912–June 16, 1977). Von Braun was a German engineer and rocket designer who, following World War II, worked in the United States on rocket technology. Although there are questions as to his role in the National Socialist (Nazi) party in Germany, eventually von Braun and his staff (including his brother, also a rocket scientist) surrendered to the Americans rather than risk capture by the Russians. He was recruited into the United States through Operation Paperclip in June 1945, and his first job at Fort Bliss, Texas, was to refurbish and assemble captured German V-2s for launch at the **White Sands** Proving Grounds. With the Korean War underway, von Braun and his staff were transferred to the Redstone Arsenal in Huntsville, Alabama, where he developed the **Jupiter** C that launched the ***Explorer 1*** on January 31, 1958. Always interested in space, von Braun authored a number of articles and essays about exploring space ("Man Will Conquer Space Soon"), a science fiction novel (*Project MARS*), and worked out a manned mission to **Mars**, though it was not developed by NASA; he even worked with Walt Disney on films about space exploration during the Cold War. However, despite these "family friendly" projects, he was at the same time working on an armed space station and on developing faster, more powerful missiles for the **U.S. Army**. He became a U.S. citizen in 1955. When **Marshall Space Flight Center** was opened in Huntsville, von Braun became its director. Its first major program, **Saturn**, was von Braun's baby, and less than a decade later, the Saturn V rocket launched the crew of ***Apollo 11*** on its way to the moon. The development of a space camp in Huntsville was his idea. In 1970, he was appointed NASA's deputy associate administrator for planning in Washington, D.C., and—in a disagreement over the future of NASA—he retired in 1972. He worked in corporate America for a while, and helped establish the National Space Institute. He was awarded the 1975 National Medal of Science, although he was too ill to attend the ceremony.

VOSTOK. The Vostok ("East") program was a **human spaceflight** project of the **Soviet Union**, comparable to some extent with the United States' Project **Mercury**. On April 12, 1961, **Yuri Gagarin** orbited Earth on board the *Vostok 1*, beating the United States into space. A total of six human spaceflights took place between 1961 and 1963, the longest of which was not quite five days. The Vostok program was followed by the Voskhod ("Ascent") program in 1964 and 1965, which again used the Vostok capsule but with a larger launch rocket. There were inevitably many comparisons between the Vostok and Mercury programs, not the least of which was that the Americans had the **Mercury Seven**, and the Soviets has the Vanguard Six.

VOYAGER 1 **AND** ***2.*** Currently farther away from the Earth than any other spacecraft have flown, *Voyager 1* and *2* (twin **probes** originally called ***Mariner*** *11* and *Mariner 12*) were launched respectively on September 1 and August 20, 1977, from **Cape Canaveral.** Impetus for the mission had begun years earlier when scientists realized that the outer planets would be aligned in such a way that a spacecraft could transit from one planet to the next without the need for an extensive onboard propulsion system. Their initial five-year mission was to explore the outer planets of **Jupiter** and **Saturn** along with their **moons**, and they provided immense amounts of data about the gas giants—including active volcanoes on Jupiter's moon Io—followed by gravitational assists to (*Voyager 1*) **Pluto** (1988) and (*Voyager 2*) **Uranus** (1986) and **Neptune** (1989), the only spacecraft to have visited those two planets. *Voyager 1* was launched after *Voyager 2*, but *1* was on a shorter (and therefore faster) trajectory that let it to a close flyby of Saturn's **moon** Titan.

Some of the surprises that scientists found during the initial and **extended mission** include:

- Jupiter's Great Red Spot is a complex storm moving in a counterclockwise direction.
- At least nine active volcanoes are found on Io, some with plumes as high as 190 miles (330 km), with debris ejected at rates near 0.65 miles per second.
- The tides on Io as the surface is pushed and pulled between the planet and the larger moons of Europa and Ganymede is about 330 feet (100 m).
- The presence of radial, spoke-like features in Saturn's rings that form and dissipate was unexpected; they are possibly fine particles moving in an electrostatic induced rhythm.
- Winds on Saturn are as strong as 1,100 mph.
- Titan's atmosphere is chemically similar to that of Earth before life evolved.

- Uranus orbits the sun on closer to a 90-degree tilt (unlike Earth's 23-degree tilt), giving it a sideways orientation.
- Neptune's Great Dark Spot (similar to Jupiter's Great Red Spot) was discovered and is roughly the size of Earth.
- Winds on Neptune are stronger than on any other planet, up to 1,200 mph (200 kph).

Both identical probes have radioactive power sources that support the electrical instruments on board, traveling outbound at about 3 AU per year; an AU—**astronomical unit**—is the distance from Earth to the sun, or 93,000,000 miles. After the Voyagers completed their planetary explorations, the mission was extended and renamed the **Voyager Interstellar Mission**. At the time of writing, *Voyager 1* is 133.34 AU from Earth, and *Voyager 2* is 109.33 AU from Earth. Both Voyagers carry greetings from the peoples of Earth to any being who encounters them, a phonographic record with images and sounds selected by a committee chaired by Dr. **Carl Sagan**, including greetings in 55 languages. Refer to appendix B for a list of their instruments.

VOYAGER GOLDEN RECORD. Phonograph records were included aboard both **Voyager** spacecraft that were launched in 1977. They contain sounds and images selected to portray the diversity of life and culture on Earth, and are intended for any intelligent extraterrestrial life form, or for future humans, who may find them. The contents of the record were selected for NASA by a committee chaired by **Carl Sagan**. Sagan and his associates assembled 116 images and a variety of natural sounds. Also included were musical selections from different cultures and periods, spoken greetings in 55 languages, and printed messages from former U.S. president Jimmy Carter and then U.N. secretary-general Kurt Waldheim.

VOYAGER INTERSTELLAR MISSION (VIM). ***Voyager*** *1* and *2* explored the outer planets, after which they continued traveling outbound. After 35 years of travel, *Voyager 1* is in interstellar space (the area between the stars, filled with material ejected by the deaths of nearby stars millions of years ago) and *Voyager 2* is in the heliosheath, the outside edges of the heliosphere, in an area where our **solar wind** still exists. *Voyager 2* will enter interstellar space in about seven years—though the exact location of the heliopause (the boundary between our solar system and interstellar space) is unknown and variable. Data are transmitted through the **Deep Space network**, and both are expected to continue transmitting data for about 10–15 years.

W

W. M. KECK OBSERVATORY. A two-telescope astronomical observatory near the summit of Mauna Kea in the U.S. state of Hawaii, Keck's two telescopes feature 33-foot (10 m) primary mirrors, currently among the largest astronomical telescopes in use. In August 2015, astrophysicists at Keck Observatory measured the farthest galaxy (EGSY8p7) ever recorded, as well as its emission when the universe was young (less than 600 million years old), providing new data with regard to how galaxies contributed to reionization. The Keck telescopes are part of NASA's **Origins Program**.

WALLOPS FLIGHT FACILITY. Part of the **National Advisory Committee for Aeronautics**, Wallops was established in 1945 then under the direction of **Langley Research Center**; it was absorbed into NASA when the latter was established by the **National Air and Space Act** in 1958 and is currently operated by **Goddard Space Flight Center**. Wallops is located on Virginia's Eastern Shore, ideal for **low-Earth orbit** rocket launches. Wallops's primary focus within NASA is for the management, development, and monitoring of suborbital research programs, particularly low-cost, next-generation orbital launch technologies. The Mid-Atlantic Regional Spaceport, located on land leased from NASA Wallops, is open to commercial launches. For example, the October 2014 launch by **Orbital Sciences** to the **International Space Station**—the third of eight contracted deliveries—took place from Wallops. Although it exploded on liftoff, the previous two had had no significant issues. Approximately 1,000 full-time civil service and **contractor** employees are at Wallops, including at the Wallops Visitor Center, which hosts public tours and **educational and public outreach** activities.

WEATHER SATELLITES. Part of the **Earth Science** Office, NASA operates a set of **satellites** for weather purposes (GOES-E) and has partnered with the **National Oceanic and Atmospheric Administration** (NOAA) for others. NASA's first series of weather satellites was the Television Infrared Observation Satellite (TIROS) series, beginning with the launch of TIROS 1 on April 1, 1960. TIROS 1 was, by today's standards, a fairly primitive satellite, measuring 42 inches in diameter and weighing 270 lbs. Powered by 9,200 solar cells, it had two television cameras with a magnetic tape recorder for each; the antennas were four rods that served as transmitters and one rod that acted as a receiver. It operated for only 78 days, but it proved that satellites could be used for weather purposes. The latest in the TIROS-N/NOAA series, nos. 14 and 15, were launched in the 1990s and are still operational.

The Nimbus series, nos. 1–7, launched between August 1964 and October 1978, studied oceans, ozone, pollution, and weather patterns, and tested meteorological sensors for varying aspects of our planet's atmosphere. Nimbus B failed on launch, so there were only six in the series. NASA launched the Synchronous Meteorological Satellites (SIMs) 1 and 2 in 1974 and 1975, which were precursors of the **Geostationary Operational Environmental Satellites** (GOES) satellites, a network of weather satellites currently orbiting Earth and with further GOES satellites scheduled for launch.

A number of other satellites currently in operation study such aspects of weather as potential for fires, hurricane prediction, and so forth, such as **Aqua**, **Aura**, **CALIPSO**, **Landsat**, and many others.

WEBB, JAMES E. (October 7, 1906–March 27, 1991). The second administrator of the newly formed NASA, Webb served from February 14, 1961, to October 7, 1968—from the **Mercury** and **Gemini** programs to just before the crewed **Apollo** spaceflights to the **moon**, and the planetary exploration programs of **Mariner** and **Pioneer**. Born in Tally Ho, North Carolina, Webb attended the University of North Carolina; after graduating, he became a pilot in the Marine Corps from 1930 to 1932. He attended law school in Washington, D.C., and then served in a variety of public service offices before reenlisting in World War II. After the war, he returned to Washington public service and to private employment in Oklahoma City, but he returned to Washington when he was asked to succeed **Keith Glennan** as NASA's administrator. Although reluctant to take the job—he had no scientific or engineering background—Webb spent much of his Washington insider energy and his political skills on successfully lobbying for support for NASA in Congress, first for President **John F. Kennedy** and then for his successor, President **Lyndon B. Johnson**, and was instrumental in creating the **Manned Spacecraft Center** (later the **Johnson Space Center**). The war in Vietnam intervened with his plans for a stronger **budget** in 1967, and just as he was overcoming those reluctant forces, the ***Apollo 1*** fire occurred. He strongly believed that NASA needed to balance **human spaceflight** with science—that one without the other would leave the agency crippled. He succeeded in taking a disparate set of facilities and personnel and creating a more or less coherent and balanced organization. The **James Webb Space Telescope** (originally the Next Generation Space Telescope) was named in his honor.

WHITE, EDWARD H., II (November 14, 1930–January 27, 1967). One of the **astronauts** killed in the ***Apollo 1*** / *Apollo-Saturn 204* training accident (along with **Gus Grissom** and **Roger Chaffee**), Ed White was the first American to walk in space. From a military family in San Antonio, Texas, White graduated from West Point and became a pilot for the **U.S. Air Force** in West Germany. He then got a master's degree in aeronautical engineering and became a test pilot, logging more than 3,000 hours of flight. He was selected as an astronaut in 1962, and as pilot of **Gemini** 4, became the first American to conduct an **extravehicular activity** on June 3, 1965. He died less than two years later during a training accident fire at **Cape Kennedy** when the hatch could not be opened from the inside. He was buried at West Point Cemetery with full honors, and he was the recipient of numerous awards and medals. In addition, the star Iota Ursae Majoris is nicknamed "Dnoces" (which was Second—as in his name—spelled backward) and the White Hill on Mars is named after him.

WHITE SANDS COMPLEX (WSC). Hosted within the overall **White Sands Test Facility**, the WSC is administered by **Goddard Space Flight Center** and is comprised of several NASA ground stations typically specialized for high-data-transmission rate Ka-band **satellite** communications. Ground stations located at the WSC include the **Tracking and Data Relay Satellite System** (TDRSS) White Sands Ground Terminal (WSGT), the TDRSS Second Tracking and Data Relay Satellite Ground Terminal, the WS1 **Lunar Reconnaissance Orbiter** ground station, and the **Solar Dynamics Observatory** ground station. A laser communication ground station, the Lunar Lasercom Ground Terminal (LLGT) was installed in 2013 and demonstrated infrared laser light communication with the LADEE spacecraft orbiting the moon.

WHITE SANDS SPACE HARBOR (WSSH). Formerly known as the Northrup strip, the WSSH supported the **space shuttle** program from 1976 to 2011 as an **astronaut** shuttle landing training facility, a backup landing site for shuttle missions, and as the primary abort-once-around landing site for shuttle launches. The Northrup strip was renamed the White Sands Space Harbor in 1982 after STS-3 landed at WSSH due to rain and harsh weather conditions at both Edwards Air Force Base and the **Kennedy Space Center**, the only time a space shuttle landed there. WSSH consists of two 35,000-foot runways and associated shuttle landing system support facilities. With the end of the space shuttle program, the WSSH was handed over to the **U.S. Army** in 2011.

WHITE SANDS TEST FACILITY (WSTF). A 94-square-mile facility located in the southwestern corner of the White Sands Missile Range, the WSTF was originally built by NASA in 1963 as a propulsion test facility in support of the **Apollo** program and later the **space shuttle** program. The WSTF is administered by **Johnson Space Center** and hosts a variety of current NASA operations, including hazardous materials testing, propulsion testing, and development that ranges from the laboratory to full-scale systems, atmospheric and vacuum testing chambers, and a curation facility for NASA lunar samples. Additionally, the WSTF administered the nearby **White Sands Space Harbor** during the shuttle program and hosts the ground stations that comprise the **White Sands Complex**.

WIDE-FIELD INFRARED SURVEY EXPLORER (WISE). WISE, launched December 14, 2009, into **low Earth orbit**, is an infrared-wavelength astronomical space telescope. Although its mission was terminated in February 2011, WISE had already identified more than 33,500 new asteroids, discovered the first Y dwarf (substellar brown dwarfs too low in mass to sustain hydrogen-1 fusion reactions in their cores) and Earth trojan asteroids (which orbit the sun on an orbital path similar to that of Earth in the vicinity of the Earth–sun **Lagrange points** L4 and L5), and located previously unknown star clusters and millions of black holes. A four-month follow-on mission, **NEOWISE** (also managed by **Jet Propulsion Laboratory**), continued a search for hazardous comets and asteroids close to Earth's orbit. A second follow-on mission was initiated (to run until some time in 2016) when the previously undetected Chelyabinsk meteor exploded over **Russia** in February 2013. Wise operated in a sun-synchronous orbit, with its telescope always pointing away from Earth (unless the **moon** was in its view), and its solar cells pointed toward the sun. In May 2015, NASA reported WISE's discovery of "the most luminous galaxy in the universe."

WIDE-FIELD INFRARED SURVEY TELESCOPE (WFIRST). Still in the proposal stages, WFIRST is a high-priority project for astronomy. If WFIRST is **budgeted** beginning in 2017, the earliest launch would be in 2024. The telescopes—possibly with one donated by the **National Reconnaissance Office**—would be used to attempt to measure the effects of dark energy. Scheduled to be part of the Joint Dark Energy Mission between NASA and the Department of Energy, this design phase is being led by **Goddard Space Flight Center**.

WILKINSON MICROWAVE ANISOTROPY PROBE (WMAP). Launched on June 30, 2001, as part of the **Explorer** (*Explorer 80*) missions, WMAP successfully provided a full-sky map of our universe (microwave background radiation) and studied the makeup of our part of the cosmos from Earth–sun **Lagrange point** 2 (L2) by measuring remnant radiant heat. WMAP followed the **Cosmic Background Explorer** space mission in 1989 and was initially simply Microwave Anisotropy Probe but was renamed in 2003 in honor of cosmologist David T. Wilkinson., establishing that the universe is 13.772 billion years old, measuring the density of atoms and nonatomic matter, and supporting the Big Bang theory of the origin of the universe. Researchers credit WMAP with converting cosmology from "a field of wide speculation" to the current Standard Model of Cosmology (the Lambda-CDM model). Among many other measurements, WMAP investigated universe inflation, and supported theories of early expansion that showed how the universe grew by more than a trillion-trillion-fold in less than a trillionth of a trillionth of a second. In addition, the composition of the universe was shown to be 4.6 percent atoms (baryonic matter), 24 percent cold dark matter, and 71.4 percent dark energy (the source of antigravity that is propelling the acceleration of universe expansion). Data was relayed daily through the **Deep Space Network**. The original two-year mission was **extended** in 2002, 2004, 2005, and 2006; after nine years of observations, the mission was terminated in October 2010. The **James E. Webb Space Telescope** is designed to complement and support these observations and the **Planck** mission (from the **European Space Agency**, launched on May 14, 2009) is designed for higher resolution and sensitivity than WMAP, indicating even more minute measurements and models; its initial data reveals an even longer age of our universe (13.798 billion years), and slightly refined measurements of the composition of our universe.

WILLIAMS, ROBIN (1951–2014). A television actor and comedian, Robin Williams woke up the crew of the space shuttle ***Discovery*** in 1988, the first flight of a **space shuttle** since the ***Challenger* tragedy** in 1986. He said, "G-o-o-d morning, Discovery! G-o-o-d morning, Discovery! Rise and shine, boys. Time to start doing that shuttle shuffle. You know what I mean." The broadcast was a reference to his *Good Morning Vietnam* role, and was followed by a ditty to the tune of "Green Acres." The crew replied with, "G-o-o-d morning, **Houston**."

WIND SATELLITE. Launched on November 1, 1994, Wind is one of two NASA missions of the Global Geospace Science initiative, which is part of the **International Solar-Terrestrial Physics** program. Wind's mission was to study the plasma, energetic particle, and magnetics fields as they related to Earth-sun behaviors and to provide baseline ecliptic plane observations for other heliospheric studies, particularly from **Ulysses**. Despite its original three-year mission, it has been operating at Earth–sun **Lagrange point** 1

(L1) since 2004 and has enough fuel on board for another 50 years. Wind has contributed significantly to our knowledge of **solar wind** and magnetosphere interactions, being the first to observe and report phenomena that had either been unknown or simply theorized. Approximately 3,000 publications have relied on data from Wind. Refer to appendix B for a list of its instruments.

See also POLAR.

X

X-RAY MULTI-MIRROR MISSION-NEWTON. The XMM-Newton is part of the **European Space Agency**'s Horizon 2000 program, for which NASA has a guest-observer facility (GOF) at **Goddard Space Flight Center**. In addition, NASA had provided input into some of the instruments. The XMM-Newton, launched on December 10, 1999, carried three high-throughput X-ray telescopes and an optical monitor to study celestial X-ray sources, including black holes and the origins of the universe. The GOF is a clearinghouse for technical information and analysis software. Refer to appendix B for a list of its instruments.

X-15. The North American X-15 rocket-powered aircraft was one of NASA's experimental aircraft series that set speed and altitude records. Originally developed for the **National Advisory Committee for Aeronautics** by **North American Aviation** (airframe) and Reaction Motors (engines), the X-15 achieved speed records that still hold true today, more than 50 years later, for human, powered flight (4518 mph) in October 1967. In addition, eight pilots achieved **astronaut** status by flying more than 50 miles above the Earth, one of them into outer space, more than 62 miles above Earth. Like many experimental aircraft, it was designed to be ferried—often by a B-52 "mother ship"—about eight miles above the Earth and then released, its rocket engine only used for a small portion of the flight to test its performance in the thin atmosphere.

X-20. Manufactured by **Boeing**, the Dyna-Soar X-20 was a two-year **U.S. Air Force** program beginning in October 1957 (in response to **Sputnik**) to develop a military hypersonic space plane that could be used for reconnaissance, bombing, maintenance, and sabotage. It was designed to be a piloted, reusable space vehicle with a payload capacity of 10,500 lbs., 35.5 feet long, and with a wingspan of over 20 feet. By the time the X-20 reached the mock-up stage, $410 million had been spent on **astronaut training** (including **Neil Armstrong**) and development. The program ran into problems with funding and with determination of the appropriate **booster** to launch Dyna-Soar into orbit. In 1963, a determination between **Gemini** and Dyna-Soar showed that the **Manned Orbiting Laboratory**, a spin-off of Gemini, was more promising, and the Dyna-Soar program was **cancelled** later that year. However, much of the research on Dyna-Soar would later resurface in development of the **space shuttle**, particularly its delta-wing shape.

X-30. An advanced technology demonstrator project designed by **Rockwell International** for the **National Aero-Space Plane** project as a single-stage-to-orbit spacecraft. The **Department of Defense** wanted it to carry a crew of two and even a small payload. These demands ultimately made X-30 larger, heavier, and more expensive than the original plan of only carrying a crew of two and, as a result, the X-30 was **cancelled** in the early 1990s.

X-33. Lockheed Martin's Skunk Works Factory's X-33, part of the **Space Launch Initiative** project in the late 1990s, was intended to be an eventual replacement for the **space shuttle** that was scheduled to be discontinued in an estimated 15 years. As a reusable launch vehicle, it would lower the costs of launching **satellites** and other spacecraft. Set in motion by NASA in July 1996 and approved for a budget of $941 million, the robotic replacement was designed to be available for orbital flight approximately every 10 days. Unfortunately, its huge 68-foot wingspan and other issues (instability, excess weight, failure of the composite liquid hydrogen tank) led to failures in pressure tests and a **cancellation** of the project in 2001—but not before NASA had spent $922 million and Lockheed Martin another $375 million in developmental costs.

XB-70. Originally designed to be a missile bomber, the XB-70 was under development when President **John F. Kennedy** cancelled it in favor of his man on the **moon** campaign. The Flight Research Center (now **Armstrong Flight Research Center**), however, was working on plans for supersonic transport, and the XB-70 was perfect for its uses, although somewhat unstable at Mach 3. At about the same time, **Ames Research Center** in California conducted wind-tunnel studies on the XB-70A, modifying it with an additional 5 degrees of dihedral on the wings, which increased its stability. NASA teamed up with the **U.S. Air Force**, conducting dozens of joint flights that resulted in studies on vibration, acceleration, stability, turbulence, and atmospheric temperature that would be immensely valuable in other programs, particularly the ongoing **Apollo** program.

XS-1. Short for "Experimental Spaceplane 1," it is a **Defense Advanced Research Projects Agency** program to develop a reusable first stage of a space transport system for access to Earth orbit. Its goal is to provide lower-cost and more approachable launches of U.S. government **satellites** while maintaining a competitive edge in the commercial launch market. The space plane must transport between 3,000 and 5,000 lbs. (1,400–2,300 kg) of payload to **low Earth orbit** for less than $5 million per flight at a rate of 10 or more flights per year. Three teams were selected to participate in Phase 1: Northrop Grumman (Scaled Composites and Virgin Galactic), Mastern Space (with XCOM Aerospace), and **Boeing** (with Blue Origin). Currently, the project is planned to have its first flight in 2017 with the first orbit in 2018.

XXM-NEWTON. *See* X-RAY MULTI-MIRROR MISSION-NEWTON.

Y

YEAGER, CHARLES "CHUCK" ELWOOD (February 13, 1923–). Chuck Yeager retired from the **U.S. Air Force** as a brigadier general in 1975, but he is chiefly known as the first test pilot to have flown faster than the speed of sound in level flight. In 1962, he became the commander of the USAF Aerospace Research Pilot School, from whose ranks many astronauts were trained. He flew at Edwards Air Force Base for NASA in its M2-F1 lifting body, and remained a consultant for NASA after his retirement. Yeager was also appointed to the presidential commission investigating the ***Challenger*** **tragedy**.

Z

ZERO GRAVITY RESEARCH FACILITY. Located in Brook Park, Ohio, the Zero-G facility was construction during the Cold War space race, but is currently being used to study the effects of a near-weightless environment for periods up to 5.18 seconds. This happens during free fall inside a 467-foot (142m) steel vacuum chamber, where pressure is reduced (a process that takes about an hour). The experiment is stopped at the bottom of a 432-foot (132 m) fall by a 20′ deep cart filled with 1/3″ (3 mm) diameter expanded polystyrene beads, which absorb the kinetic energy. At its peak, the experiment vehicle travels about 113 mph.

Appendix A: FY 2015 Budget Request

National Aeronautics and Space Administration

FY 2015 President's Budget Request Summary

Budget Authority ($ in millions)	Actual 2013	Enacted 2014	Request 2015	Notional 2016	Notional 2017	Notional 2018	Notional 2019
NASA Total	**16,865.2**	**17,646.5**	**17,460.6**	**17,635.3**	**17,811.5**	**17,989.7**	**18,169.7**
Science	**4,781.6**	**5,151.2**	**4,972.0**	**5,011.7**	**5,071.9**	**5,122.6**	**5,173.9**
Earth Science	1,659.2	1,826.0	1,770.3	1,815.5	1,837.6	1,861.9	1,886.3
Planetary Science	1,274.6	1,345.0	1,280.3	1,304.9	1,337.1	1,355.7	1,374.1
Astrophysics	617.0	668.0	607.3	633.7	651.2	696.8	933.0
James Webb Space Telescope	627.6	658.2	645.4	620.0	569.4	534.9	305.0
Heliophysics	603.2	654.0	668.9	647.6	676.6	673.3	675.5
Aeronautics	**529.5**	**566.0**	**551.1**	**556.6**	**562.2**	**567.8**	**573.5**
Space Technology	**614.5**	**576.0**	**705.5**	**712.6**	**719.7**	**726.9**	**734.2**
Exploration	**3,705.5**	**4,113.2**	**3,976.0**	**4,079.9**	**4,049.4**	**4,107.7**	**3,673.4**
Exploration Systems Development	2,883.8	3,115.2	2,784.4	2,863.3	2,905.9	2,982.1	3,106.6
Commercial Spaceflight	525.0	696.0	848.3	872.3	791.7	730.9	172.0
Exploration Research and Development	296.7	302.0	343.4	344.3	351.8	394.7	394.7
Space Operations	**3,724.9**	**3,778.0**	**3,905.4**	**3,951.9**	**4,062.8**	**4,085.6**	**4,601.8**
Space Shuttle	38.8	–	–	–	–	–	–
International Space Station	2,775.9	–	3,050.8	3,126.5	3,266.9	3,290.3	3,818.6
Space and Flight Support	910.2	–	854.6	825.4	795.9	795.3	783.2
Education	**116.3**	**116.6**	**88.9**	**89.8**	**90.7**	**91.6**	**92.6**
Cross Agency Support	**2,711.0**	**2,793.0**	**2,778.6**	**2,806.4**	**2,834.4**	**2,862.8**	**2,891.4**
Center Management and Operations	1,991.6	–	2,038.8	2,059.2	2,079.7	2,100.5	2,121.6
Agency Management and Operations	719.4	–	739.8	747.2	754.7	762.3	769.8
Construction and Environmental Compliance and Restoration	**646.6**	**515.0**	**446.1**	**379.0**	**382.7**	**386.6**	**390.4**
Construction of Facilities	589.5	–	370.6	302.7	305.7	308.8	311.8
Environmental Compliance and Restoration	57.0	–	75.5	76.3	77.0	77.8	78.6
Inspector General	**35.3**	**37.5**	**37.0**	**37.4**	**37.7**	**38.1**	**38.5**
NASA Total	**16,865.2**	**17,646.5**	**17,460.6**	**17,635.3**	**17,811.5**	**17,989.7**	**18,169.7**

Note: As reflected in the August 2013 Operating Plan, FY 2013 includes rescissions per P.L.113-6 Division G, Section 3001(b)(1)(B) and Division G, Section 3004(c)(1) and reductions due to sequestration per BBEDCA Section 215A.

FY 2014 reflects funding amounts specified in P.L. 113-76, Consolidated Appropriations Act, 2014, including amounts noted in the Explanatory Statement. Where amounts were not specified, no amount is shown in the budget table.

Funds associated with out-year estimates for programmatic construction remain in programmatic accounts.

BUD-1

FY 2015 Budget Request Source: FY 2015 President's Budget Request Summary, available at www.nasa.gov/sites/default/files/files/508_2015_Budget_Estimates.pdf.

Appendix B: List of NASA Instruments, by Mission

Table 26.1. List of NASA Instruments, by Mission

Mission/Satellite	Class of Equipment	Instrument	Purpose
ACE	Spectrometer	CRIS	A spectrometer investigating the isotopic composition of cosmic rays. The instrument uses charge-coupled detector camera assemblies, silicon detectors, and a scintillating optical fiber hodoscope to detect particles and determine their trajectories.
	Spectrometer	EPAM	Detects electrons, protons, and alpha particles using the following telescopic instruments: two low-energy foil spectrometers, two low-energy magnetic spectrometers, and a composition aperture that measures ion composition.
	Magnetometer	MAG	Uses two triaxial fluxgate magnetometers on booms to measure local interplanetary magnetic field strength and direction.
	Charged Particle Detector	SEPICA	Measures energetic particle distributions and ionic charge state, kinetic energy, and nuclear charge of energetic ions.
	Charged Particle Detector	SIS	Measures the isotopic composition of energetic particles from helium to nickel in high resolution, typically for particles streaming from the solar corona but also from galactic cosmic rays during quiet solar periods.
	Charged Particle Detector	SWEPAM	Uses electrostatic analyzers with fan-shaped fields of view to measure solar wind plasma and ion rates of flow.
	Spectrometer	SWICS and SWIMS	Measure chemical and isotopic composition of solar and interstellar matter using time-of-flight mass spectrometers and electrostatic analyzers.
	Spectrometer	ULEIS	A high-resolution mass spectrometer that measures composition, energy spectra, and flux rates of energetic solar particles.
AIM	Imaging	CIPS	Images polar mesopheric clouds using four cameras arranged in a cross pattern that provides a 2000 × 1000 km coverage and a spatial resolution that ranges from 1 × 2 km to 5 × 5 km at the edges of the picture.

	Spectrometer	SOFIE	Measures atmospheric temperature and gases such as carbon dioxide, methane, nitric oxide, ozone, polar mesospheric clouds, and aerosols using a solar occultation technique by measuring solar energy passing through the atmosphere at sunset and sunrise relative to the spacecraft.
	Charged Particle Detector	CDE	Measures the influx of cosmic dust into the Earth's mesosphere by measuring impact signals on thin polarized polyvinylidene fluoride films mounted on the satellite on the zenith side of the spacecraft. Two detectors are covered in order to provide a control and to measure background noise.
Apollo	Various	Examples include imaging in multiple wavelengths, hand-held photography, solar wind / Gamma-ray / Alpha-particle spectrometers, micrometeorite detection, radiation detection, suprathermal ion detection	Typical instruments were focused on studies of lunar atmosphere, surface, soil, and geology through the use of instruments such as spectrometers, imaging, and radar. Other studies monitored the radiation environment, solar wind, and micrometeorite detection. Of note was direct sampling and return of lunar soil and rock.
Aqua	Imaging	AIRS	Studies the atmosphere in the infrared and near infrared in order to make highly accurate temperature and humidity measurements throughout the layers of the atmosphere.
	Radiometer	AMSU-A	Uses a microwave radiometer to measure atmospheric temperature profiles in the upper atmosphere.
	Radiometer	CERES	Two broadband scanning radiometers measuring reflected solar radiation, emitted terrestrial radiation, and total thermal radiation.
	Radiometer	HSB	A Brazil-provided microwave sounder that profiled humidity in the atmosphere.
	Spectroradiometer	MODIS	A spectroradiometer measuring visible and infrared radiation in medium resolution in order to derive measurements of vegetation coverage, ocean chlorophyll fluorescence, snow and ice cover, and related phenomena.

	Radiometer	AMSR-E	Uses a passive microwave radiometer to measure precipitation, cloud water, water vapor, sea surface winds, sea surface temperatures, soil moisture, and ice and snow coverage.
Aquarius	Radiometer	Aquarius	A NASA-built instrument on board the Argentinian SAC-D satellite. The instrument measures sea surface salinity using a radiometer to observe ocean surface thermal emissions and a radar scatterometer to correct for ocean wave action.
Ariel	Magnetic Fields	Radio Frequency Capacitance Probe	A capacitance probe used to measure the density and temperature of electrons.
	Magnetic Fields	Electron Temperature Gauge	An electron temperature sensor, complementary to other temperature sensors on the satellite.
	Charged Particle Detector	Cosmic Ray Detector	Used an omnidirectional Cerenkov counter and Anton Geiger tube detector to measure cosmic ray spectra.
	Charged Particle Detector	Ion Mass Sphere	A rhodium-plated sphere detector used to measure ion concentration and change of atomic hydrogen, helium, and oxygen.
	Charged Particle Detector	Lyman Alpha Gauge	Three flush-mounted detectors for monitoring the intensity of ultraviolet radiation.
	X-ray Detector	X-ray Detector	Experiment to measure and study solar X-rays from the 3 to 12 Angstrom band.
ARTEMIS	Two Lunar Satellites	See Themis spacecraft instruments entry for instrument list.	Two spacecraft originally used in the THEMIS mission repurposed to study lunar interaction with the Earth's magnetotail, magnetic interactions in the lunar wake, as well as interactions with the solar wind.
ATTREX	Various	Global Hawk UAV Platform	Carries a variety of instruments that can be interchanged. Examples include cloud droplet profilers, cloud LIDARS, hygrometers, and solar radiometers.
Astro-E2	X-ray Detector	X-ray Spectrometer	Uses an instrumentation system cooled to approximately 0.06 Kelvin (−273°C/−460°F) in order to detect the temperature rise of an X-ray photon hitting a silicon target.
	X-ray Detector / Imaging	X-ray Imaging Spectrometer	Consists of four X-ray imaging spectrometers using an X-ray sensitive 1024 × 1024-pixel CCD.

	X-ray Detector	Hard X-ray Detector	Detects high-energy X-rays and determines directional origin by comparing detections between the primary detector and a secondary detector that acts as a collimator.
	X-ray Detector	X-ray Telescope	Thin-foil, gold-coated, grazing incidence telescope.
Astro-1	Imaging	Ultraviolet Imaging Telescope (UIT)	An ultraviolet imaging telescope using a 38-cm f/9 mirror within a Ritchey Chretien design and two 70-mm film cameras for recording images. The UIT recorded in a range from 1200 to 3000 Angstrom and a diffraction grating could be used for low-resolution spectroscopy.
	Imaging	Hopkins Ultraviolet Telescope (HUT)	The HUT ultraviolet telescope used 0.9-m f/2.0 primary mirror and spectrograph, and operated in a range from 425 to 1850 Angstroms with an emphasis on the 900–1200 Angstroms.
	Imaging / Polarimeter / Spectrometer	Wisconsin Ultraviolet Photopolarimetry Experiment (WUPPE)	A 0.5-m f/10 Cassegrain telescope operating in the 1400 to 3300 Angstrom range that could take simultaneous photometry and polarization measurements using a spectropolarimeter and spectrometer. In addition, the WUPPE experiment used a blue sensitive zero-order detector (ZOD) camera for guiding.
	X-ray	Broad Band X-ray Telescope (BBXRT)	A broadband imaging telescope using cryogenically cooled spectrometers and X-ray mirrors, the BBXRT was primarily focused on studying the bright supernova SN1987A, with a secondary mission of studying other galactic and extragalactic X-rays.
Astro-2	Telescopes	UIT, HUT, WUPPE	Same instrumentation as the Astro-1 but without the BBXRT. In addition, the HUT received a new optical coating, which nearly doubled its performance.
Aura	Radiometer	HIRDLS	An infrared radiometer scanning the Earth's upper atmosphere to determine concentrations of ozone, carbon dioxide, methane, nitrous acid, CFCs, water vapor, aerosols, and various atmospheric gases, as well as stratospheric cloud tops and locations.
	Radiometer / Microwave / Spectrograph	Microwave Limb Sounder	The microwave limb sounder measures a variety of atmospheric gases and temperatures from the lower stratosphere to the upper troposphere. The MLS measures ozone and associated compounds in the middle and upper stratosphere for global stratospheric study, water vapor ice

			content, and temperature in the upper troposphere, and measures temperature in the lower stratosphere as well as concentrations of atmospheric gases such as water, ozone, chlorates, hydroxyls, and other gases.
	Imaging	OMI	Uses hyperspectral imaging to observe solar backscatter radiation from the visible to the ultraviolet in order to monitor air quality indicator compounds, dust, smoke, and sulfates, and to monitor and map ozone.
	Spectrometer	TES	A pointable infrared imaging Fourier transform spectrometer that can use both natural infrared emissions and reflected sunlight for observations. TES focuses on tropospheric chemical compounds, tropospheric and stratospheric exchanges, acidic compounds, and ozone, water vapor, carbon dioxide, and other key tropospheric chemical compounds.
BARRELL	X-ray	NaI Scintillator	A detector that indirectly measures relativistic electrons entering the Earth's atmosphere by measuring Bremsstrahlung X-rays, also known as braking radiation, that emit detectable light when impacting the sodium iodide scintillator crystal.
	Magnetometer	DC Magnetometer	Measures the Earth's magnetic fields local to the instrument.
Big Bird/Keyhole	Radio	Atmospheric sensor	(Nonclassified NASA payload) Doppler beacons used to map upper-atmospheric densities.
Biosatellite 1, 2, 3	Biomedical	Medical Instrumentation	This series of satellites carried insects, seedlings, bacteria, and a pig-tailed monkey. Besides basic imaging observation of various experiments, most instrumentation and study was done on Earth after recovery, with the exception of the monkey, which was instrumented with various monitors to study brain waves, urine, fecal, and blood biochemical analysis, and other medical measurements such as cardiovascular, temperature, and related biomonitors.
Cassini	Spectrometer	CIRS	An infrared spectrometer comprised of two infrared interferometers, a reference interferometer, a moving scan mechanism, and thermal control equipment.
	Imaging	Imaging Science Subsystem	This imaging instrument consists of a narrow- and wide-angle camera using 1024 × 1024 charge-coupled devices.

Spectrometer	UVIS	An ultraviolet imaging spectrometer consisting of both an extreme and far-ultraviolet channel.
Spectrometer	VIMS	The VIMS instrument is a spectrometer operating in both the visible and infrared, recording measurements over 352 wavelengths generally between 0.35 to 5.1μ to study surface, atmospheric, and ring chemistry of Titan and Saturn.
Spectrometer	CAPS	The Cassini plasma spectrometer consists of an ion beam spectrometer, an electron spectrometer, and an ion mass spectrometer to study the energy, temperature, flows, and density of ions, electrons, plasma, and solar wind interactions in Saturn's magnetosphere.
Charged Particle Detector	CDA	The cosmic dust analyzer uses a hemispherical target, an ion collector, four charge pickup grids, electron multiplier, and sensor electronics to record dust, ice, and positively charged plasma ions.
Spectrometer	INMS	An ion and neutral mass spectrometer is a quadrupole mass spectrometer that studies gases and ions encountered in Saturn's ring environment and Titan's upper atmosphere.
Magnetometer	MAG	A dual technique magnetometer that uses a fluxgate magnetometer and vector/scalar helium magnetometer in order to study the magnetic fields of Saturn and Titan in strength, magnitude, and direction.
Magnetic Fields, Spectrometer	MIMI	The magnetospheric imaging instrument uses three sensor heads, a low-energy magnetospheric measurement system using magnetic deflection and solid state detectors, a charge-energy-mass spectrometer using electrostatic deflection and time of flight, and an ion and neutral camera recording oxygen and hydrogen energetic neutral atoms, in order to measure energy, mass, and species of energetic neutral atoms for magnetospheric imaging, hot plasmas, sources of plasmas and energetic ions, and other magnetospheric interactions and associated particle studies.
Radio	RPWS	The radio and plasma wave science instrument consists of an electric field sensor antenna, magnetic search coil, and Langmuir probe in order to measure Saturn's radio waves,

			magnetic fields, radio signals from lightning, plasma waves and magnetospheric plasma, electron densities, and related magnetospheric phenomena.
	Radar	Microwave Radar	Cassini's radar uses a five-beam Ku-band synthetic aperture imager, radar altimeter, and radiometer to map Titan's surface beneath cloud layers as well as to image Saturn, Saturn's rings, and other Saturn moons.
	Radio	RSS	Cassini's radio science experiment uses radio waves transmitted to NASA's deep space network through Saturn's rings and atmosphere to study their composition or non-obstructed radio transmissions in order to measure spacecraft changes in velocity to derive gravitational interactions with the spacecraft.
Cassini-Huygens Probe	Radiometer	Aerosol Collector and Pyrolyzer	Analyzed chemical composition of Titan's atmosphere by taking a series of samples during descent and subsequently heating these for analysis as well as feeding vaporized material to the GCMS instrument.
	Radiometer, Imaging	Descent Imager and Spectral Radiometer	Took descent images and spectral measurements from the infrared to the ultraviolet. Also profiled Titan's atmosphere during descent by using a solar occultation technique to view the sun and analyze aerosols through light scattering.
	Spectrometer, Chromatograph	Gas Chromatograph and Mass Spectrometer	A gas chromatograph and spectrometer for analyzing Titan's atmosphere.
	Meteorology	Doppler Wind Experiment	Used synced ultrastable oscillators and radio transmissions to examine winds during descent by performing radio-Doppler analysis between the probe and Cassini orbiter.
		Huygens Atmospheric Structure Instrument	Used a variety of sub-instruments such as accelerometers, temperature sensors, atmospheric pressure, and an electric fields sensor array.
		Surface Science Package	A variety of instruments for analyzing the surface. Accelerometers, tilt sensors, a thermal properties probe, electrodes for measuring fluid permittivity, a fluid density sensor, a refractive index prism, photodiode sensor, and acoustical properties sensors.

CALIPSO		Cloud Aerosol Lidar with Orthogonal Polarization	A cloud and aerosol profiling LIDAR that measures laser backscatter and orthogonally polarized backscatter components. Two-channel (512-nm & 1024-nm) CO_2 laser. Lasers were built by Composite Optics.
		Imaging Infrared Spectrometer	An infrared spectrometer using a microbolometer detector array and filter wheel for multiple channel measurements. Developed by Sodern in Paris, France, and uses microbolometers from Boeing.
		Wide Field Camera	A straightforward imaging camera using an off-the-shelf modified Ball Aerospace star tracker camera, delivered through CNES.
Chandra	Telescope	Mirror Assembly	Chandra's high-resolution mirror assembly for X-rays, unlike a traditional telescope, consists of a series of 2.7-m nested paraboloid and hyperboloid mirrors in a barrel-like configuration and is accurate to within 1.3μ.
	Imaging	HRC	A high-resolution camera using microchannel plates consisting of lead oxide glass tubes that excite when struck by X-rays.
	Imaging	ACIS	An array of charge-coupled devices tuned for recording X-rays.
	Spectrometer	HETG and LETG	Two high-resolution spectrometers used for high and low energies, these spectrometers use gold transmission gratings and Chandra's overall telescopic system to produce X-ray spectrographs.
CINDI	Magnetic Fields, Charged Particle Detector	IVM/NWM	The ion velocity meter and neutral wind meter instruments measure densities and velocities of ionized and neutral particles in the ionosphere in order to study plasma bubbles, their structures, and precursor conditions.
Clementine	Imaging	UV/Vis	An ultraviolet and visible light camera assembly using a 46-mm aperture catadioptric telescope design and charge-coupled device (CCD) camera. The CCD camera was equipped with a six-wavelength filter wheel and could take bursts of 12 images in 1.3 seconds.
	Imaging	NIR	A near-infrared camera using a 29-mm aperture catadioptric design and a mechanically cooled charge-coupled device camera.

	Imaging	LWIR	A long-wavelength infrared camera using a catadioptric design with a 131-mm aperture and a cooled charge-coupled device camera.
	Imaging	HIRES	A 131-mm high-resolution telescope with an image intensified charge-coupled device camera and a six-position filter wheel for selecting optional wavelengths. This telescope/camera assembly had a narrow field of view so was used to study specific targets; also the telescope itself was shared with the LIDAR.
	Imaging	Star Tracker Cameras	Two wide-field, CCD cameras used to image stars to determine spacecraft attitude, some basic lunar and Earth imaging could also be performed.
	Laser	LIDAR	A laser image detection and ranging instrument that provided altimetry information using laser pulses bounced off the lunar surface and the HiRES telescope and a silicon avalanche photodiode to detect laser return.
	Radar	Bistatic Radar Experiment	This experiment transmitted S-band radio signals through the spacecraft's high-gain antenna at lunar targets, typically lunar polar craters in the search for ice, that were then detected on Earth for analysis by NASA's Deep Space Network.
	Radio	Gravity Experiment	The spacecraft transmitted pings using an S-band transponder and S-band antennas that were sampled and analyzed on Earth for Doppler shifts in order to calculate subtle changes in velocity and infer lunar density. Accuracy of velocity data varied between 0.3 mm/s and 3 mm/s; data could not be collected on the moon's far side.
	Charged Particle Detector	Charged Particle Telescope	A telescope and silicon detector assembly used to record the flow and spectra of protons, electrons, and some heavier ions.
CloudSat	Radar	Cloud Profiling Radar	A 94-GHz downward-looking radar that detects clouds and density of water and ice.
Cluster	Magnetometer	FGM	A fluxgate magnetometer mounted on 5-m booms, used to measure Earth's planetary magnetic field strength and direction.

	Magnetic Fields	EDI	This instrument shoots beams of electrons into space and measures magnetic and electrical field strengths in the vicinity of the spacecraft by taking advantage of the fact that electrons travel in circular motions within magnetic fields, allowing field strengths to be determined from time to return and drift.
	Magnetic Fields	ASPOC	An experimental method for reducing spacecraft electrical charge buildup by emitting charged ions through a needle into space.
	Magnetometer	STAFF	A boom-mounted magnetometer measuring rapid magnetic field changes, or magnetic waves, in both low and high frequencies.
	Magnetic Fields	EFW	Studies electrical fields, plasma waves, and convection at up to a rate of 36,000 samples per second using 50-m wire booms.
	Magnetic Fields	DWP	A computing control experiment that coordinates the onboard wave experiments in terms of power and telemetry and correlates local electron readings with wave instruments data.
	Radio, Magnetic Fields	WHISPER	Emits radio pulses using two 50-m antennas and measures the echoes in order to study electron densities.
	Radio, Magnetic Fields	WBD	Records high-resolution data within specific radio bands to study magnetic and electrical fields at the Earth's magnetic poles.
	Magnetic Fields, Charged Particle Detector	PEACE	Studies velocities and counts of incoming electrons using high- and low-frequency, half-sphere electrostatic analyzer sensor heads.
	Magnetic Fields, Charged Particle Detector	CIS	Uses a hot ion analyzer and time-of-flight ion composition and distribution function analyzer to record study mass, composition, and 3D distributions of ions in the spacecraft vicinity.
	Spectrometer	RAPID	A dual sensor spectrometer for analyzing ion and electron velocities and mass.
CGRO	Spectrometer, Gamma-Ray Detector	BATSE	Detected gamma-ray bursts using NaI scintillation detectors—produces light when struck by specific particles— a spectroscopy detector, and a large area detector (LAD). The

			BATSE could switch to a fast acquisition mode when a burst was detected.
	Spectrometer, Gamma-Ray Detector	OSSE	A scintillation spectrometer that could be directed to observe targets of interest, used four NaI scintillation detectors to observe a target and background for sensitive observations of gamma-ray bursts, pulsars, neutron stars, solar flares, and other sources of radiation within the instruments range.
	Gamma-Ray Detector	COMPTEL	Used a liquid scintallator and NaI scintallator to detect gamma-rays. Could use both detectors to operate in a double-scatter mode by comparing detector data.
	Gamma-Ray Detector, Calorimeter	EGRET	Detected energetic gamma-rays in the range from 20 megaelectronvolts (MeV) to 30 gigaelectronvolts (GeV) using a spark chamber, NaI calorimeter, and absorption shower counter.
COBE	Radiometer	DIRBE	A cryogenically cooled radiometer used to measure diffuse infrared radiation.
	Radiometer	DMR	Used three passively cooled, differential radiometers that used pairs of horn antennas to research fluctuations in cosmic microwave background radiation at frequencies 31.5, 53, and 90 GHz.
	Interferometer	FIRAS	A helium-cooled polarizing Michelson interferometer for surveying the galactic far-infrared.
CHIPSat	Spectrometer	CHIPS Instrument	A whole spacecraft spectrometer instrument designed to study interstellar plasma.
CubeSats	Various	Various	CubeSats originate from a variety of sources and instrumentation varies. Typically testing off-the-shelf equipment complexity varies from simple radio instruments to X-ray detection, magnetometers, and spectrographs.
Dawn	Imaging	FC	A pair of redundant 150-mm visible and near-infrared light cameras.
	Spectrometer	GRAND	A spectrometer that measures gamma-rays and neutrons using an array of 21 sensors and will use the interaction of galactic cosmic rays with the surface of the target asteroids in order to determine the abundance of major elements on their surfaces.

	Spectrometer	VIR	An Offner spectrometer paired with a Shafer telescope and cryogenically cooled in order to map spectra in the ultraviolet, visible, and infrared regions.
	Radio	Gravity Science	Will transmit radio waves from the spacecraft's radio antenna; bounced signals will be analyzed on Earth for information on rotation, gravity, and mass.
Deep Impact Spacecraft	Spectrometer, Imaging	HRI	A high-resolution 30-cm aperture telescope with an infrared spectrometer, charge-coupled device camera, and a special imaging module that will allow the camera to record in white light, color, and infrared.
	Imaging	MRI	A medium-resolution 12-cm aperture telescope, images in visible light, similar in design to the HRI instrument.
Deep Impact Impactor	Imaging	ITS	Nearly identical to the MRI, the ITS targeting camera was used for impactor navigation; a secondary objective is to return images before destruction during descent.
EO-1	Imaging	ALI	A multispectral, wide-field camera using silicon carbide optics, a proof-of-concept design to reduce complexity, weight, and power.
	Imaging	LEISA/LAC	A medium-resolution hyperspectral system with an atmospheric corrector that corrects primarily for water vapor variation.
	Imaging	Hyperion	A hyperspectral imaging system with a high resolution and capability of resolving 220 spectral bands with 30-m resolution.
ERBS	Radiation Detector	ERBE Scanner	An instrument package that measures shortwave, long-wave, and total waveband radiation.
	Radiation Detector	ERBE Non-scanner	Measures radiation from the Earth in short wavelengths and total wavelength energies, as well as monitoring solar energy.
	Spectrometer	SAGE II	Measured compounds and gases in the stratosphere such as ozone, water vapor, and nitrogen dioxide by imaging the sun through the atmosphere with a spectrometer.
EPOXI	Various	See Deep Impactor	Same instrument package as the Deep Impactor. A continuation of the Deep Impact mission.
Euclid	Imaging	Visual Imager (VIS)	A visible-light, charge-coupled device camera that uses Euclid's 1.2-m aperture Cassegrain telescope.

	Spectrometer, Photometer	Near Infrared spectrometer and photometer (NISP)	An instrument that uses Euclid's 1.2-m telescope to take both near-infrared spectrometry and near-infrared photometry using a Teledyne TIS H2RG integrated circuit detector and filter wheel assembly.
Explorer 1	Charged Particle Detector	Geiger-Müller tube	A detector to study the flux of energetic particles.
	Temperature Sensor	Temperature sensors	Basic internal and external temperature sensors.
	Charged Particle Detector	Acoustic detector	A micrometeorite impact microphone.
	Charged Particle Detector	Wire Grid micrometeorite detector	A wire-grid-mounted externally on the satellite that would record breaks in the wire by micrometeorites.
Extreme Ultraviolet Explorer	X-ray Detector	Grazing incidence telescopes	Wolter-Schwarzschild grazing-incidence telescopes; this type of telescope uses very low angles of reflection or incidence.
	Spectrometer	Three EUV Spectrometers	Extreme ultraviolet spectrometers with a resolution of approximately 1–2 Angstroms.
FAST	Magnetic Fields, Charged Particle Detector	Electrostatic Analyzer (ESA)	Sixteen electrostatic analyzers with 180-degree fields of view mounted around the spacecraft using ion or electron spectrographs in order to measure flux, energy, angles, and time resolution of ions and electrons, typically associated with auroras.
	Electric Fields/ Magnetic Fields Detector	Electric Fields Sensors	Spherical sensors deployed on booms to measure electrical fields and plasma density.
	Spectrometer	TEAMS	A time-of-flight mass spectrograph that can record a 3D distribution of ions, energy, and velocities.
	Magnetometer	AC/DC Magnetometers	A DC fluxgate magnetometer for measuring magnetic field strength, direction, and an AC search-coil magnetometer for measuring waves in the extremely low and very low frequencies.

Fermi Gamma-Ray Telescope	Gamma-Ray Detector	Large Area Telescope	A gamma-ray telescope made up of four major components: a tracker that uses silicon strip detectors, a calorimeter using caesium iodide hodoscope design, an anticoincidence shield used to detect and subtract stray gamma-rays, and a data acquisition system.
	Gamma-Ray Detector	GLAST Burst Monitor	A gamma-ray burst monitor that uses 12 sodium iodide (NaI) scintallators (a scintallator reacts in the presence of specific radiation wavelengths to produce light) and two bismuth germinate scintillators.
FUSE	Spectrometer	FUSE Instrument	Comprised of four telescopes and four spectrographs for imaging in the far-ultraviolet spectrums. The four-channel design allowed mirror coatings to be optimized for different specific wavelengths of study.
Galex	Imaging, Spectrometer	Richey-Chretien f/6 telescope	A 50-cm Ritchey-Chretien telescope that could image in the near and far ultraviolet; either images or spectrographs could be captured.
Galileo Spacecraft	Imaging	Solid state imager	A mission-optimized CCD camera capable of taking multispectral, high-resolution images.
	Spectrometer	NIMS	A near-infrared spectrometer system radiatively cooled to temperatures as low as 64 Kelvin and consisting of a 228-mm aperture Ritchey-Chretien telescope and 17 individual detectors.
	Spectrometer	UVS/EUV	An ultraviolet and extreme ultraviolet set of spectrometers, the UVS used a Cassegrain telescope design and the EUV was a spare extreme ultraviolet spectrometer from the Voyager missions that was modified for Galileo.
	Imaging, Radiometer	PPR	A photopolarimeter radiometer using a Cassegrain Dall-Kirkham 10-cm aperture telescope that recorded intensity and linear polarization of sunlight reflected from Jupiter and its satellites and could measure emitted infrared radiation.
	Charged Particle Detector	Dust Detector System	An impact ionization detector for measuring velocities, mass, and charge state of impacting dust particles.

	Charged Particle Detector	EPD	The energetic particles detector system was composed of two instruments: a low-energy magnetospheric measurement system using a double-ended telescope and silicon detectors, and a composition measurement system using a time-of-flight telescope and delta-E × E solid-state detector telescope.
	Charged Particle Detector	HIC	Used two solid-state detector telescopes to monitor the flux of heavy ions.
	Magnetometer	MAG	Two triaxial fluxgate magnetometers on 11.03-m booms for monitoring interplanetary and Jovian system magnetic field strength and direction.
	Magnetic Fields	PLS	A plasma study instrument consisting of a set of four spherical electrostatic analyzers and three magnetic mass spectrometers for studying magnetospheric plasma in the Jovian system.
	Magnetic Fields, Radio	PWS	Studied plasma and radio waves in the Jovian system using two search coil magnetic antennas and 6.6-m electric dipole antenna mounted on the magnetometer boom.
	Radio	RS	A radio science experiment that used radio waves transmissions between the spacecraft and Earth to study Jupiter's upper atmosphere, study Doppler shifts for clues of local mass and gravity variations of Jupiter and its satellites, search for very low-frequency gravitational waves, and other radio-based investigations of the Jovian system.
Galileo Atmospheric Probe	Imaging, Interferometer	HAD	A double-path-length interferometer, filter, and photodetector array for measuring atmospheric helium.
	Meteorology	ASI	Equipped with temperature, pressure, and acceleration sensors for studying the atmospheric structure of Jupiter during descent.
	Spectrometer	NMS	A quadrupole mass spectrometer with a dual-channel inlet system designed for studying the isotopic and chemical composition of Jupiter's atmosphere.
	Radiometer	NFR	A multichannel radiometer for studying net flux solar energy during descent, detect cloud layers, and study clouds and aerosols in the infrared.

	Nephelometer	NEP	A nephelometer (detects suspended particles using a light beam) using a gallium-arsenide LED light source and solid state detectors for studying particles in Jupiter's clouds in terms of density, structure, and size distribution.
	Radio, Electromagnetic	LRD	A lightning and radio wave detector using an electromagnetic sensor with an antenna and preamplifier and an optical sensor with a photodiode and fisheye lens.
	Charged Particle Detector	EPI	An energetic particles investigation using two optical sensors, a particle sensor, and an RF antenna to measure protons, electrons, alpha particles, and ions.
Gemini	Various	Examples: imaging in multiple wavelengths, electron / proton / uv spectrometers, ion and electron detection, radiometers, magnetometers, micrometeorite detection, radiation detection	Instrumentation varied between missions. Typical instruments included imaging in the visible and infrared; spectrometry of Earth's clouds, atmosphere, and surface; radiation exposure, and micrometeorite detection. An instrument study of note was an ion and electron wake detection study for spacecraft attitude sensing.
Genesis	Particle Collection	Gold Foil Collector	A solar wind collector constructed of gold foil.
	Charged Particle Collection	Solar Wind Concentrator	Used high-transparency grids and high voltages adjusted using data from the ion monitor to reject protons and accelerate ions into the target.
	Particle Collection	Wafers and Collectors	Solar wind collector arrays consisting of wafers constructed of high-purity materials such as sapphire, aluminum, gold, germanium, and silicon, that were designed to capture a variety of solar wind particles.
	Spectrometer, Magnetic Fields	Ion and Electron Monitors	Consisted of two instruments, an ion and electron spectrometer that monitored the solar wind in real time in order to adjust voltages in the concentrator and change the deployed collector array for optimal collection.
Geotail	Electromagnetic	EFD	Studies electrical fields in the Earth's magnetotail using a pair of probes on wire booms.

	Magnetometer	MGF	Studies magnetic field strength and direction, specifically in frequencies below 50 Hz, using dual triaxial fluxgate magnetometers and a three-axis search coil magnetometer.
	Spectrometer, Magnetic Fields	LEP	A set of three sensors, a quadrispherical electrostatic analyzer, a spherical electrostatic analyzer, and an ion mass spectrometer, used to monitor velocities of plasma, electrons, and ions, and ion compositions.
	Magnetic Fields	CPI	Three sets of quadrispherical analyzers for measuring hot plasmas, solar wind electrons, ions, and ion composition measurements.
	Charged Particle Detector	HEP	Comprised of five instruments for studying high-energy particles, a low-energy particle detector consisting of three ion mass spectrometers, a burst detector using three delta-E × E solid-state telescopes, and medium- and high-energy isotope detectors using silicon semiconductor telescopes.
	Charged Particle Detector, Spectrometer	EPIC	An instrument system for the study of energetic particles and ion compositions, it is composed of two subcomponents, a suprathermal ion composition spectrometer using a quadrispherical analyzer with a solid-state detector time-of-flight telescope, and an ion-composition instrument using a solid-state detector electron telescope.
	Magnetic Fields	PWI	A plasma investigation instrument constructed of triaxial magnetic search coils using two electric dipole antennas.
GLAST			See Fermi Gamma-Ray Telescope
GPM	Radar	Dual Frequency Precipitation Radar	Ku-band and Ka-band precipitation radars that use microwave radio waves to map precipitation in 3D and estimate rainfall.
	Radiometer	GPM Micro Imager	A 13-channel microwave radiometer using a 1.2-m antenna for detailed observations of rain and snowfall.
Glory	Imaging	Aerosol Polarimetry Sensor (APS)	Designed to scan visible, near-infrared, and shortwave infrared reflected from clouds and other atmospheric aerosols using the polarization of reflected light.
	Imaging	Cloud Cameras	Used to track clouds to support APS analysis.
	Radiometer	Total Irradiance Monitor (TIM)	Four radiometers used to monitor the sun to record total solar radiation.

GOES N-P Series	Radiometer	Imager	An imaging radiometer capable of capturing data in five channels in order to image the Earth's surface for water vapor, cloud cover, fog, ice, and land and sea temperatures.
	Radiometer, Imaging	Sounder	Measures radiation or spectra in 19 bands using a scan mirror and rotating filter wheel that sweeps the Earth's surface, optimized for weather-related measurements such as temperature, water vapor, ozone, and reflected solar radiation.
	Magnetic Fields	Magnetometer	Two boom-mounted magnetometers monitoring the Earth's magnetic field in both in strength and direction; besides data collection, these magnetometers provide alerts of sudden changes in geomagnetic activity.
	X-ray Detector	X-ray and EUV sensor	Consists of a solar X-ray telescope that monitors the sun for solar flares from 0.05 to 0.3 nm for calculating potential risk and interruptions of satellite and radio communications, and an extreme ultraviolet sensor monitoring solar energy impacting the upper atmosphere that can cause significant changes in the density of the ionosphere, which can impact low-Earth-orbiting satellites. (See the Skylab entry for an extreme example.)
	X-ray Detector	Solar X-ray Imager	An X-ray grazing incidence mirror telescope to monitor the solar X-ray emission and capture full-disk images of the sun from 0.6 to 6.0 nanometers and can image features such as coronal holes, loops, flares, and coronal mass ejections.
	Charged Particle Detector	Energetic Particles Sensor	A set of solid-state detector instruments designed to monitor protons, electrons, and alpha particles in order to asses risks to satellites and orbiting astronauts.
	Radio / Rescue	Search and Rescue emergency beacon listeners, Data Relay	A suite of instruments involved in a variety of satellite communication assistance missions such as relaying data between ground-based transponders or sources, detecting and relaying information on search and rescue beacons, and relaying emergency weather information.
GOES R-series	Radiometer	ABI	A multi-channel imaging radiometer that can image the Earth in 2 visible channels, 4 near-infrared, and 10 infrared, providing imagery or data of atmospheric data such as aerosols, clouds, fog, vegetation burn scars, winds, snow, water vapor, rainfall, dust, ozone, air temperatures, ash, and cloud heights.

	Imaging	GLM	A near-infrared detector tuned for sensing lightning activity; GLM will provide detection of both cloud-to-ground and cloud-to-cloud activity.
	X-ray Detector, Imaging	EXIS	Consisting of an extreme ultraviolet sensor and X-ray sensor, this pair of instruments will monitor solar activity such as solar flares and X-ray flare events for calculating potential risk to satellites and interruptions of radio communications and to monitor solar heating of the upper atmosphere that can cause significant changes in the density of the ionosphere, which can impact low-Earth-orbit satellites. (See the Skylab entry for an extreme example).
	Magnetic Fields	MAG	A magnetometer monitoring Earth's magnetic field in both strength and direction, the primary purpose of this instrument is to provide alerts of sudden changes in geomagnetic activity such as magnetic storms and solar wind shocks.
	Charged Particle Detector, Magnetic Fields	SEISS	A set of four instruments for monitoring flux of protons, electrons, and heavy ions, the suite consists of an energetic heavy ion sensor that indicates energetic particle activity surrounding the Earth, high and low electron and proton flux sensors that also can monitor in the medium energies, and a solar and galactic proton sensor that provides data crucial to solar radiation storm warnings.
	Imaging	SUVI	An extreme ultraviolet solar telescope that can provide full disk solar images and can resolve features such as solar flares, solar filaments, and coronal holes, the SUVI instrument provides an upgrade in resolution and spectral coverage compared to its predecessor the SXI on the GOES-N–P series.
	Radio, Rescue	UPS	A suite of instruments involved in a variety of satellite communication assistance missions such as relaying data between ground-based transponders or sources, detecting and relaying information on search-and-rescue beacons, and relaying emergency weather information.
GRAIL A & B	Radar	LGRS	A ranging system that used Ka-band and S-band two-way transmissions between the two GRAIL spacecraft in order to measure micron-level changes in distance used to derive local variations in the moon's gravity

	Imaging	MoonKam	MoonKam, Moon Knowledge Acquired by middle school students, was a visible light camera and the first planetary instrument dedicated to outreach and public education.
Gravity Probe B	Gyroscopes, Relativity	Gyroscopes and a reference telescope	Four fused-quartz gyroscopes cooled cryogenically to 1.8 Kelvin and machined to unprecedented circular precision in order to measure the warping of space time due to frame dragging. A 14-cm aperture star-tracking telescope was used to maintain alignment with a guide star. The gyroscopes, star-tracking telescope, and mounting assembly were constructed with fused-quartz, a material with minimal and uniform expansion and contraction properties.
GRACE	Radio	KBR	K- and Ka-band communication system used between the GRACE satellites for ranging information.
	Clock, Radio	USO	An ultra-stable clock reference used in conjunction with ranging and communications.
	Imaging	SCA	Star-guide cameras that provide reference for both spacecraft attitude control and reference for the ranging experiment.
	Accelerometer	ACC	Accelerometers that record nongravitational accelerations such as drag from the upper atmosphere, solar radiation, and attitude corrections in order to subtract these forces from the gravitational study.
Herschel	Photometer, Spectrometer	PACS	An instrument package with two components, a photometer using two bolometer arrays—a detector sensitive to heating by radiation—and an integral field-line spectrometer using two germanium/gallium photoconductor arrays.
	Photometer, Spectrometer	SPIRE	Contains an imaging photometer that can record in three wavelength bands using arrays of bolometers in a spider-web configuration, and an imaging Fourier transform spectrometer using a Mach-Zehnder design.
	Spectrometer	HIFI	A heterodyne high-resolution spectrometer that translates signals to a lower frequency for observation and covers seven wavelength bands.
HETE-2	Spectrometer, Gamma-Ray Detector	FRECATE	A French-built omnidirectional spectrometer designed to detect and record spectrometry data on gamma-ray bursts.

	X-ray detector	WXM	A wide-field X-ray monitor using an aluminum and gold coded mask and two position-sensitive proportional counters (PSPCs). Each PSPC is divided into an upper cell with carbon fiber anode wires for X-ray detection and a lower cell with tungsten wires that acts as an anticoincidence shield; the entire PSPC is filled with xenon and carbon dioxide gases.
	X-Ray Detector	SXC	A set of two soft X-ray CCD cameras employing coded masks similar to the WXM and that complement the WXM instrument with some nonoverlapping range sensitivity and a superior angular resolution.
Hinode	Imaging	SOT	A solar optical 50-cm aperture telescope with a Gregorian design, collimator lens, and a thermal-removing mirror at prime focus. The SOT contains broad- and narrow-band imagers and a spectrapolarimeter for studying polarization of spectral lines for magnetic field vector studies.
	X-ray Detector	XRT	A grazing incidence X-ray telescope with a high resolution and wide coronal-temperature coverage with a purpose of observing solar magnetic fields and magnetic field features such as flares and coronal mass ejections.
	Spectrometer	EIS	An extreme-ultraviolet spectrometer using an off-axis parabolic primary mirror with high-reflectance coatings and a toroidal diffraction grating; observations are recorded with a pair of CCD cameras. The off-axis design and coatings allow high spatial resolutions and high sensitivity.
Hubble	Telescope	Telescope	A 2.4-m aperture Ritchey-Chretien Cassegrain telescope constructed of ultra-low expansion glass. After an initial installation of corrective optics for a spherical aberration, all instruments were later upgraded to include built-in correction.
	Imaging	Advanced Camera for Surveys (ACS)	A camera package consisting of three instruments, a wide-field visible and near-infrared camera, a high-resolution camera, and a solar blind channel for imaging faint objects in the far ultraviolet.
	Spectrometer	Cosmic Origins Spectrograph (COS)	A highly light-efficient far-ultraviolet and near-ultraviolet spectrograph for studying large-scale galactic structures such as filaments and intergalactic gases.

	Spectrometer, Imaging	Near Infrared Camera/ Multi-Object Spectrometer (NICMOS)	A near-infrared camera and spectrometer with an electric cooler added during a repair mission.
	Spectrometer, Imaging	STIS	An imaging and spectrograph instrument package that covers wavelengths from ultraviolet to the near-infrared.
	Imaging	WFC3	The last generation of Hubble primary cameras, a wide-field and wide-wavelength imaging system using both solid-state detector and charge-coupled devices to image in the visible, ultraviolet, and near-infrared light. It was installed in 2009.
Huygens			See Cassini-Huygens Probe
IBEX	Charged Particle Detector	IBEX-Hi	An energetic neutral atom detector capable of detecting and determining velocities of particles in energy ranges from 300 electronvolts to 6,000 electronvolts. Uses a collimator and electrostatic analyzer to block undesired particles and direct energetic neutrals to detectors.
	Charged Particle Detector	IBEX-Lo	An energetic neutral atom detector with a time-of-flight instrument and capability of detecting and determining velocities of particles in energy ranges from 10 electronvolts to 2,000 electronvolts. This high-energy instrument uses a collimator and electrostatic analyzer to block undesired particles and direct energetic neutrals to detectors.
ICESAT	Laser	GLAS	A laser-ranging altimeter system, or LIDAR, that transmits short laser pulses in the infrared and in a visible green; return reflections are recorded by the spacecraft for determining distances from the spacecraft to Earth surface features such as ice and terrain and includes significant cloud layers.
ICESAT 2	Laser	ATLAS	Similar to the GLAS LIDAR on IceSat, but upgraded to use a multibeam micro-pulse approach that will greatly improve elevation measurements in sloped terrain, rough surfaces, and other surfaces that were difficult to parse in the previous generation LIDAR.
ICON	Spectrometer, Interferometer	MIGHTI	A Michelson interferometer using a spatial heterodyne spectroscopy design (a technique that eliminates the need for moving parts by using diffraction gratings, each groove acting as an interferometer mirror) and a cubic beam splitter that

			provides a reduction in size compared to previous designs, to monitor high-altitude thermospheric winds and temperatures through the Doppler shift of atomic oxygen and spectral shape of oxygen molecules.
	Spectrometer	EUV	An extreme ultraviolet spectrograph utilizing a straightforward imaging spectrograph design for monitoring ionospheric density.
	Spectrometer	FUV	A far-ultraviolet spectrograph using a Czerny-Turner design with a collimating mirror and far-ultraviolet grating that helps to filter unwanted wavelengths, designed to monitor thermospheric composition and ion density.
	Charged Particle Detector	IVM	An instrument for measuring the velocities, temperatures, and densities of ions. The instrument has two apertures with semitransparent conducting grids facing the incoming plasma stream that are used by two subcomponent instruments, a retarding potential analyzer and an ion drift meter. All together, the three components use ion interaction with the instrument's electrical currents to derive measurements.
IMAGE	Charged Particle Detector	LENA, MENA, HENA	Energetic neutral-atom imagers that operated in the low-, medium-, and high-energy ranges, these energetic particle detectors were used to essentially image the Earth's inner magnetosphere, the ring current of the magnetosphere, and other hot plasmas. The velocity, time-of-flight, and energy of incoming particles were calculated to form an image.
	Spectrometer, Imaging	FUV	A far-ultraviolet instrument system that consisted of a broadband ultraviolet imaging camera, a narrow-band spectrographic imager, and a geocoronal emission photometer that detected Lyman-alpha spectra of hydrogen.
	Imaging	EUV	An extreme-ultraviolet imager that used three sensor heads, giving the instrument a 30 × 84 degree field of view, combined with the rotation of the spacecraft the instrument could sweep the sky in an 84 × 360 degree fashion.
	Radar	RPI	A low-power radar with 250-m and 10-m wire antennas that observed plasma densities in the magnetosphere.
InSight	Seismometer	SEIS	Seismometer for monitoring quakes.

	Temperature	HP	The heat flow and physical properties instrument will insert a probe 5 m into Martian surface with heat sensors every 5 cm along the instrument.
	Radio	RISE	An X-band radio that will broadcast transmissions to be received and monitored on Earth in order to discern oscillations, or wobble, in the rotation of Mars.
	Imaging	Black and White Camera for Instrumentation monitoring	Black and white cameras that will monitor instruments on the lander.
INTEGRAL	Spectrometer, Gamma-Ray	SPI	A sensitive gamma-ray spectrometer with a high resolution, uses high-purity germanium detectors cooled to 85 Kelvin and using a bismuth germinate oxide anticoincidence shield (a detector used to subtract detections not originating from the target). The spectrometer uses a hexagonal coded mask constructed of tungsten and configured into 127 hexagonal elements, the coded mask is used by several of the gamma X-ray instruments on board.
	Gamma-Ray	IBIS	A gamma X-ray imager using two detector planes, a layer of cadmium telluride for low-energy gamma rays, and a layer of cesium iodide for high-energy gamma rays. The combination of two layers, a lead shield, and anticoincidence detector enables the detector to focus on a target. IBIS makes use of the coded mask described in the SPI instrument.
	Charged Particle Detector	IREM	A radiation environment monitor using three particle detectors for straightforward localized or background radiation monitoring.
	Gamma-Ray	JEM-X	A gamma X-ray monitor using high-pressure gas chambers filled with methane and xenon and uses the coded mask described in the SPI instrument.
	Imaging, Gamma-Ray	OMC	A multi-wavelength visible light camera used to image gamma X-ray targets and uses a Johnson V-Filter, a CCD cooled to −80°C for imaging, and a 5-cm lens.
ISS	Various	Various	The ISS can accommodate modular instrumentation additions.
	Laser	CATS	A LIDAR for studying clouds and aerosols such as smoke, pollution, and dust.

	Imaging	ISERV	Uses a small telescope and off-the-shelf digital camera technology.
IRIS	Spectrometer	High Frame Rate Ultraviolet Imaging Spectrometer	A high-resolution ultraviolet telescope and spectrographic system. The telescope has a 20-cm aperture that will image fine solar disc features as small as 240 km, and the spectrograph is a slit-jaw or slit-based imaging spectrograph.
James Webb Space Telescope	Imaging	NIC	The primary camera for the James Webb telescope, NIC is a near-infrared camera imaging in wavelengths between 0.6 to 5 μ.
	Spectrometer, Imaging	NIRSPEC	A near-infrared spectrograph with an unusual micro-shutter array capability allowing it to observe over 100 objects simultaneously. This is especially helpful for imaging many long-exposure faint objects and selectively blocking interfering sources.
	Imaging	MIRI	An instrument built for the mid-ranges of the infrared spectrum, the instrument consists of a broadband, wide-field infrared camera, and medium-resolution spectrograph. The instrument uses a two-stage cooling system, a heat exchanger, and a cryocooler assembly with a pulse tube design.
	Spectrometer, Imaging	NIRISS	A near-infrared broadband imaging and slit-less spectrograph that can perform either wide-field or single-target spectroscopy using grating-prisms, called grisms, and interferometry using a masking aperture.
Jason-1, 2 & 3	Radar	Radar Altimeter	Emits two radio pulses on two frequencies, which allows the instrument to determine sea level, wind speeds, atmospheric electron content, and wave heights.
	Radiometer	AMR/JMR	A radiometer using three frequencies in order to derive atmospheric water vapor and content such as rain or snow.
	Radio	DORIS	A dual-frequency instrument using Doppler measurement to determine the spacecraft's velocity and, combined with GPSP signals, the satellite's position can be determined accurately.
	Radio	GPSP	A global positioning and tracking system to support spacecraft location measurements; data are combined with DORIS measurements.

	Laser	LRA	An array of mirrors used as targets for ground-based laser bounces that can be used for calibrating the satellite's on-board location tracking.
Juno	Radiometer	MWR	A microwave radiometer instrument system consisting of six radiometers for studying microwave radiation emitted from six different cloud layers.
	Spectrometer, Imaging	JIRAM	An infrared camera and spectrometer designed to study Jupiter's auroras; will be able to study convection and compounds contained in the processes of Jupiter's upper atmosphere.
	Magnetic Fields	Magnetometer	Two redundant flux gate magnetometers for mapping Jupiter's magnetic fields in strength and direction.
	Radio	Gravity Science	Uses an X-band radio transponder and a Ka-band radio transmitter to communicate with monitoring stations on Earth in order to precisely measure Jupiter's gravity field using Doppler shifts and spacecraft velocities.
	Magnetic Fields	JADE / JEDI	A plasma and energetic particles detector for monitoring Jupiter's magnetosphere, polar particle streams, and ions emitted from volcanoes on Jupiter's moon Io.
	Magnetic Fields	Waves	Uses an electric dipole antenna and a magnetic search coil antenna to study radio and plasma waves and magnetic fluctuations.
	Spectrometer, Imaging	UVS	An ultraviolet imager and spectrograph that will image Jupiter's auroras in the far and extreme ultraviolet.
	Imaging	JunoCam	A pushbroom imaging camera that will take visible light images of Jupiter during close flybys. The camera is not expected to survive more than 7 orbits due to high-energy particles around Jupiter.
Kepler Space Telescope	Imaging	Primary Mirror, Camera	A .95-m aperture Schmidt telescope with a 1.4-m primary mirror, and a photometer consisting of an array of 21 imaging modules each containing two 2200 × 1024 CCDs, which makes this the largest camera NASA has launched into space to date.

LADEE	Spectrometer	Neutral Mass Spectrometer (NMS)	A quadrupole neutral mass spectrometer that identified and monitored abundances of ions and molecules up to 150 atomic mass unit such as carbon dioxide, helium, hydrogen, water, xenon.
	Spectrometer	UV-Vis Spectrometer	An ultraviolet and visible light spectrometer that analyzed dust, gases, and other particles close to the lunar surface.
	Charged Particle Detector	Lunar Dust EXperiment (LDEX)	A dust analyzer using a hemispherical impact target, ions, and electrons emitted after collision with the target were analyzed to measure particle sizes, density, and distribution.
	Laser	LLCD	A laser communication and proof-of-concept experiment conducted from lunar orbit. The spacecraft used a gimbal-mounted optical module consisting of an infrared laser and 10-cm telescope to transmit and receive infrared laser communications with a ground station on Earth at the WSC.
LAGEOS 1 & 2	Laser	Laser Reflector	The satellite is the instrument, acting as a laser target for ground-based ranging experiments and consists of 426 cube corner reflectors constructed primarily of fused silica glass, with four reflectors constructed of germanium for infrared laser reflection.
Landsat 1, 2, 3	Imaging	Return Beam Vidicon	A set of three cameras that imaged in blue-green, yellow-red, and near-infrared using vidicon camera technology, a form of pre-CCD cathode ray tube technology.
Landsat 1, 2, 3, 4, 5	Imaging	Multispectral Scanner	A multispectral scanner-telescope that investigated Earth's climate factors, such as snow cover, pollution, and ocean currents.
Landsat 4, 5, 6, 7	Imaging	Thematic Mapper	A seven-spectral-band radiometer for Earth studies.
Landsat 6, 7	Imaging	Enhanced Thematic Mapper	Improvements on the original thematic mapper instrument, such as a panchromatic channel and multispectral imaging.
Landsat 7	Imaging	High-Resolution Multispectral Stereo Imager	A multispectral stereo imaginer that used pushbroom imaging techniques and a silicon detector array.
Landsat 8	Imaging	Operational Land Imager	A visible, near-infrared, and shortwave infrared imager that uses a pushbroom imaging technique.

		Thermal Infrared Sensor	An infrared sensor for measuring surface temperatures in two thermal bands, uses Quantum Well infrared photodetectors specialized for thermal infrared and with a cost savings compared to traditional infrared detectors.
LCROSS	Imaging	VIS	A visible light video camera.
	Imaging	NIR	Cameras optimized for recording in near-infrared wavelengths.
	Imaging	MIR	Cameras optimized for recording in mid-range infrared wavelengths.
	Imaging	TLP	A photometer that measured total luminance or visible light intensity data.
	Spectrometer	VSP	A visible light spectrometer constructed of a modified commercial model, similar in design to a spectrometer used on the MSL.
	Spectrometer	NSP	Near infrared spectrometers with narrow and wide fields of view, NSP 1 has a 1 degree field of view and NSP 2 has a 130 degree field of view.
Lunar Reconnaissance Orbiter	Spectrometer	Cosmic Ray Telescope for the Effects of Radiation (CRaTER)	Measures spectra of galactic and solar cosmic rays. Also is testing models of long-term space radiation exposure to humans by taking spectra behind a plastic that simulates human tissue.
	Photometers / Infrared	Diviner Lunar Radiometer Experiment (DLRE)	Day and night lunar surface and subsurface temperature mapping. Data has a wide range of use, from selecting landing sites to identifying potential water-ice.
	Photometers	Lyman Alpha Mapping Project (LAMP)	Instrument that maps the lunar surface in the far ultraviolet to search for water-ice and image permanently shadowed polar regions. This instrument also contains a new night-vision experiment using stellar starshine in the ultraviolet to see in the dark.
	Charged Particle Detector	Lunar Exploration Neutron Detector (LEND)	Sensitive hydrogen and hydrogen compound bearing detector that generates high-resolution mapping of these particles in the search for water-ice. Was built at the Space Research Institute of the Russian Academy of Sciences.

	Lasers	Lunar Orbiter Laser Altimeter (LOLA)	Uses a laser diffraction and backscatter detection technique to create high-resolution 3D surface elevation maps, and quantify surface roughness and surface reflectance.
	Imaging	Lunar Reconnaissance Orbiter Camera (LROC)	A system of two narrow-angle cameras and one wide-angle camera that creates high-resolution lunar surface pictures.
	Radar	Mini-RF Technology Demonstration	Two radar instruments mapping the lunar surface, especially in the polar regions and searching for radar signatures of water-ice. The instrument demonstrates lightweight radar technology and an alternative spacecraft to Earth communication technique.
Magellan	Radar	Synthetic Aperture Radar (SAR)	The SAR module used the spacecraft's 3.7-m high-gain antenna to transmit and receive microwave pulses. This data combined with altimeter data creates a 3D map of Venus's surface. Resolution of the SAR instrument was approximately 150 m.
	Radar	Altimeter Antenna	A fan-beam antenna that sent pulses in between SAR bursts to record altimeter data that would be combined with SAR data. Resolution of the altimeter was approximately 30 m.
	Radiometer	Radiometer	This instrument recorded radiated thermal microwave energy from the Venusian surface by operating Magellan's high-gain antenna in a passive listening mode in between SAR bursts.
	Radio	Whole Spacecraft, High-Gain Antenna	Speed changes in the spacecraft's orbit were monitored to map Venus's gravitational field. This was accomplished by aerobraking the craft into a nearly circular orbit and recording detailed telemetry from the high-gain antenna to calculate Doppler shifts.
Magnetospheric Multiscale Mission	Magnetic Fields	Fast Plasma Instrument (FPI)	This instrument will record a 3D picture of the ion plasma every 150 ms and of the electron plasma every 30 ms for the study of sudden changes in the magnetic field, especially for magnetic field reconnection events and related.
	Magnetic Fields, Spectrometer	Hot Plasma Composition Analyzer (HPCA)	This instrument will quantify ions of hydrogen, helium, and oxygen using a novel radio frequency mass filter technique to prevent the spectrometer from being saturated with solar wind protons.

	Charged Particle Detector, Magnetic Fields	Fly's Eye Energetic Particle Sensor (FEEPS)	This sensor will collect whole-sky snapshots of electron energies and directions using solid-state detectors. Two FEEPS sensors per craft will result in 18 simultaneous data collection views. This multiple viewpoint is the origination of the "fly eye" name.
	Charged Particle Detector, Magnetic Fields	Energetic Ion Spectrometer (EIS)	The EIS sensor is similar to FEEPS; however, EIS quantifies mass of energetic ions in addition to direction. The combination of data between the EIS and FEEPS will provide information on ions at a rate faster than one third of spacecraft spin for better resolution during fast-changing magnetic reconnection events.
	Magnetic Fields	Analog Fluxgate (AFG) and Digital Fluxgate (DFG)	These two magnetometers will collect similar sets of magnetic field and current structure data.
	Charged Particle Detector, Magnetic Fields	Electron Drift Instrument	This instrument will measure electric and magnetic field strength by sending a beam of electrons into space and measuring time and drift of the return flight with its electron gun detector unit. This device utilizes the fact that electrons travel in nearly circular orbits.
	Magnetic Fields	Spin-plane double probe (SDP) and Axial Double Probe (ADP)	These instruments measure electrical fields by measuring the voltage between two electrodes separated as widely as possible. This will be accomplished by mounting the sensors on long booms that are extended away from the main body of the craft. The SDP will use four 200-foot wire booms and the ADP will use two 30-foot antennas.
	Magnetic Fields	Search Coil Magnetometer (SCM)	The SCM will use an induction magnetometer technique to measure magnetic field changes through induced current in a wire coil.
Mariner Missions carried similar instruments	Radiometer	Radiometers	Radiometers for measuring radiation in various wavelengths, typically in the infrared.
	Spectrometer	Spectrometers	Various spectrometers, operating in the ultraviolet, extreme ultraviolet, infrared. *Mariner 2* contained a solar wind plasma spectrometer

	Charged Particle Detector	Cosmic Ray and Energetic Particle Detectors	Various cosmic ray detector, particle detectors, and telescopes for measuring flux and spectra of particles, radiation, plasma, and ions. Often solid-state detectors made of silicon or Geiger-Mueller tubes, Faraday cups, solid-state detectors, or scintillating materials (those that emit light when struck by certain wavelengths of radiation).
	Magnetic Field	Magnetometers	Magnetometers, typically triaxial fluxgate magnetometers, for measuring magnetic field strength and direction in the vicinity of the spacecraft.
	Imaging	Cameras	Cameras were carried on most missions and were typically vidicon tube television cameras, a form of cathode ray tube technology used before charge-coupled devices (CCD).
	Radio	Radio Science	Some radio experiments were included on some missions, *Mariner 4*, *5*, *6*, *9* included occultation experiments in which radio signals from the spacecraft would be monitored to derive properties of the upper atmosphere and ionosphere of the planet under study. *Mariner 5* included a radio experiment for studying interplanetary solar wind and electrons.
	Charged Particle Detector	Dust Detectors	Cosmic dust detectors, typically sensitive microphones mounted to sensor plates.
Mars Climate Orbiter	Radiometer	PMIRR	A scanning, narrow-band infrared radiometer built to profile Martian temperatures, carbon dioxide content, dust, water vapor, and clouds.
	Imaging	MARCI	Consisted of wide- and medium-angle color cameras using a push-frame imaging technique and CCD detectors.
Mars Curiosity	Mars Curiosity	Mars Curiosity	Mars Curiosity—see Mars Science Laboratory
Mars Exploration Rover Spirit and Opportunity (Instrumentation is identical)	Imaging	Pancam	Two high-resolution color CCD cameras with multispectral capabilities including some solar imaging capability. The camera head is mounted to a mast assembly that can swing the cameras through 180 degrees of vertical motion and can rotate 360 degrees horizontally allowing photos of panoramic landscapes to be taken.

Imaging	Engineering Cameras	A set of two types of camera systems: navigation cameras and hazardous avoidance cameras. The hazardous avoidance cameras are a set of four black and white cameras mounted in the lower front and rear that are interpreted by the rover's software for avoiding hazards. The navigation cameras are a stereo pair mounted on the mast assembly that take visible light panoramic images.
Spectrometer	Miniature Thermal Emission Spectrometer (Mini-TES)	A miniature infrared spectrometer that uses a viewing port and scanning mirrors on the mast assembly to capture and direct light to instrument. The Mini-TES can identify target mineralogy of targets through variation in thermal emissions as well as collect atmospheric temperature, water vapor, and dust density data.
Spectrometer	Mössbauer Spectrometer	An iron Mössbauer spectrometer mounted on the rover arm used to study iron-bearing rock, the spectrometer avoids sample preparation through a backscatter technique with a Cobalt isotope source. The sensor requires 6–12 hours against a target for analysis.
Spectrometer, X-ray	Alpha Particle X-ray Spectrometer	An alpha particle X-ray spectrometer mounted on the rover's arm for analyzing elements in rock and soil, the sensor is placed over a sample, and alpha particle and X-ray data is collected over a 10-hour time period.
Charged Particle Detector, Magnetic Fields	Magnets	Three sets of magnetics are in use by the rover(s), one set is used with the RAT to collect magnetic samples, a second set collects magnetic particles on the front of the rover that are in reach by the arm instruments, and a final set is mounted on top of the rover near the panoramic camera in order to deflect wind-borne magnetic dust.
Imaging	Microscopic Imager	A black and white microscopic imaging CCD camera mounted on the rover's arm and designed to take high-resolution close-up images.
Sample Collection	Rock Abrasion Tool (RAT)	A powerful rock grinder using three electric motors and two high-speed grinding wheels, the RAT is capable of creating a 4.5-cm hole to a depth of 5 mm

Mars Express	Spectrometer	Visible and Infrared Mineralogical Mapping Spectrometer (OMEGA)	A visible light and near-infrared spectrometer with an objective of creating a global mineralogical map using emitted infrared radiation and reflected sunlight.
	Spectrometer	Ultraviolet and Infrared Atmospheric Spectrometer (SPICAM)	An ultraviolet and infrared spectrometer built for profiling ozone, carbon dioxide, water content, and other atmospheric gases through solar occultation (viewing the sun through the Mars atmosphere), surface-reflected sunlight, and after-glow types of emissions.
	Radar	Sub-Surface Sounding Radar Altimeter (MARSIS)	Mars Advanced Radar for Subsurface and Ionospheric sounding is a radar instrument employing 20-m antennas for transmitting and receiving radar sounder signals for profiling the upper crust of Mars in terms of roughness, water, ice, and structure as well some sounding of the ionosphere.
	Spectrometer	Planetary Fourier Spectrometer (PFS)	An infrared Fourier-transform spectrometer, a design covering a wide spectral range, employing a double pendulum interferometer design for profiling water vapor, carbon dioxide, and other atmospheric gases, as well as monitoring surface temperatures and any mineralogical data that can be determined.
	Spectrometer	Analyzer of Space Plasmas and Energetic Atoms (ASPERA)	A space plasma and energetic analyzer system that uses an ion and electron spectrometer, a neutral particle imager, and two energetic neutral atom detectors. The instrument system is intended to study solar wind interaction with the Martian ionosphere and the plasma and neutral gas environment in the upper atmosphere.
	Imaging	High-Resolution Stereo Camera (HRSC)	The high-resolution stereoscopic camera operates in the visible and near-infrared and has two photometric channels. The camera provides color stereo images and uses a pushbroom imaging technique.
	Radio	MaRS	An experiment that will monitor Doppler shifts to study Martian gravity and will study the Martian ionosphere with occultation techniques where radio waves are sent through the Mars atmosphere to Earth's Deep Space Network ground stations.

Mars Global Surveyor	Imaging	Mars Orbiter Camera (MOC)	The Mars Orbiter Camera had a black and white narrow-angle camera, and wide-angle high-resolution red and blue cameras. The MOC imaged to a resolution of approximately a meter with the narrow angle camera and the wide-angle views were used to monitor large-scale global atmospheric changes and to provide context or targets for narrow-angle targets.
	Laser	Mars Orbiter Laser Altimeter (MOLA)	An orbital laser altimeter that used a neodymium yttrium aluminum garnet (Nd:YAG) laser for conducting global surveying of the Martian topography, seasonal changes in the polar ice caps, and water and carbon dioxide cloud tops.
	Spectrometer	Thermal Emission Spectrometer (TES)	The thermal emission spectrometer was designed to study surface and atmosphere composition and conditions and was constructed of a spectrometer with a Michelson interferometer design, a bolometer channel for broadband infrared radiance, and a reflectance channel for broadband visible and near-infrared.
	Magnetic Fields	Magnetometer (MAG/ER)	Consisted of a pair of triaxial fluxgate magnetometer for measuring magnetic field strengths and direction, and an electron reflectometer, which is a quadraspherical electrostatic analyzer used for measuring electron flux.
	Radio	Ultrastable Oscillator for Doppler measurements (USO/RS)	A radio science experiment that used an ultra-stable oscillator as a reference clock or frequency reference, and the spacecraft's radio transmitters. Studies included gravity measurements of Mars by measuring radio signal Doppler shifts, and ionospheric and upper atmosphere studies using radio occultation, which is a technique of transmitting through the Martian atmosphere to NASA's Deep Space Network receivers.
	Radio	Mars Relay	The orbiter relayed data from the Mars Exploration Rovers to Earth.
Mars Odyssey (also 2001 Mars Odyssey)	Imaging	Thermal Emission Imaging System (THEMIS)	A thermal emission imaging system using an infrared imaging spectrometer and high-resolution camera for mapping Martian mineral distribution by studying the characteristic signatures of spectral bands emitted by rock and soil. The visible imaging

			portion of the experiment could resolve to 18 m and is used to study surface geology and scout locations for following Mars missions such as the Mars Exploration rover.
	Gamma-Ray	Gamma Ray Spectrometer (GRS)	A gamma-ray spectrometer consisting of a high-energy neutron detector built by the Space Research Institute of Russia, a gamma-ray subsystem with a boom-mounted detector that uses high-purity germanium crystal, and a neutron spectrometer designed to detect thermal, epithermal, and fast neutron energy bands. The purpose of the GRS system is to identify and study elements present on the Martian surface and water- or ice-bearing rock and soil.
	Spectrometer, Charged Particle Detector	Mars Radiation Environment Experiment (MARIE)	Mars Radiation Environment Experiment is an experiment to study potential astronaut radiation hazards using an energetic particle spectrometer to measure spectra and count of incoming protons, neutrons, and galactic cosmic rays.
Mars Observer	Various	MOC, MOLA, TES, MAG/ER, RS, MBR	Due to the loss of the mission all the instruments were reincorporated into later Mars missions, primarily on the Mars Global Surveyor.
		PMIRR	See Mars Climate Orbiter
		GRS	See Mars Odyssey
Mars Pathfinder (Lander)	Imaging	Imager for Mars Pathfinder IMP	Imager for Mars Pathfinder was a two-camera stereo-imaging system with filters for multispectral imaging that could acquire panoramic, multispectral, and narrow-band solar images. A macro or close-up lens was available for imaging any magnetic dust particles attached to a small magnet near the camera.
	Meteorology	Atmospheric Structure Instrument/Meteorology Package (ASI/MET)	An atmospheric and meteorological system that measured temperatures, pressures, and wind speed and direction using a wind sock experiment. During landing, atmospheric density and temperatures were recorded including accelerometers for studying atmospheric turbulence and wind.
	Radio	RS	A radio science experiment that studied Mars's rotational precession and libration, or wobble, by measuring radio signal Doppler shifts by Earth-based receivers.

Mars Pathfinder –Sojourner (Rover)	Imaging	Imaging System	Consisted of two forward-facing black and white cameras and one rearward facing color camera. The cameras could take close-up images of targets designated for sampling, aid in navigation, and examine results of abrasion tests.
	X-ray, Spectrometer	Alpha Proton X-ray spectrometer (APXS)	An alpha particle X-ray spectrometer mounted on the rover's single-actuator deployment mechanism. The APXS was used for analyzing elements in rock and soil and was operated by placing the sensor over a sample for collection of alpha particle and X-ray data.
	Charged Particle Detector	MAE	The material adherence experiment used a solar cell sensor under a moveable glass plate for measuring the rate of dust accumulation. The glass dust collector could be moved temporarily to make comparisons between the dust-free sensor and accumulated dust.
	Materials Test	WAE	The Wheel Abrasion Experiment was a test of a variety of metals attached to a rover wheel and at times the wheel was spun without moving the rover to simulate harsher conditions. Materials tested were aluminum, nickel, and platinum, and measurements were taking by studying reflectivity changes over time.
Mars Polar Lander	Imaging	Mars Descent Imager (MARDI)	A camera that was mounted on the bottom of the lander and was to be used to image the Martian terrain during landing.
	Imaging	Stereo Surface Imager (SSI)	A multispectral, stereoscopic imaging camera that had two cameras and filters allowing study of mineralogy, the Martian atmosphere, imaging of the sun directly, and other multispectral experiments.
	Laser	Light Detection and Ranging (LIDAR)	A laser image detection and ranging sounder that would have been used to study atmospheric dust, aerosols, and ice in the atmosphere up 2 or 3 km above the lander. The LIDAR used a gallium-aluminum-arsenic laser diode and had a passive detection mode that could use the sun as a light source for collecting data.
	Sample Collection	Robotic Arm	A 2-m arm that could collect soil samples and take surface temperature measurements. The arm included an arm camera for taking close-up images.

	Meteorology	Meteorological Package (MET)	A meteorological instrument package mounted on the lander to measure wind, pressure, and humidity.
	Spectrometer	Thermal and Evolved Gas Analyzer (TEGA)	This instrument contained a spectrometer, electrochemical cell, scanning calorimeter, and ovens. The robotic arm would deposit samples in the ovens for heating for vapor analysis by the spectrometer and electrochemical cell. The TEGA instrument was intended to study water-ice, carbon dioxide, and other minerals with volatiles detectable by the instrument.
	Microphone	Mars Microphone	A microphone with electronics similar to a hearing aid that would have recorded and transmitted sounds within range of the lander, such as blowing sand, winds, and possible electrical discharges in dust clouds.
Mars Reconnaissance Orbiter	Imaging	High-Resolution Imaging Science Experiment camera (HiRISE)	A high-resolution imaging telescope with a 0.5-m aperture and filters for red, blue-green, and near-infrared. HiRISE uses a CCD with a pushbroom imaging technique and is capable of 30-cm resolution.
	Imaging	Context Camera (CTX)	The context camera provides coverage in medium resolutions and takes overlapping images with HiRISE in order to provide overviews and context of the high-resolution images provided by HiRISE.
	Imaging	Mars Color Imager (MARCI)	A wide-angle color imaging camera that acquires daily global images of Mars in five visible and two ultraviolet wavelengths. MARCI's pictures will be used to study global weather, dust storms, polar ice-cap changes, frost, and other global weather phenomena.
	Imaging, Spectrometer	Compact Reconnaissance Imaging Spectrometer for Mars (CRISM)	A visible and near-infrared imaging spectrometer for studying surface minerals on Mars. CRISM has two detectors for the visible and infrared and uses cryocoolers to cool the infrared sensor heads.
	Radiometer	Mars Climate Sounder (MCS)	The MCS is a radiometer, or climate sounder, operating in nine infrared channels from the medium to far infrared and in one broadband channel in the visible and near infrared. The MCS is used to profile the Martian atmosphere in 3D in terms of water vapor, temperatures, dust, and other atmospheric conditions.

	Radar	Shallow Subsurface Radar (SHARAD)	The shallow subsurface radar uses radar to search for water deposits up to 1 km in depth in the Martian crust.
	Radio	Radio Science	A gravity experiment that will monitor Doppler shifts from the spacecraft's radio signals to study Martian gravity.
	Accelerometer	Atmospheric Structure Investigation	Accelerometers on board the spacecraft will provide data on changes to the spacecraft's motions due to interactions with the Martian upper atmosphere.
Mars Science Laboratory - Aeroshell	Accelerometer, Temperature	MEDLI	A landing system instrument suite that measured the performance of the entry heat shield, atmospheric conditions during descent, and aerodynamic forces and accelerations during Mars entry.
	Imaging	MARDI	A CCD color camera designed to take a series of pictures at five frames per second during the landing sequence to be used for location and geological context analysis after landing.
Mars Science Laboratory - Curiosity Rover	Imaging	MASTCAM	A set of two fixed-focal-length color cameras that can image in the visible and infrared and that include filter wheels for multispectral imaging.
	Imaging	MAHLI	A CCD color camera on the end of the robotic arm that includes white light LEDs for nighttime imaging, ultraviolet LEDs to search for fluorescent materials, and can also capture high-definition video.
	Spectrometer	CHEMCAM	An instrument package consisting of a laser-induced breakdown spectrometer and a remote micro-imager that can use laser pulses and spectrometry analysis to remotely analyze elements of a target.
	Spectrometer	APXS	An X-ray spectrometer for analyzing elements in rock and soil that operates by placing the sensor head over a sample and recording X-ray spectra over approximately 3 hours per target.
	Spectrometer, X-ray	CHEMIN	An X-ray diffraction and X-ray fluorescence instrument system that analyzes rock and soil samples by transmitting a collimated X-ray beam through a sample and recording diffracted or fluoresced X-rays on a CCD. Samples are prepared into a powder by a sample and processing handling system.

	Spectrometer	SAM	A sample analysis suite consisting of a quadrupole mass spectrometer, a gas chromatograph, and a tunable laser spectrometer. Uses a sample manipulation system and chemical separation and processing laboratory to prepare samples for analysis.
	Charged Particle Detector, Radiation Detector	RAD	A charged particle telescope using solid-state detectors and a cesium-iodide calorimeter, or scintillator, for analyzing energetic particle radiation during the cruise to Mars and especially on the Mars surface. One of the purposes of this instrument is to study radiation exposure for astronauts on the surface of Mars.
	Spectrometer	DAN	A neutron spectrometer that can operate in either a passive acquisition mode or an active pulsing mode using a neutron source. DAN is used for analyzing hydrogen and hydrogen-bearing minerals on the surface and shallow surface layer of Mars.
	Meteorology	Rover's Environmental Monitoring Station (REMS)	The rover's environmental monitoring station records winds, atmospheric pressure, humidity, air temperatures, ground temperatures, and ultraviolet radiation.
MAVEN	Spectrometer	Imaging Ultraviolet Spectrograph (IUVS)	An imaging ultraviolet spectrograph will study neutrals and ions and is optimized for profiling composition of the Martian middle atmosphere, the upper atmosphere, and exosphere, covering a range from approximately 30 to 4500 km. The IUVS is one of the most advanced interplanetary spectrographs to date and employs a modified Czerny-Turner plane-grating spectrograph design, a toroidal camera mirror for astigmatism correction, and an ability to scan in a side or downward view. The instrument will use stellar occultation for taking measurements in the middle Martian atmosphere—a technique similar to solar occultation by measuring missing spectra absorbed by atmospheric gases. IUVS is also able to take global imagery in multiple-ultraviolet bands and with atomic deuterium and hydrogen studies will contribute to measuring historic lifetime atmospheric water-vapor.

	Spectrometer	Neutral Gas and Ion Mass Spectrometer (NGIMS)	Spectrometer designed to study the composition and abundance of neutral atoms, thermal and suprathermal ions, and isotope ratios to create upper atmospheric profiles. The NGIMS uses a Bayard-Alpert pressure gauge, quadrupole mass analyzer, and ion detector, and operates in either a neutral or ion collection mode.
	Magnetic Fields, Imaging	Langmuir Probe and Wave (LPW)	The experiment studies heating of the ionosphere and extreme ultraviolet irradiance of the ionosphere and upper atmosphere. It consists of an extreme ultraviolet sensor with three EUV channels, Lyman-Alpha, solar corona and hot corona, and two 0.5-m titanium-nitride cylindrical whip antennas mounted on 7-m booms for measuring electron density and temperature.
	Charged Particle Detector	Suprathermal and Thermal Ion Composition (STATIC)	The instrument uses an ion analyzer and deflector, a 15-kilo-volt accelerator, carbon foils, and micro-channel plate detector for studying thermal and energetic ions escaping from the atmosphere of Mars.
	Charged Particle Detector	Solar Energetic Particle (SEP)	An instrument employing two solid-state telescopes with silicon detectors. The telescopes are orthogonal and double-ended, effectively giving the instrument an ability to observe in four directions. The SEP measures energetic protons and electrons in the upper atmosphere for studying the relationship between solar wind interactions and escaping particles.
	Charged Particle Detector	Solar Wind Ion Analyzer (SWIA)	A hemispherical electrostatic analyzer for measuring the input of solar wind ions to Martian atmosphere and is mounted in solar-facing view.
	Charged Particle Detector	Solar Wind Electron Analyzer (SWEA)	A hemispherical electrostatic analyzer mounted on a 1.75-m boom to minimize spacecraft charge interference. The SWEA measures ionospheric electrons and solar wind particles.
	Magnetic Fields	MAG	Two boom mounted fluxgate magnetometers for measuring magnetic field strength and direction in the Martian ionosphere.
MESSENGER	Imaging	Mercury Dual Imaging System (MDIS)	An imaging system using a wide- and narrow-angle imager with filter wheels to image in specific wavelengths. The wide-angle system uses an achromatic lens system (nearly distortion free) and contains a filter wheel. The narrow-angle imager uses a

		25-mm aperture, off-axis Ritchie-Chretien telescope. Both imaging setups share the spacecraft's special passive cooling system for operations so near the sun.
Spectrometer, Gamma-Ray Detector	Gamma-Ray Spectrometer (GRS)	A gamma-ray and neutron spectrometer instrument system for measuring gamma-ray emissions emitted due to radioactive decay of surface minerals on Mercury for creating global mineralogy maps. The gamma-ray spectrometer is a 31-cm cylinder containing a germanium semiconductor scintallator (a type of crystal material that emits light in the presence of specific radiation), and a bismuth germanate anticoincidence shield that is used to detect and subtract stray gamma-rays from results. The neutron spectrometer consists of two lithium-glass scintillators with wedges of neutron absorbing plastic.
Spectrometer, X-ray Detector	X-ray spectrometer (XRS)	An X-ray spectrometer for mapping abundances of magnesium, aluminum, calcium, sulfur, titanium, and iron. The XRS uses an upgraded design of the XRS on the NEAR-Shoemaker spacecraft and consists of gas-filled proportional counters, anticoincidence wire detectors, and passive X-ray protective shielding for stray X-rays.
Magnetic Fields	Magnetometer (MAG)	A three-axis fluxgate magnetometer for measuring Mercury's magnetic field strength and direction.
Laser	Mercury Laser Altimeter (MLA)	An orbital laser altimeter similar to the MOLA instrument on the Mars Global Surveyor, that uses a chromium-aluminum-garnet (Cr:Nd:YAG) laser, primarily for surveying the northern hemisphere of Mercury, where MESSENGER makes it closest approaches.
Spectrometer	Mercury Atmospheric and Surface Composition Spectrometer (MASCS)	An instrument used for profiling iron-bearing surface minerals and searching the exosphere for species of particles within range of the instrument capabilities. The system consists of an ultraviolet-visible spectrometer (UVVS) and an infrared-visible spectrograph (VIRS) with a shared Cassegrain 250-mm aperture telescope. The UVVS uses an Ebert-Fastie diffraction grating and the VIRS uses a fiber optic bundle to feed a visible detector using a silicon line array and the infrared detector using an indium-gallium-arsenide line array.

	Spectrometer	Energetic Particle and Plasma Spectrometer (EPPS)	Spectrometer using time-of-flight technique, and solid silicon detector to measure energies and classify ions and electrons in the ionosphere and exosphere of Mercury.
	Radio	Radio Science	A Radio Science experiment using X-band transponders to monitor the spacecraft's movement through radio Doppler analysis; combined with MLA measurements and radio occultation, this experiment will analyze libration or wobble of Mercury's rotation.
Mini-RF	Radar	Mini-RF	Two miniature radio-frequency, or radar, lunar mapping instruments on two different lunar spacecraft: the LRO mission and India's *Chandrayaan-1*. The mini-RFs use a synthetic radar aperture technique and demonstrate a lightweight version of this technology.
Moon Mineralogy Mapper	Spectrometer	Moon Mineralogy Mapper	An imaging spectrometer using a pushbroom technique that can operate in a wide-global mode or a higher resolution targeted mode. The intent of this instrument is to study and map the mineral composition of the lunar surface. This instrument was built by NASA and is carried aboard India's *Chandrayaan-1*.
NSCAT	Radar	Radar scatterometer	A microwave wavelength radar scatterometer that measured sea surface wind speeds.
NEAR	Imaging	MSI	A multi-spectral imaging system for analyzing asteroid surface composition or mineralogy using an optical telescope, filter wheel, and CCD detector, and imaged in broadband, specific color wavelengths, and the near infrared.
	Spectrometer	NIS	Near-infrared diffraction-grating spectrometer for studying asteroid surface composition and mineralogy used a 32-element germanium and a 32-element indium-gallium-arsenide line-array detectors.
	X-ray Detector, Gamma-Ray Detector, Spectrometer	XRS-GRS	A Gamma and X-ray fluorescence spectrometer for studying surface composition. The gamma-ray detector consisted of a sodium-iodide scintillator crystal (emits light in the presence of specific radiation) and bismuth-germanate anticoincidence detector for subtracting stray rays. The X-ray fluorescence spectrometer used gas-proportional counters.

	Magnetic Fields	MAG	A three-axis fluxgate magnetometer used for measuring interplanetary magnetic field strength and direction as well as any magnetic fields associated with Eros or bow waves into the solar wind.
	Laser	NEAR Laser Rangefinder (NLR)	The rangefinder used a yttrium aluminum garnet (Nd:YAG) laser and receiver for analyzing the surface of Eros and performing navigation.
	Radio	RS	A Radio Science experiment that used X-band transponders to monitor the spacecraft's movement through radio Doppler analysis for studying the gravity fields of Eros and Mathilde.
New Horizons	Spectrometer	ALICE	An ultraviolet mapping spectrometer for profiling the atmosphere of Pluto in terms of specific molecules, temperature profiles, and mapping specific spectral lines of Hydrogen. ALICE can gather data in either an airglow mode or using occultation techniques where a star or the sun is observed through the target area.
	Imaging	LORRI	A 20.8-cm aperture panchromatic narrow-angle imager, primarily for long-range imaging and planning leading up the encounter with Pluto and any other objects of opportunity.
	Imaging	RALPH	A multispectral visible light camera and a linear etalon imaging spectral array for high-resolution color mapping and to study surface composition. Both instruments share a 6-cm aperture telescope with color filters and one near-infrared filter.
	Spectrometer	PEPSSI	An energetic particle spectrometer for analyzing ions, protons, and electrons and consists of a time-of-flight segment and silicon solid-state detector array.
	Radio	Radio Science Experiment (REX)	A radio studies experiment that will use radio for various studies including listening for weak radio emissions, sending radio waves through Pluto's ionosphere to Earth for study, and for gravitational studies through the use of radio Doppler techniques.

	Dust Detector	Venetia Burney Student Dust Counter	An outreach, student-designed dust counter using a detector assembly constructed of a surface-mounted polyvinlidine fluoride film and detector electronics. It is named *Venetia* in honor of Venetia Burney Phair, who suggested the name Pluto in 1930 at the age of 11.
	Charged Particle Detector	SWAP	A solar wind analyzer using plasma toroidal electrostatic analyzer and retarding potential analyzer will be used mostly for interplanetary studies during cruise.
NEOWISE	Imaging	See the WISE instrument.	Same instrument as WISE; however, this did not use the active cryogenic cooling system.
NOAA-N and N-Prime / NOAA-18 and 19	Radiometer	AVHRR	A radiometer able to image in the visible, near-infrared, and infrared for conducting studies of atmospheric conditions as well as some surface features such as clouds, rain, snow, and vegetation.
	Radiometer, Spectrometer	HIRS	A high-resolution infrared sounder for temperature and moisture that uses a 20-channel visible and infrared spectrometer.
	Radiometer, Microwave	Advanced Microwave Sounding Units A & B (AMSU-A & B)	Units designed to profile temperatures and water vapor using microwave radiation waves and detection. AMSU B is only included on NOAA-18.
	Radiometer	Solar Backscatter Ultraviolet (SBUV)	An ultraviolet radiometer monitoring solar backscatter ultraviolet radiation. The instrument maps atmospheric ozone and can detect cloud cover.
	Radiometer, Microwave	MHS	A 5-channel microwave radiometer for profiling atmospheric water vapor.
	Radiation Detector	Space Environment Monitor (SEM)	Monitors radiation flux in the satellites vicinity.
NPP			See Suomi-NPP
NuSTAR	X-rays	X-ray spectrometer	An X-ray telescope employing the X-ray grazing-incidence mirror design using nested paraboloid and hyperboloid mirrors in a barrel-like configuration. The instrument used a 10-m deployable mast in order to extend the optics and achieve necessary focal length.

OSTM/Jason-2	Radar	Poseidon-3 radar altimeter	A radar altimeter providing distance to the sea surface.
	Microwave	Advanced Microwave Radiometer (AMR)	Radiometer that measures radiation from the Earth's surface for monitoring water vapor.
	Radio	Doppler Orbitography and Radio-positioning Integrated by Satellite (DORIS)	A radio positioning system that analyzes Doppler shift using radio signals transmitted from ground stations to subtract gravity and atmospheric effects from the radar altimetry data.
	Radio	Global Positioning System Payload (GPSP)	A global positioning tracking system that uses GPS signals to determine the satellite's location in orbit.
	Laser	Laser Retroreflector Array (LRA)	An array of mirrors acting as laser targets from ground stations for supplementing and calibrating the satellite's altitude.
	Radiation Detector	CARMEN-2	An instrument for studying the impact of radiation on satellite components.
	Laser	T2L2	A laser link instrument for synchronizing remote ground clocks.
	Radiation Detector	Light Particle Telescope (LPT)	A telescope for studying radiation in the satellite's environment.
Operation Ice Bridge	Aircraft	Aircraft Platforms	Primarily NASA's P-3B and DC-8 aircraft, as well as a C-130, King Air B-200, HU-25C Falcon, and Gulfstream G-V. Some smaller chartered aircraft have been used as well.
	Laser	Laser Altimeters	Various laser altimeters used onboard aircraft.
	Radar	Radars	Various sources, aircraft borne radars.
	Gravimeter & Magnetic Fields	Gravimeter and Magnetometer	A gravitation monitor studying seawater cavities in glaciers, and magnetometers for studying magnetic field strength and direction.
	Imaging / Photometry	Imaging	Various sources, imaging below aircraft as well as some radiometric temperature sounding.
OCO-2	Spectrometer	Spectrometer	Three high-resolution near-infrared spectrometers for measuring carbon dioxide and oxygen.
OSIRIS-REx	Imaging	OCAMS	A visible light camera suite consisting of an 8-inch telescope, a camera for imaging in four specific wavelengths to search for outgassing plumes, and a continuous imager for documenting the mission.

	Laser	OLA	A laser altimeter that uses laser light and return detection for ranging and creating topographic maps.
	Spectrometer	OVIRS	A visible and infrared spectrometer, sensitive from the visible blue spectrum to the near-infrared for identifying and mapping mineralogy and organic material.
	Spectrometer	OTES	A thermal emission spectrometer for identifying mineralogy and collecting temperature data using an infrared spectrometer.
	Spectrometer, X-ray Detector	REXIS	Regolith X-ray imaging spectrometer will map surface minerals by detecting X-ray fluorescence, which are X-rays that have been emitted by surface minerals absorbing solar wind energy.
	Sample Collection	TAGSAM	A sampling head and articulated arm that will collect soil samples from the asteroid.
Phoenix Mars Lander	Imaging	RAC	The robotic arm camera provided close-up images of targets of interest within range of the robotic arm, had its own illumination system, and was similar in design to a 35-mm CCD camera.
	Imaging	Surface Stereo Imager (SSI)	A stereo high-resolution imaging system that could take panoramic pictures. The SSI could image in visible and infrared light and with filters could image in multiple spectrums.
	Spectrometer	Thermal and evolved gas (TEGA) analyzer	The thermal and evolved gas analyzer instrument used eight, single-use only, miniature furnaces for heating samples, and a mass spectrometer for studying isotope ratios of samples. In addition, a scanning calorimeter was used to monitor phase changes of materials as they heated.
	Imaging	Mars Descent Imager	(was not used due to mission risk)
	Wet Chemistry Laboratory	Microscopy, Electrochemistry, and Conductivity Analyzer (MECA)	Analyzer contained a wet chemistry laboratory, optical and atomic force microscopes, and thermal and electrical conductivity probes attached to the end of the robotic arm for testing soils.
	Meteorology	Meteorological station (MET)	A meteorological station for monitoring atmospheric pressure and temperatures, and used a LIDAR to sound dust in the atmosphere above the lander.
Pioneer 1	Imaging	Scanning Television	A scanning television camera, a form of cathode ray tube technology used before charge-coupled devices (CCD).

	Charged Particle Detector, Gamma-Ray Detector	Ionization chamber	An aluminum-walled gas-filled ion chamber for studying cosmic rays, protons, gamma-rays, and any other particles detectable by the instrument.
	Dust Detector	Micrometeorite Detector	A detector consisting of a diaphragm and microphone for recording impacts.
	Magnetic Fields	Magnetometer	A single-axis magnetometer instrument able to detect fields perpendicular to the spacecraft spin-axis, used a nonlinear amplifier and series of switched magnetometers of varying sensitivity for detecting cruise and lunar magnetic fields.
Pioneer 2	Charged Particle Detector	Cosmic Ray Telescope	A telescope that used seven triple coincidence proportional counters as detectors.
	Charged Particle Detector, Gamma-Ray Detector	Ionization Chamber	An aluminum-walled gas-filled ion chamber for studying cosmic rays, protons, gamma-rays, and any other particles detectable by the instrument.
	Dust Detector	Micrometeoroid Detector	A detector consisting of a diaphragm and microphone for recording impacts.
	Magnetic Fields	Magnetometer	A search coil magnetometer able to detect both magnitude and direction of magnetic fields.
	Imaging	Earth-Moon TV Scanner	Used a spherical mirror, tape recorder, electronic amplifier, and transmitter.
Pioneer 3 - 4	Radiation Detector	Radiation Detector	Two Geiger-Mueller tubes for studying radiation.
Pioneer 5	Charged Particle Detector, Radiation Detector	Proportional Counter Telescope	A radiation detector for studying trapped radiation in the near Earth environment, it was a triple coincidence omnidirectional proportional counter telescope.
	Magnetic Fields	Search-Coil Magnetometer	A single-axis search coil for measuring magnetic fields perpendicular to the spacecraft's spin axis. Similar to the magnetometer on *Pioneer 1*.
	Charged Particle Detector	Ion Chamber and Geiger-Mueller Tube	A Neher-type ion chamber and Geiger-Mueller tube for detecting protons, electrons, and ions.
	Dust Detector, Spectrometer	Micrometeorite spectrometer	Two diaphragm microphones intended to measure the count and momentum of micrometeorite impacts.

Pioneer 6 Only	Radio	Radio Science	Used radio transmissions to seek any relativistic effects by studying radio Doppler shifts. Relativistic effects of the sun were too minimal for detection; however, it produced the first measure of solar coronal noise on radio Doppler shifting.
Pioneer 6 & 7	Charged Particle Detector	Electrostatic Analyzer	A quadrispherical electrostatic analyzer with eight collectors for studying solar wind electrons and ions.
Pioneer 6, 7, 8	Magnetic Fields	Uniaxial Fluxgate Magnetometer	A single-axis fluxgate magnetometer for detecting magnetic field strength, it used the spin of the spacecraft to determine field direction.
	Charged Particle Detector	Solar Wind Plasma Faraday Cup	A Faraday cup consisting of two collectors for detecting solar wind ions and electrons.
	Charged Particle Detector	Cosmic Ray Telescope	A charged-particle telescope that used four solid-state detectors for detecting protons and alpha particles.
		Two-Frequency Beacon Receiver	A radio experiment that used transmissions between the spacecraft and ground stations to determine ionospheric electron content.
	Charged Particle Detector	Cosmic Ray Anisotropy	A galactic and solar cosmic ray detection instrument that used three solid-state telescopes arranged in a 23-degree fan and caesium-iodide CsI scintillator crystal (lights in the presence of specific radiation) with a plastic anticoincidence shield.
Pioneer 8 Only	Dust Detector	Cosmic Dust	Measured cosmic flux density, particle speed, and gradient using a film-grid sensor array and acoustical impact plate.
	Charged Particle Detector	Electrostatic Analyzer	A truncated hemispherical electrostatic analyzer with three current collectors for studying solar wind electrons and ions.
Pioneer 9	Magnetic Fields	Magnetometer	A three-axis fluxgate magnetometer used to measure local interplanetary magnetic field strength and direction.
	Charged Particle Detector	Electrostatic Analyzer	A truncated hemispherical electrostatic analyzer with three current collectors for studying solar wind electrons and ions.
	Charged Particle Detector	Solar Plasma Detector	A truncated hemispherical electrostatic analyzer with three current collectors for studying solar wind electrons and ions.
	Radio	Two-Frequency Beacon Receiver	A radio experiment that used transmissions between the spacecraft and ground stations to determine ionospheric electron content.

	Dust Detector	Cosmic Dust Detector	Equivalent to *Pioneer 8* instrument. Measured cosmic flux density, particle speed, and gradient using a film-grid sensor array and acoustical impact plate.
	Charged Particle Detector	Cosmic-Ray Anisotropy	A galactic and solar cosmic ray detection instrument that used three solid-state telescopes arranged in a 23-degree fan and cesium-iodide CsI scintillator crystal (lights in the presence of specific radiation) with a plastic anticoincidence shield.
	Charged Particle Detector	Cosmic Ray Gradient	A particle detector telescope with five solid-state sensors, a Cerenkov detector, and an anticoincidence shield.
	Electric Fields, Magnetic Fields	Electric Field Detector	Used an electric dipole antenna and collected data in three telemetry channels for measuring electrostatic and electromagnetic plasma waves.
Pioneer 10 & 11	Magnetic Fields	Magnetometer	A three-axis helium magnetometer used to measure local interplanetary magnetic field strength and direction. *Pioneer 11* carried an additional triaxial fluxgate magnetometer.
	Charged Particle Detector	Charged Particle Instrument	Used telescopes and detectors to study solar and galactic particle composition, and detect electron currents, gamma-rays, and proton fluxes.
	Charged Particle Detector	Asteroid/Meteoroid Astronomy	Used four nonimaging, particle-detector telescopes using a Ritchey-Chretien design.
	Dust/Particle Detector	Meteoroid Detectors	Comprised of 234 cells pressurized with an argon-nitrogen mixture; loss of gas due to micrometeorite holes was measured to determine particle impact size.
	Charged Particle Detector, Radiation Detector	Jovian Trapped Radiation	An array of five particle detectors consisting of a Cerenkov counter, an electron-scatter counter, a minimum ionization counter, and two scintillator detectors.
	Photometer	Ultraviolet Photometry	A broadband photometer for analyzing the solar wind during cruise, and for studying Jupiter's aurora.
	Photometer, Radiometer	Imaging Photopolarimeter	Used a 2.54-cm Maksutov catadioptric telescope and photopolarimeter detector arrangement of prisms, coatings, and photocathodes for conducting radiometric and polarimetric measurements.

	Radiometer	Infrared Radiometers	A 2-channel infrared radiometer using a Cassegrain telescope and a pair of 88-channel bimetallic thermocouples for studying irradiance of Jupiter's atmosphere and cloud-surface.
	Radio	S-Band Occultation	A radio experiment for profiling the Jovian ionosphere and determining Jupiter's diameter by studying radio waves transmitted through the ionosphere from the spacecraft.
	Charged Particle Detector	Geiger Tube Telescope	Comprised of seven Geiger tubes in three arrays to study proton and electron fluxes.
	Charged Particle Detector	Cosmic Ray Spectra	Consisted of one high-energy and two low-energy solid-state telescopes for detecting cosmic-ray particles.
	Charged Particle Detector	Plasma Analyzer	Dual quadrispherical electrostatic analyzer with 26 particle detectors and five current collectors for measuring incident plasma, protons, ions, and electrons.
	Photopolarimeter, Photometer	Zodiacal-Light Two-Color Photopolarimetry	An imaging photopolarimeter for mapping the zodiacal light in blue and red wavelengths.
Pioneer Venus Orbiter	Charged Particle Detector	OETP	A pair of Langmuir probes used to measure electron density and temperature.
	Radar	ORAD	A radar altimeter that used S-band pulses to study the orbiters altitude, surface topography of Venus, and derive other measurements such as surface temperature and interior structures.
	Radio	OGPE	Used S-band and X-band radio signals to study the Venusian atmosphere, ionosphere, and solar wind flow and interaction with the Venusian atmosphere.
	Gamma-Ray Detector	OGBD	An omnidirectional gamma-ray burst detector using scintillator spectrometers.
	Imaging, Photometer	OCPP	A form of imaging photopolarimeter (see similar instruments on *Pioneer 10* and *11*) for mapping cloud cover, haze, and cloud particles.
	Charged Particle Detector	ORPA	A Langmuir-probe retarding-potential analyzer that measured electron densities and temperatures, ions, and energy of ambient photoelectrons.
	Spectrometer	ONMS	A quadrupole mass spectrometer for analyzing particles and molecules in the upper atmosphere of Venus.

	Magnetic Fields	OMAG	A triaxial, or three-axis, fluxgate magnetometer for measuring Venusian magnetic field strength and direction as well as interplanetary magnetic fields during cruise.
	Electric Fields	OEFD	An electrical field detector similar to the instrument on *Pioneer 8* and *9*.
	Spectrometer	OUVS	Used a 125-mm Cassegrain telescope and 125-mm Ebert-Fastie spectrometer that used ultraviolet wavelengths for studying the thermosphere and ionosphere.
	Radiometer	OIR	An eight-channel infrared radiometer for sounding temperatures of the upper atmosphere.
	Spectrometer	OIMS	A Bennett radio-frequency mass spectrometer that studied composition and densities of ions.
	Charged Particle Detector	OPA	A quadrispherical electrostatic analyzer with five current collectors for measuring incident plasma, protons, ions, and electrons, and was based on similar instruments in previous Pioneer missions.
	Radio	Radio Science	A variety of radio experiments were performed using S-band and X-band radio signals for measuring properties of the Venusian atmosphere such as refractivity, electron densities, studying the gravity and spin vectors of Venus, internal densities, and atmospheric height and density.
Pioneer Venus Multiprobe Bus	Spectrometer	Ion Mass Spectrometer	An ion mass spectrometer used for studying the solar wind and upper Venusian atmosphere.
	Spectrometer	Neutral Mass Spectrometer	A neutral particle mass spectrometer used for studying the upper Venusian atmosphere.
Pioneer Venus Multiprobe - Large Probe	Accelerometer, Temperature	Atmospheric Structure	Used a three-axis accelerometer, pressure and temperature sensors for profiling the atmosphere during descent.
	Nephelometer	Nephelometer	A cloud particle measuring device that used a gallium-arsenide laser diode and backscatter detector.
	Spectrometer	Cloud Particle Size Spectrometer	Measured particle size and distribution densities using a laser and optical sensors.
	Chromatograph	Gas Chromatograph	A gas chromatograph used to analyze the atmosphere during descent.

	Radiometer	Infrared Radiometer	A four-channel infrared radiometer that profiled atmospheric thermal flux and cloud layers.
	Spectrometer	Neutral Mass Spectrometer	Used a double-focusing magnetic deflection mass spectrometer.
	Radiometer	Solar Flux Radiometer	A solar flux radiometer that used six narrow-field-of-view detectors for measuring scattered light.
	Nephelometer	Differential Long Base-Line Interferometer	Used radio signals and very long baseline interferometry techniques for studying wind speeds in the lower atmosphere.
Pioneer Venus Multiprobe - Small Probes	Accelerometer, Temperature	Small Probe Atmospheric Structure	Used a three-axis accelerometer, pressure and temperature sensors for profiling the atmosphere during descent.
	Nephelometer	Small Probe Nephelometer	A cloud particle measuring device that used a gallium-arsenide laser diode and backscatter detector.
	Radio, Interferometer	Differential Long Base Line Interferometer	Used radio signals and very long baseline interferometry techniques for studying wind speeds in the lower atmosphere.
	Radiometer	Net Flux Radiometer	A small net flux radiometer for profiling solar energy absorbed by the Venusian atmosphere.
	Radio	Atmospheric Turbulence	Used telemetry data from the probe's descent to study turbulence and winds in the atmosphere during descent.
	Radio	Atmospheric Propagation	Used descent telemetry to study communications through the Venusian atmosphere for application on later missions.
Planck Space Observatory	Radiometer	High-Frequency Instrument (HFI)	The high-frequency instrument uses 54 bolometer detectors, sensitive infrared-heat detectors, cryogenically cooled, and using a spider web of tiny thermistors.
	Radiometer	Low Frequency Instrument (LFI)	A microwave radiometer instrument consisting of 11 feed horns and makes use of microwave amplifiers.
Polar Satellite	Magnetic Fields	Plasma Waves Investigation (PWI)	Uses orthogonal electric dipole antennas, a magnetic loop antenna, and a triaxial magnetic search coil antenna.
	Magnetic Fields	HYDRA	A collection of eight pairs of electrostatic analyzer heads (aka Hydra the many-headed Greek serpent and constellation) for studying magnetic events such as reconnections, study field-

		aligned currents in the aurora, and quantify and measure with high-time-resolution magnetic interactions between the magnetosheath and plasmas.
Magnetic Fields	MFE	Magnetic fields experiment that used a pair of triaxial magnetometers with flippers for studying magnetic events in the polar cusp (a region where the magnetosphere interacts with the ionosphere), study magnetic reconnection events, and study field-aligned currents.
Spectrometer	TIMAS	A 128-channel toroidal ion mass spectrometer for studying, with an emphasis on mass studies, solar wind plasma interactions in the magnetosphere, hot plasmas, cool plasmas, and other magnetosphere events.
Electric Fields	EFI	Electric fields investigation that used three orthogonal double probes mounted on booms for studying electrical fields.
Charged Particle Detector	TIDE	A thermal ion dynamics experiment that used two sensors mounted on opposite sides of the satellite for studying low-energy ions in the magnetosphere and determining their transport, origin, and loss.
Imaging	UVI	An ultraviolet imaging camera that has a near-ultraviolet and vacuum-ultraviolet channel, as well as a visible imaging ability. UVI imaged global auroras at several wavelengths.
Imaging	VIS	A visible imaging system consisting of three cameras, one of which can image in the far ultraviolet, intended for taking imagery of nighttime polar auroras in high time resolution and high spatial resolution.
X-ray Detector	PIXIE	Polar ionospheric X-ray imaging system that used a pinhole camera system with a position-sensitive proportional counter detector and signal processing electronics that identifies X-ray events and derives energy.
Charged Particle Detector	Charge and Mass Magnetospheric Ion Composition Experiment (CAMMICE)	Experiment that used a magnetospheric ion composition sensor and heavy ion telescope, CAMMICE studied composition of the Earth's plasmas, sources, and transport systems.

	Charged Particle Detector	Comprehensive Energetic Particle Pitch Angle Distribution (CEPPAD)	Instrument system that consisted of an imaging proton sensor, an imaging electron sensor, a high-sensitivity telescope, and an energetic particle spectrometer. The instrument measured pitch angle measurements of energetic particle fluxes, high-energy ion composition, and resolution between ions and electrons.
QuickSCAT	Radar	Seawinds	A Ku-band scatterometer that used microwave pulses and backscatter reflections to collect data on surface ocean winds, study vegetation changes, and measure sea ice edge movements at Arctic and Antarctic ice packs.
Ranger 1–2 (Block I)	Imaging	Lyman-alpha telescope	A telescope for imaging the Earth in the Lyman-alpha spectra region using a parabolic mirror, ionization chamber, and radiation shielding.
	Magnetic Fields	Magnetometer	A rubidium-vapor magnetometer inside a fiberglass shell for measuring electromagnetic field strength and direction.
	Charged Particle Detector	Electrostatic analyzers	A pair of electrostatic analyzers for measuring the flux of electrons and ions.
	Charged Particle Detector	Medium-energy range particles detectors	A set of charged particle detecting modules using photoconductive materials for detecting protons, electrons, and X-rays.
	Radiation Detector	Two Triple Coincidence telescopes	An assembly of seven proportional-counter tubes for studying high-energy radiation.
	Charged Particle Detector	Cosmic Ray Integration ion chamber	An ion chamber using a spherical argon gas chamber for detecting cosmic ray particles.
	Dust Detector	Cosmic Dust Detectors	Used a light-flash impact detector and a crystal transducer that is basically a microphone.
	X-ray Detector	X-ray Scintillation counters	A pair of X-ray scintillation detectors provided by the Atomic Energy Commission project Vela. Used a type of crystal that emitted photons when encountering specific radiation.
Ranger 3–5 (Block II)	Imaging	Vidicon television camera	Vidicon tube television camera, a form of cathode ray tube technology used before charge-coupled devices (CCD). The optical system used a Cassegrain telescope design.

	Spectrometer, X-ray Detector	Gamma-Ray Spectrometer	Used a cesium-iodide crystal scintillator (emits photons upon detection) for detecting gamma-rays and a plastic anticoincidence scintillator shield for subtracting stray rays.
	Radar	Radar altimeter	A pulse-type radar altimeter that was intended for ranging the lunar surface for spacecraft operations.
	X-ray Detector	Seismometer	A magnetic seismometer built with a coil, spring-suspended magnet, and internal calibration. The seismometer would become active upon impact with the lunar surface and was designed to return data for a limited time.
Ranger 6–9 (Block III)	Imaging	Television cameras	Six vidicon TV cameras for transmitting pictures of the lunar surface, the camera setup was built with maximum reliability design and resolution. Vidicon was a form of cathode ray tube technology used before charge-coupled devices (CCD).
RapidScat (also ISS RapidScat)	Radar	Radar Scatterometer	A pencil-beam scatterometer for measuring ocean surface winds and is very similar to and derives a design heritage from the SeaWinds scatterometer. Primarily an ISS technology demonstration.
RHESSI	X-ray Detector, Gamma-Ray Detector	X-ray and Gamma-Ray Spectrometer	An X-ray and gamma-ray spectrometer and imager primarily intended for solar flare study. RHESSI uses a spectrometer with nine germanium detectors and a cryogenic cooler, and an imager using a Fourier-transform imaging technique and nine rotating collimators.
ROSAT	X-ray Detector	X-ray telescope (XRT)	An X-ray telescope that consisted of a 2.4-m assembly of four nested grazing incidence paraboloid and hyperboloid mirrors in a barrel-like configuration. The mirrors were constructed of a ceramic glass with a gold coating for X-ray reflectivity.
	Imaging, X-ray Detector	High Resolution Imager (HRI) / Position Sensitive Proportional Counter	A high-resolution X-ray imager consisting of cascading microchannel plates and a grid position system that converted X-rays to images using a photocathode.
	Imaging	Wide Field Camera (WFC)	An ultraviolet wide-field camera that used three nested grazing incidence mirrors and a micro-channel plate detector.
	Photometer	PSPC	A position sensitive proportional counter with a gas counter that detected incoming photons.

Rosetta (An ESA mission, NASA contributed 3 instruments listed)	Spectrometer	ALICE	A far-ultraviolet light imaging spectrograph for studying the cometary nucleus for noble gases such as krypton, helium, neon, and argon.
	Spectrometer, Radiometer	Microwave Instrument for Rosetta Orbiter (MIRO)	Consists of a spectrometer and radiometer for detecting temperatures, molecules, isotope ratios, outgassing, and gases.
	Charged Particle Detector	IES	An electrostatic analyzer and ion analyzer for measuring cometary plasma and solar wind ions and electrons.
RXTE	Imaging, X-ray Detector	All Sky Monitor	An X-ray sky monitor of known and transient X-ray sources that uses a multiple pinhole camera and collimator design.
	X-ray Detector	Proportional Counter Array	Large area X-ray proportional counter detects X-rays using five proportional counters.
	X-ray Detector	High Energy X-ray Timing Experiment	High-energy X-ray timing experiment consists of X-ray scintillation detectors for detecting X-rays from sources within and outside the galaxy.
SAMPEX	Charged Particle Detector	Heavy Ion Large Telescope (HILT)	Used an ion drift chamber to detect heavy ions from solar and cosmic rays. In certain cases, ionic charge states were inferred by using the Earth's magnetic field, itself as a kind of massive magnetic mass spectrometer. Was built at the Max Planck Institute for Extraterrestrial Physics.
	Spectrometer	Low-Energy Ion Composition Analyzer (LICA)	Measures solar and magnetospheric ions using a time-of-flight mass spectrometer.
	Spectrometer	Mass Spectrometer Telescope (MAST)	Measures isotopic composition of a range of elements from Li to Ni, and He.
	Charged Particle Detector	Proton Electron Telescope (PET)	Measures energy spectra, composition of protons and helium nuclei from the sun, from other solar system bodies, the galaxy, and man-made sources in low Earth orbit.
SEAC^{4}RS	Satellites	CloudSat, CALIPSO, GOES	CloudSat studies clouds and precipitation using radar; CALIPSO and GOES provide various meteorological data such as water vapor, solar radiation, cloud and weather analysis, and temperature

	Aircraft	DC8, ER-2, Learjet	Aircraft carry a wide variety of instrumentation related to the mission of studying atmospheric feedback cycles. Example instrument studies include aerosols, molecules, gases, cloud particles, water vapor, and temperature.
	High-Altitude Balloons	High-Altitude Balloons	A series of balloons containing a frost-point hygrometer for measuring water vapor from the surface to the stratosphere, and ozonesondes for studying ozone distribution.
	Ground Station, Photometer	Aerosol Robotic Network (AERONET)	Worldwide network of aerosol detection equipment, typically using photometers.
	Ground Station, Laser	Micropulse LIDAR Network	Measures aerosol and vertical cloud structure using lasers (LIDAR) colocated with AERONET sensors.
Shuttle Radar Topography Mission	Radar	Interferometric Synthetic Aperture Radar (SAR)	A synthetic aperture radar system that used a main antenna in the shuttle's cargo bay for transmitting and receiving, and an antenna mounted on a 60-m mast for receiving.
Skylab	Apollo Telescope Mount (ATM)	White Light Coronagraph	Used solar occulting disks, lenses, and folding mirrors to image the solar corona in visible wavelengths. Images were recorded either with a film camera or TV camera for live operator viewing.
	ATM	X-Ray Spectrographic Telescope	Produced broadband images of solar flares and the solar disk in the soft X-ray spectrum. Used grazing-incidence mirrors, grating filters for spectroscopy, and a film camera.
	ATM	Ultraviolet Scanning Polychromator Spectroheliometer	Measured intensity of far-ultraviolet wavelengths from targeted solar locations and created spectroheliograms using seven atomic spectral lines. Used a telescope assembly and spectrometer to collect data using photomultiplier tubes.
	ATM	X-Ray Telescope	Used grazing-incidence mirrors and six filters for producing solar X-ray images on film. Proportional counters using thin-metallic gas-filled chambers were associated with imaging for recording X-ray intensity and to aid crew in selecting filters.
	ATM	Extreme Ultraviolet Spectroheliograph	Used an optical assembly consisting of a grating, heat-rejection mirrors, an aluminum filter, and film camera to image the solar chromosphere and corona.
	ATM	Spectrograph and Extreme Ultraviolet Monitor	An assembly consisting of a film camera, TV camera, grating spectrograph, and extreme ultraviolet monitor. The spectrograph recorded spectra in select solar regions and XUV

			monitor provided real-time, full-solar disk video for identification of targets of interest.
	ATM	Two Hydrogen-Alpha Telescopes	Hydrogen-alpha wavelength telescopes that used a Vidicon video camera for real-time monitoring, a film camera, and could image either the full solar disc or zoom in to specific features.
	Various	Examples: Microwave radiometers, multispectral scanners, cameras, medical instrumentatio n	In addition to the ATM solar observatory, Skylab carried various instruments such as visible, infrared, and multispectral cameras for imaging and studying the Earth in multiple spectrums; microwave radiometers for studying temperatures, surface terrain, and ocean features; and a full suite of medical instruments for monitoring and studying astronaut physiology in the space environment.
SOFIA	Telescope	Cassegrain telescope	A reflecting telescope with a 2.7-m (2.5 useful) mirror. Instruments can be interchanged; primarily will be used to study light in the infrared.
	Aircraft	Boeing 747 special performance (SP)	747 with a short body designed for ultra-long endurance flight.
SOHO	Spectrometer	Global Oscillations at Low Frequencies (Golf)	An instrument that uses an optical resonance technique within a sodium spectral line for detecting Doppler shifts. The intent is to determine internal solar structure by measuring low-frequency solar oscillations. A second objective is monitoring the average line-of-sight solar magnetic field.
	Photometer, Radiometer	Variability of Solar Irradiance and Gravity Oscillations (VIRGO)	Uses radiometers, a three-channel photometer, and luminosity oscillation imager to measure total solar irradiance and variation, and solar frequencies, amplitudes, and oscillations.
	Interferometer	Michelson Doppler Imager (MDI) (Superseded by the Helioseismic and Magnetic Imager (HMI) on the SDO)	Images the full disc of the solar surface with narrow-band filters and a tunable Michelson interferometer to measure oscillations in the solar magnetic field.
	Spectrometer	Solar UV Measurement of Emitted Radiation (SUMER)	An ultraviolet spectrometer that measures profiles intensities and spectral Doppler shifts as well as images of the sun in the extreme ultraviolet.

	Spectrometer	Coronal Diagnostic Spectrometer (CDS)	Detects detailed emission line spectra in the extreme ultraviolet in order to study the makeup and actions of hot plasmas in the corona.
	Imaging	Extreme UV Imaging Telescope (EIT)	Imaged solar features such as coronal holes, prominences, and filaments in the inner corona in the extreme ultraviolet in wide full-disk or subfield images. Full operations of this instrument ended in July 2010 in order to transfer useful bandwidth to the LASCO instrument and to optimize cost and science return.
	Spectrometer	Ultraviolet Coronagraph and Spectrometer (UVCS)	Consisting of a spectrometer assembly, three reflecting telescopes with solar occultation, and a visible light polarimeter, UVCS directly collects solar wind particle data from 1.3 to 11.8 solar radii.
	Spectrometer	Large Angle Spectrometer Coronagraph (LASCO)	A telescope that uses an occulting disc to block the sun's disc that allows wide-field imaging and study of the corona. This method is similar to viewing a solar eclipse on Earth.
	Charged Particle Detector	Solar Wind Anisotropies (SWAN)	Monitors large-scale structure and movement of the solar wind by detecting backscatter of solar wind photons off of interstellar neutral hydrogen that passes through the solar system.
	Charged Particle Detector, Spectrometer	Charge, Element, Isotope Analysis (CELIAS)	Consisting of three sensors, the combined instrument package studies the composition of the solar wind in terms of low- and high-energy ions, their mass, composition, and abundance.
	Charged Particle Detector	Comprehensive Suprathermal and Energetic Particle Analyzer (COSTEP)	Consists of two instruments, an electron proton helium instrument (ephin) and a low-energy ion and electron instrument (LION) that detect and measure energetic particles, protons, electrons, and isotopes of hydrogen and helium by using semiconductor detectors.
	Charged Particle Detector	Energetic Analyzer (ERNE)	Detects energetic particles and quantifies isotopic and elemental abundance from hydrogen (H) to iron (Fe), and energies of fast-moving particles from various solar ejections and eruptions.
SORCE	Radiometer	Total Irradiance Monitor	Measures total solar irradiance (TSI) in watts per square meter with a high absolute accuracy using four electrical substitution radiometers equipped with robust noise and thermal protection technology and thermal background subtraction.

	Spectrometer	Spectral Irradiance Monitor	Uses a dual Fery Prism spectrometer to measure visible and near-infrared solar irradiance primarily from 310 to 2,400 nm and 200–300 nm as a secondary target wavelength.
	Spectrometer	Solar Stellar Irradiance Comparison Experiment	A pair of redundant two-channel grating spectrometers that are calibrated against ultraviolet radiation from blue stars with known constant energy. This instrument captures daily solar spectral measurements in the 115–310 nm range.
	Photometer	X Ray Ultraviolet (XUV) Photometer System (XPS)	Uses a filter wheel and set of 12 silicon photodiodes to measure solar soft X-ray radiation in wavelengths between 1 and 34 nm and the solar bright hydrogen Lyman-alpha emission between 121 and 122 nm.
Solar Probe Plus	Charged Particle Detector	Solar Wind Electrons and Protons Investigation	Will use a Faraday cup, a conductive metal cup, to catch and count charged particles.
	Imaging	Wide-Field Imager	A wide-field telescope that will take 3D images of the solar wind as the craft passes through it within the corona.
	Spectrometer	Integrated Science Investigation of the Sun	Will use a mass spectrometer to inventory, weigh, and catalog ions.
	Magnetic Fields	Fields Experiment	Measure magnetic and electrical fields, radio emissions, and shock waves in the solar atmospheric plasma. Will also inventory basic particles, dust, by monitoring voltage changes on the probe's antenna.
Space Technology 5	Magnetic Fields	Magnetometers	Used to measure the magnetic fields of current sheets associated with aurora near the poles.
Spitzer	Imaging	IRAC	A four-detector array near- and mid-infrared camera imaging system with a large field of view. All instruments on Spitzer are cryogenically cooled and use the 0.85-m telescope.
	Spectrometer	IRS	A set of infrared spectrographs designed to detect redshifts of target objects. All instruments on Spitzer are cryogenically cooled and use the 0.85-m telescope.
	Photometer	MIPS	The multiband imaging photometer is a far-infrared photometer operating in three broad spectral bands and uses three detector arrays. MIPS has multiple modes of operations and is able to operate in a large scanning mode, a super resolution and

			photometry mode, a spectral energy distribution mode, and a total power-absolute brightness mode. MIPS is cryogenically cooled and uses the 0.85-m telescope.
Stardust- NExT	Imaging	Imaging and Navigation Camera	Used a Petzval lens design and filter wheel for multispectral imaging. The filter wheel became stuck in broadband, which still allowed it to complete most of its primary mission.
	Spectrometer	Cometary and Interstellar Dust Analyzer	A time-of-flight mass spectrometer designed to study chemical composition of encountered particles.
	Dust Detector	Dust Flux Monitor Instrument	Used two polyvinylidene fluoride disks with mounted electrodes to detect dust impacts, partial flux, intensity, and mass distribution
		Stardust Sample Collection	Used a low-density aerogel made of an inert silica-based substance for capturing and storing encountered cometary particles and interstellar dust. The aerogel sample collector was returned to Earth using a small reentry capsule and recovery system.
STEREO A and B	Imaging	Sun Earth Connection Coronal and Heliospheric Investigation (SECCHI)	Composed of two white-light coronagraphs, an extreme ultraviolet imager, and a heliospheric imager. Used to study coronal mass ejections.
	Various, Spectrometer, Magnetic Fields	In-Situ Measurements of Particles and CME Transients (IMPACT)	A suite of instruments: a suprathermal electron telescope consisting of silicon semiconductor detectors, a solar wind electron analyzer using hemispherical electrostatic analyzers, a triaxial fluxgate magnetometer, a solar electron proton telescope, a suprathermal ion telescope that uses an ion mass spectrometer, and high- and low-energy telescopes using solid-state detectors.
	Spectrometer	Plasma and Suprathermal Ion Composition (PLASTIC)	A time-of-flight/energy mass spectrometer using a time-of-flight chamber, an electrostatic analyzer, and solid-state detectors.
	Radio	STEREO/WAVES	Uses radio receivers to study accelerated electrons, wave intensity, angular size, as well as radio and plasma waves associated with coronal mass ejections.

SWAS	Microwave	Submillimeter radiometer instrument	Uses a 69 × 53-cm off-axis Cassegrain telescope and an acousto-optical spectrometer for studies in submillimeter wavelengths, generally between infrared and microwave wavelengths.
Suomi NPP (NASA contributed instruments)	Microwave	Advanced Technology Microwave Sounder (ATMS)	A cross-track scanner for profiling temperatures, water vapor, and atmospheric pressure, it has 22 observation channels.
	Radiometer	Visible Infrared Imaging Radiometer Suite (VIIRS)	A scanning radiometer in the visible and infrared for studying clouds, aerosols, the ocean surface, and surface temperatures.
	Spectrometer	Cross-Track Infrared Sounder (CRIS)	A Fourier-transform spectrometer with 1305 spectral channels for profiling temperature, pressure, and moisture.
	Spectrometer, Imaging	Ozone Mapping Profiler Suite (OMPS)	A set of redundant instruments consisting of three hyperspectral imaging spectrometers for profiling ozone.
	Radiation Detector	Clouds and Earth's Radiant Energy System (CERES)	A three-channel radiometer for measuring emitted and reflected solar energy for long-term monitoring of Earth's radiation budget.
Surveyor (All)	Imaging	Television Camera	Vidicon tube television cameras that were pre-CCD-era cathode ray technology detectors.
	Soil Test	Strain Gauge, various	Used to monitor landing forces to determine the hardness of the lunar surface.
Surveyor 3–7	Sample Collection	Soil Surface Sampler	Surface-trenching tool that could extend approximately 1.5 m and was used to study the lunar surface.
Surveyor 5–7	Charged Particle Detectors	Alpha-scattering instrument	Instrument that used an alpha source to irradiate lunar soil samples and measured energy spectra of alpha particles and protons to quantify major elements. In a passive mode, was also used to study cosmic rays and solar protons.
Suzaku			See Astro E-2.
Swift	Gamma-Ray detector	Burst Alert Telescope	A gamma-ray detection telescope that uses a coded aperture mask and wide field of view for detecting gamma-ray bursts.
	X-ray Detector	X-ray Telescope	Uses X-ray focusing grazing incidence mirrors and measures spectra of X-rays, light curves, fluxes, and afterglows emanating from gamma-ray burst targets.

	Imaging	Ultraviolet Optical Telescope	A 12-inch Ritchey-Chretien telescope for imaging afterglow of gamma-ray burst targets.
TDRS	Radio	Ku, Ka, S band, radio communications	Used for communications with several space platforms such as the ISS, other satellites, aircraft, balloons, and ground stations.
Terra	Radiometer	ASTER	A thermal emission and reflection radiometer operating in 14 multispectral channels from the visible to the infrared.
	Radiometer	Clouds and Earth's Radiant Energy System (CERES)	System uses two broadband-scanning radiometers for studying Earth's radiation budget.
	Spectrometer, Radiometer	Multi-Angle Imagine Spectroradiometer (MISR)	The instrument studies cloud tops, surface reflectance, and airborne particles using nine pushbroom cameras using four spectral bands.
	Spectrometer	Moderate Resolution Imagine Spectrometer (MODIS)	Spectrometer using a cross-scan mirror, collecting optics and detectors for studying Earth's surface and lower atmosphere with measurements such as surface temperatures, ocean surface colors, chlorophyll fluorescence, and airborne particles.
	Spectrometer	Measurement of Pollution in the Trophosphere (MOPITT)	A scanning spectrometer instrument system studying reflected and emitted infrared specifically sourced from carbon monoxide and methane.
TESS	Imaging	Four identical wide field cameras	Cameras will survey target stars with the intent of detecting transiting exoplanets.
THEMIS (On Spacecraft)	Electric Fields	EFI	Electric field instrument consisting of spin-plane spherical sensors mounted on cables and two axial tubular sensors mounted on 5-m stacer (spiral tube and actuator for controlled extension/retraction) booms.
	Magnetic Fields	FGM	A triaxial fluxgate magnetometer mounted on a 2-m boom for measuring magnetic fields strengths and direction.
	Magnetic Fields	SCM	A triaxial search coil magnetometer.
	Charged Particle Detector	ESA	An electrostatic analyzer that monitors electron and ion flux.
	Particle Detector	SST	Solid-state, double-ended telescope with silicon detectors for measuring supra-thermal electron and ion profiles.

THEMIS (Ground Based)	Imaging	All Sky Imager	All Sky White light imager array is a set of 20 ground-based cameras located in the far Northern Hemisphere from Alaska and through Canada that image the aurora with a 1-km resolution.
	Magnetic Fields	Ground Based Magnetometer Array	Twenty-eight ground-based, fluxgate magnetometers located in the far Northern Hemisphere from Alaska and through Canada. Eleven of the sensors are installed in schools as part of an outreach program.
TIMED	Imaging, Spectrometer	GUVI	A far-ultraviolet, cross-track scanning imaging spectrograph for mapping the auroral oval, energy fluxes, temperature information, and ionospheric electron density profiles.
	Radiometer	SABER	A 10-channel infrared radiometer for profiling the atmosphere in components such as ozone, carbon dioxide, water vapor, temperature and other gases or molecules in detection range.
	Spectrometer, Photometer, X-ray Detector	SEE	A spectrometer and set of photometers for studying solar irradiance. The EUV is a grating spectrograph and the photometers use silicon photodiodes and measure soft solar X-rays (XUV).
	Interferometer	TIDI	Consisting of four telescopes, a Fabry-Perot interferometer, and a CCD detector for studying the mesospheric and lower-thermosphere winds, noctilucent clouds, and auroral emission.
TOMS / EP	Spectrometer	Total Ozone Mapping Spectrometer	A spectrometer that measures ozone through both incoming solar energy and ultraviolet light backscattered off the Earth's atmosphere.
TOPEX Poseidon	Radar	Dual-Frequency Topex Radar Altimeter-ALT	A radar altimeter using the Ku-Band and C-band for measuring sea height, wave heights, and wind speeds.
	Radiometer	Topex Microwave Radiometer (TMR)	A microwave nonscanning radiometer with three channels for measuring water vapor over the oceans.
	Laser	LRA	Laser reflectors used by ground-station lasers for precise calibration of the satellite's orbit and altimeters.
	Radio	GPSDR	Global positioning system demonstration receiver that used GPS satellites for orbit determination.

	Radio	Coppler Orbitography and Radio-Positioning Integrated by Satellite (DORIS)	DORIS used ultra-stable microwave frequencies and an ultra-stable oscillator to communicate with ground stations for precise satellite position tracking using Doppler techniques.
	Radar	SSALT	A single-channel Ku-band radar altimeter used to measure sea surface height and satellite to surface ranging.
TRACE	Imaging	EUV/ UV Telescope	A 30-cm extreme ultraviolet telescope for imaging the solar corona and solar transition region.
TRMM	Radar	Precipitation Radar (PR)	Radar using a single 14-GHz frequency for profiling precipitation heights.
	Radiometer	TMI	A nine-channel passive microwave radiometer for profiling rainfall and water vapor.
	Radiometer	VIRS	A five-channel scanning radiometer operating in the near-infrared, mid-infrared, and thermal infrared for monitoring sea and land surface temperatures and atmospheric moisture content.
	Radiation Detector, Thermal	Clouds and the Earth's Radiant Energy System (CERES)	CERES measures long-wave and total-wavelength radiation energies.
	Imaging	LIS	An optical lightning detection sensor that can detect both cloud-to-ground and cloud-to-cloud lightning.
Ulysses	Particle Detector	BAM solar wind plasma experiment	Solar wind plasma studies instrument that used ion and electron spherical electrostatic analyzers.
	Radio	Coronal Sounding	A radio science investigation for studying coronal plasma that studied Doppler shifts in spacecrafts radio transmissions.
	Spectrometer	GLG/Solar Wind Ion Composition Spectrometer (SWICS)	SWICS used an electrostatic analyzer, time-of-flight chamber, and a solid-state detector for analyzing temperatures, composition, and speed of solar ions.
	Gravity Waves	Gravitational Waves	A search for gravity waves using radio Doppler studies.
	Magnetic Fields	Magnetic Fields	A triaxial fluxgate magnetometer and a helium magnetometer for measuring magnetic fields strength and direction.

	Magnetic Fields	EPAC	An energetic particles study using four particle detection telescopes at multiple angles providing a wide-angle field of view for studying energy, composition, and distribution of particles.
	Spectrometer	LAN	Low-energy ions and electrons instrument for studying solar flares and consisted of two double-ended solid state detectors that used foil and magnetic spectrometers.
	Particle Detector	SIM/Cosmic Ray and Solar Particles instrument (COSPIN)	COSPIN used six, charged-particle telescopes covering high energies, low energies, electron, and proton flux.
	Magnetic Fields	Unified Radio and Plasma (URAP)	URAP Waves studied solar radio bursts and used a set of three antennas.
	X-ray Detector, Gamma-Ray Detector	HUS/GRB	Solar X-rays and Cosmic Gamma-Ray Bursts experiment that used two hemispherical cesium-iodide scintallator crystals and solid-state detectors.
	Particle Detector	Cosmic Dust	A multicoincidence plasma impact detector with speed, mass, and direction measurement capabilities.
	Photometer	Ulysses Interstellar Neutral Gas Experiment	Studied interstellar neutral helium gas using a pinhole camera design concept for the detection of neutral helium and ultraviolet photons.
UARS	Radiometer	ACRIM2	Used three active cavity radiometer pyrheliometers for measuring solar energy in the far infrared.
	Spectrometer	Cryogenic Limb Array Etalon Spectrometer (CLAES)	CLAES analyzed atmospheric composition and temperature using spectrometers with a solid-etalon Fabry-Perot design, telescope, and solid-state linear detector array.
	Radiometer	Halogen Occultation Experiment (HALOE)	HALOE profiled upper-atmosphere molecules using occultation techniques and two four-channel radiometers.
	Imaging	High-Resolution Doppler Imager (HRDI)	A high-resolution Doppler imager that studied light spectra of emitted or scattered light using a telescope and Doppler imaging, interferometer with a three-etalon Fabry-Perot design.
	Radiometer	ISAMS	An infrared radiometer that used a pressure-modulated radiometry technique for profiling molecules, water vapor, and nitrogen bearing molecules.

	Spectrometer	Microwave Limb Sounder (MLS)	Used a spectrometer, three-mirror antenna system, optics, and radiometer.
	Particle Detector, magnetic Fields	PEM	Studied particle interactions and molecules in the Earth's atmosphere such as ozone, nitric oxide, and energetic particle interaction with the upper atmosphere. The monitor used electrostatic analyzers, solid-state charged-particle telescope detectors, X-ray detectors, and triaxial fluxgate magnetometer.
	Spectrometer	Solar Stellar Irradiance Comparison Experiment (SOLSTICE)	SOLSTICE studied solar variability using a three-channel Ebert-Fastie spectrometer, photomultiplier tubes, and spectral grating.
	Spectrometer	Solar Ultraviolet Spectral Irradiance Monitor (SUSIM)	SUSIM was a dual dispersion spectrometer for monitoring in the ultraviolet.
	Interferometer	Wind Imaging Interferometer (WINDII)	Interferometer using a Michelson interferometer design and CCD camera.
Van Allen Probes	Magnetic Fields	Electric Field and Waves (EFW)	EFW suite analyzes electric fields using a set of four spin-plane electric field antennae and a set of two spin-axis stacer booms with mounted metal spheres.
	Magnetic Fields	EMFISIS	Electric and Magnetic Field Instrument Suite for studying plasma waves and magnetic fields comprised of a triaxial magnetometer, triaxial search coil magnetometer, and sweep frequency receiver.
	Particle Detector	ECT	Energetic Particle, Composition, and Thermal Plasma instrument suite for analyzing particles, electrons, and ions in the near-Earth environment and is made up of a magnetic electron ion spectrometer, a helium-oxygen proton electrostatic analyzer, and a relativistic electron proton telescope.
	Particle Detector	Radiation Belt Storm Probes Ion Composition Experiment (RBSPICE)	RBSPICE studies the "ring current" around Earth using a time-of-flight and energy spectrometer for studying protons, electrons, and ions as well as vectors of detected particles.
	Spectrometer	Relativistic Proton Spectrometer (RPS)	Spectrometer for measuring radiation belt protons using a Cherenkov detector and silicon detectors.
Viking 1 & 2 (Spacecraft)	Imaging	VIS	A pair of high-resolution scanning television cameras.

	Spectrometer	MAWD	An infrared grating-spectrometer that searched for water vapor by analyzing reflected sunlight from the Martian surface.
	Radiometer	IRTM	A multichannel radiometer that measured atmospheric and surface temperatures using a system of four telescopes with infrared detectors.
	Radio	Radio Science	A radio science experiment using the orbiters transmitters to analyze the gravity of Mars through radio Doppler analysis, and other measurements possible with radio occupation techniques such as the Martian atmosphere and the radius of Mars.
(Lander)	Imaging	Imaging	Two high- and low-resolution scanning cameras that could acquire images in color, infrared, and black and white for high resolution.
	Spectrometer	Gas Chromatograph Mass Spectrometer (GCMS)	Spectrometer that was used for analyzing the atmosphere and soil samples near the lander. Atmospheric samples bypassed the gas chromatogram.
	Seismometer	Seismometer	A seismology instrument that used a fixed magnet and moving coil for sensing seismic activity.
	Spectrometer	X-ray Fluorescence Spectrometer (XRFS)	Spectrometer that detected X-rays emitted from irradiated samples using gas-filled proportional counters.
	Biology	Biology Laboratory	A biology laboratory experiment suite that had a gas exchange experiment, a pyrolytic release experiment, and a labeled release experiment.
	Meteorology	Meteorology	A meteorological package that recorded temperatures, wind speed and direction, and atmospheric pressure.
	Magnets	Magnetic Properties	Used samarium-cobalt magnets on the lander and sampling arm to capture iron-bearing particles; the magnets were examined by the lander's camera.
	Particle Detector	Retarding Potential Analyzer (RPA)	Instrument that was capable of detecting plasmas, electrons, and ions for analyzing the ionosphere in terms of composition, temperature, and structure.
	Meteorology	Atmospheric Structure	A meteorological package that consisted of accelerometers, radar altimeters, temperature and pressure sensors, for profiling the Martian atmosphere during descent.

	Spectrometer	Neutral Mass Spectrometer (NMS)	Spectrometer used to profile the Martian atmospheric composition during descent.
	Radio	Radio Science	A Radio Science experiment using the lander's transmitters to analyze precession and rotation of Mars through radio Doppler analysis.
Voyager	Imaging	Imaging Science	A narrow-, wide-, and long-focal-length set of cameras that used pre-CCD era vidicon cathode ray technology detectors.
	Particle Detector	Cosmic Ray Detector	A high- and low-energy solid-state telescope system for detecting cosmic-ray particles and an electron telescope using solid-state detectors for electron detection.
	Spectrometer	Infrared Spectrometer	Infrared interferometer spectrometer instrument is a combined Michelson interferometer and radiometer using a 0.5-m Cassegrain telescope and was used to study atmospheric compositions, temperature profiles, and large particle studies such as Saturn's rings.
	Spectrometer	Ultraviolet Spectrometer (UVS)	The spectrometer could use either an airglow or solar occultation mode for analyzing atmospheres of encountered planets.
	Spectrometer	Plasma Spectrometer	Comprised of two Faraday cup detectors for recording plasma ions and their velocities, densities, and pressure.
	Particle Detector	Low-Energy Charged Particles (LECP)	Detector that could record particle data in a planetary mode consisting of six solid-state detectors, or an interplanetary mode that could use a particle telescope, solid-state detectors, and an anticoincidence shield.
	Magnetic Fields	MAG	A pair of triaxial fluxgate magnetometers for measuring magnetic field strength and direction of Jupiter and Saturn's magnetosphere, as well as interplanetary magnetic fields and any interstellar magnetic fields encountered at the solar wind boundary and beyond.
	Particle Detector, Magnetic Fields	Plasma Wave System	Provided electron-density profiles and plasma wave information using a 16-channel receiver and a low-frequency waveform receiver.

	Photon Detector	Photopolarimeter System	Used a 20-cm telescope, eight band filters, and a photomultiplier tube for analyzing surfaces, composition, and atmospheric scattering of Jupiter and Saturn.
	Radio	Planetary Radio Astronomy	A radio astronomy experiment using a radio receiver and monopole antennas for studying Jupiter and Saturn radio emissions.
	Radio	Radio Science	A study of spacecraft radio signals for deriving data on planetary ionospheres, gravity fields, and specific occultation experiments such as material in Saturn's rings.
Wide Field Infrared Explorer	Imaging	Infrared Telescope	A highly sensitive wide-field infrared telescope operating in four infrared channels, used a 40-cm telescope and active cryogenic cooling system.
WIND spacecraft	Particle Detector	3DP	A 3D plasma analyzer using semiconductor detectors with telescopes, quadrispherical electrostatic electron and proton analyzers with design features enabling measurements at low flux levels.
	Particle Detector	TGRS	A gamma-ray spectrometer for recording transient sources, the instrument uses a cryogenically cooled n-type Germanium scintillator crystal.
	Magnetic Fields	MAG	Magnetic fields instrument is a boom-mounted triaxial fluxgate magnetometer for measuring magnetic field strength and direction.
	Magnetic Fields	WAVES	Measures solar wind plasmas using an antenna array, triaxial search coil, and radio wave analysis electronics.
	Particle Detector	Solar Wind Experiment (SWE)	Solar Wind Experiment includes a high-time-resolution, six-axis ion and electron spectrometer and a pair of Faraday cups for 3D measurements of ions.
	Particle Detector	EPACT	An energetic particle acceleration, composition, and transport instrument using a series of particle detection telescopes.
	Particle Detector	SWIC/STICS	A solar wind and supra-thermal ion composition study using time-of-flight sensors and proton/alpha particle telescope.
	Particle Detector	KONUS	A particle detector using a pair of KONUS gamma-ray detectors.

Wilkinson Microwave Anisotropy Probe	Radiometer	Microwave Radiometers	Pseudo-correlation radiometer instrument is a pair of differential radiometers operating in five frequency bands using 1.5-m Gregorian reflectors.
XMM-Newton	Imaging	EPIC	A set of three photon imaging X-ray CCD cameras; two cameras use X-ray telescopes that house a reflection grating typically used in spectrometers and one camera with an unobstructed telescope.
	Spectrometer	Reflection Grating Spectrometer (RGS)	A reflection grating spectrometer for acquiring X-ray spectra.
	Imaging	OM	Optical monitor instrument is a visible light 30-cm Ritchey-Chretien telescope for identifying sources of X-ray acquisition.

Appendix C: NASA Employees, by Location, 2015

Location	No. of Full-Time, Professional Employees
Ames Research Center	1,120
Armstrong Flight Research Center	522
Glenn Research Center	1,508
Goddard Space Flight Center	3,106
Johnson Space Center	2,885
Kennedy Space Center	1,914
Langley Research Center	1,745
Marshall Space Flight Center	2,291
Stennis Space Center	286
Headquarters	1,089
NASA Shared Services Center	129
Total	16,595

Appendix D: Partial List of NASA Active Missions and Their Assigned Directorates or Programs

Mission (Acronym)[1]	Launch Date	Directorate/ Program	Managed by
ACE	25-Aug-97	Heliophysics, Astrophysics, Explorer	GSFC
AIM	25-Apr-07	Heliophysics	GSFC
AirMOSS	Mar-12	Earth Sciences	JPL
Aqua	4-May-02	Earth Sciences	GSFC
Aquarius	10-Jun-11	Earth Sciences	JPL, GSFC
ARTEMIS	Feb-07	Heliophysics, Explorers	GSFC
Asteroid Redirect Initiative	Future	Near Earth Object	JPL
Astro-H	Sch. 2015	Astrophysics Explorers, Mission of Opportunity	GSFC
ATTREX	Dec-11	Earth Sciences	ARL
Aura	15-Jul-04	Earth Sciences	GSFC
CALIPSO	28-Apr-06	Earth Sciences	LaRC,GSFC
CARVE	1-Mar-12	Earth Sciences	JPL, LaRC
Cassini-Huygens	15-Oct-97		JPL
Chandra X-ray Observatory	23-Jul-99	Physics of the Cosmos	MSFC
CINDI	16-Apr-08	Heliophysics	GSFC
CLARREO	Future	Earth Sciences	LaRC
CloudSat	28-Apr-06	Earth Sciences	JPL, GSFC
Cluster II	16-Jul-00	Heliophysics	GSFC
Curiosity	26-Nov-11	Mars Exploration	JPL
CYGNSS	Sch. Oct. 2016	Earth Sciences	ARL

Dawn	27-Sep-07	Discovery	JPL, MSFC
Deep Impact	Jan-05	Discovery	JPL, MSFC
ECOSTRESS	Sch. Sep. 2014	Earth Sciences	JPL, LaRC
Euclid	Scheduled 2020	Physics of the Cosmos	GSFC, JPL
FAST	21-Aug-96	Small Explorers	GSFC
Fermi Gamma-ray Space Telescope	11-Jun-06	Physics of the Cosmos	GSFC, MSFC
GEDI	Sch. Sept. 2014	Earth Sciences	LaRC
Geotail	24-Jul-92	Heliophysics	GSFC
GPM	Feb-14	Earth Sciences	GSFC
GOES	Variable	Earth Sciences	GSFC
GRACE	17-Mar-02	Earth Sciences	JPL, GSFC
GRAIL	10-Sep-11	Discovery	JPL, MSFC
HS3	Aug-12	Earth Sciences	ARL, GSFC
Hinode (Solar-b)	23-Sep-06	Heliophysics	GSFC, MSFC
Hubble	25-Apr-90	Cosmic Origins	GSFC
IBEX	19-Oct-08	Heliophysics, Explorers	GSFC
ICEsat -2	Sch. 2017	Earth Sciences	GSFC
InSight	Sch. March 2016	Discovery	JPL
INTEGRAL	17-Oct-02	Astrophysics, Physics of the Cosmos	GSFC

IRIS	27-Jun-13	Heliophysics, Small Explorer	GSFC, ARL
James Webb Space Telescope	Sch. 2018	Cosmic Origins	GSFC, MSFC
Jason-1, -2, -3	2001–2015	Earth Sciences	JPL
Juno	5-Aug-11	New Frontiers Program	JPL, MSFC
Kepler	6-Mar-09	Exoplanet Exploration	ARL, MSFC
LADEE	6-Sep-13	Science Mission Directorate	ARL, GSFC, MSFC
LAGEOS 1 and 2	4-May-76	Earth Sciences	GSFC
Landsat 7	15-Apr-99	Earth Sciences	GSFC
LDCM/LandSat 8	11-Feb-13	Earth Sciences	GSFC
LISA	Sch. 2020	Physics of the Cosmos	GSFC
LRO	18-Jun-09	Science Mission Directorate	GSFC
Mars Exploration Rovers Spirit and Opportunity	10-Jun-03	Mars Exploration	JPL
Mars Odyssey	7-Apr-01	Mars Exploration	JPL
Mars Reconnaissance Orbiter	12-Aug-05	Mars Exploration	JPL
Mars Science Laboratory	26-Nov-05	Mars Exploration	JPL, ARL
MAVEN	18-Nov-13	Mars Exploration	GSFC, KSC
MESSENGER	4-Aug-04	Discovery	GSFC, MSFC
NEOWISE	Dec-09	Near Earth Object	GSFC
New Horizons	19-Jan-06	New Frontiers Program	GSFC
NICER	Sch. 2016	Astrophysics	JPL

NOAA-N	20-May-05	Earth Sciences	GSFC
NPOESS	Sch. March 2014	Earth Sciences	GSFC
NuSTAR	13-Jun-12	Astrophysics Explorer	JPL, GSFC
Operation Ice Bridge	15-Oct-09	Earth Sciences	GSFC, ARL
OCO-2	Sch. July 2014	Earth Sciences	JPL
OCO-3	Sch. Dec. 2016	Earth Sciences	JPL
PIPER	Various	Astrophysics	GSFC
POES	Various	NOAA/Meteorology	GSFC
QuikSCAT	19-Jun-99	Earth Sciences	JPL
RadarSat	4-Nov-95	Earth Sciences	JPL
RHESSI	5-Feb-02	Heliophysics	GSFC
Rosetta	Mar-04	European Space Agency	JPL
Sage III-ISS	Sch. 2014	Earth Sciences	LaRC
SDO	11-Feb-10	Heliophysics	GSFC
SEAC4RS	Various	Airborne Sciences	ARL
SeaWinds (ADEOS II)	14-Dec-02	Earth Sciences	JPL
SERVIR	Various	Earth Sciences	MSFC
SI	Future	Astrophysics	GSFC
SMAP	Nov-14	Earth Sciences	JPL
SOFIA	N/A	Cosmic Origins	ARL, AFRC
SOHO	2-Dec-95	Heliophysics	GSFC
Solar Probe Plus	Future	Heliophysics, Living with a Star	GSFC
Solar Orbiter Collaboration	Sch. 2017	Heliophysics, Living with a Star	KSC
SORCE	25-Jan-03	Earth Sciences	GSFC
Spitzer	25-Aug-03	Cosmic Origins	JPL

Stardust-NExT	7-Feb-99	Discovery	JSC, MSFC
STEREO	25-Oct-06	Heliophysics, Astrophysics	JPL, GSFC
Suomi NPP	25-Oct-11	Earth Sciences	GSFC
Suzaku	Jul-05	Astrophysics Explorer	GSFC
Swift	2004	Astrophysics Explorer	GSFC
SWOT	Sch. 2020	Earth Sciences	JPL
TDRS	Various	Space Communications	GSFC, KSC
TEMPO	Future	Earth Sciences	LaRC
Terra	18-Dec-99	Earth Sciences	GSFC
TESS		Astrophysics	JPL
THEMIS	7-Feb-07	Heliophysics	GSFC
TIMED	7-Dec-01	Heliophysics	GSFC
TWINS A&B	13-Mar-08	Heliophysics	GSFC
Van Allen Probes	Aug-12	Heliophysics	GSFC
Voyager	5-Sep-77	Heliophysics	JPL
WIND	1-Nov-94	Heliophysics, Astrophysics	GSFC
XMM Newton	Dec-99	Physics of the Cosmos	GSFC

[1]Not all missions here are included in the dictionary, and some missions might have become decommissioned or activated since this table was created. In addition, many programs are handled (at least in part) by more than one of NASA's field centers and offices.

Bibliography

I. INTRODUCTION

The best documentation of NASA's space program is from, naturally enough, NASA itself. Its website (www.nasa.gov) maintains a complete listing of past and present missions, wonderfully presented and generally positive in tone. (The consistency and quality of its websites have not gone unnoticed; see, for example, Shaunacy Farro's article from December 2014.) The sites are full of dates and details of accomplishments, lists of instruments and goals, photos, and links. The write-ups, of course, present NASA in the best possible light. There are few other repositories of information on NASA that were not published by NASA, leaving a reader limited choices to try and find an "objective" authority. For that reason, much of the information in this historical dictionary has been gleaned from NASA's website, cross-checked against various other references as appropriate. We have chosen not to list each NASA site we used, as they would be easily found by anyone clicking on the Search function, as we ourselves certainly did, and we have treated some of the information NASA presents as factual, without verifying each and every launch date or launch site, vehicle weight, mission goal, and so forth. This is especially true with regard to the instruments used on each mission; where possible, we verified the instruments with NASA's partners, but for the most part we relied on NASA's own lists of instruments and their capabilities.

NASA does an excellent job of documenting its own history. A listing of e-books, reports, and memoirs can be found at http://history.nasa.gov/series95.html#ebooks. Many of the listed publications are out of print but available digitally, many of them at no charge.

The listing below includes all the other sources we used in trying to confirm or complete the information we found on NASA's website. One of the best sources for information on NASA is John M. Logsdon's *Exploring the Unknown: Selected Documents in the History of the U.S. Civil Space Program*, projected to be in eight volumes (and currently at vol. 7, part 2). For those readers who want to delve into NASA's history through the review of primary documents, there is no better collection. Another exceptional book—and one much more readable—on NASA's history is Walter A. McDougal's ". . . The Heavens and the Earth: A Political History of the Space Age." And from the perspectives of many of NASA's administrators and proponents, an excellent review of NASA's activities and personnel is *NASA's First 50 Years: Historical Perspectives*. Finally, Malcolm Plant's *Dictionary of Space* (1986), though nearly 30 years old, is still a valuable resource.

A number of books cover the early NASA programs, less so the more-recent missions with the exception of the tragedies involved. Probably one of the best books—or at least one of the most accessible—on Project Mercury is the ever-popular *The Right Stuff* by Tom Wolfe. NASA's own history of Mercury is documented in *This New Ocean: A History of Project Mercury*, an extensive and exhaustive account of early spaceflight.

For a comprehensive look at NASA's own perspective of where they are heading, see *NASA Strategic Plan 2014*. Although it is, in essence, a rather vague document ("constant attention to safety," "expand the frontiers of knowledge," "strategic global partnerships," and so forth), it does provide insight into the directions that NASA sees itself following in the coming years.

NASA's future direction can also be inferred from its budget, and the Space Foundation prepared a comprehensive analysis of NASA's proposed 2015 budget and its implications. A copy of the report can be found at http://www.spacefoundation.org/sites/default/files/downloads/Update%204%20FY%202015%20NASA%20Budget.pdf. In addition, for an extremely comprehensive and detailed look at NASA's development, complete with schematics, tables, budgets, and more information than one person could possibly need, refer to the *NASA Historical Data Books*, volumes 1–8, recently reprinted from its initial volume 1 in 1973.

Because much of science—especially with regard to astronomy and the space sciences—is changing rapidly, print editions are nearly out of date by the time they are published. Hence, much of what follows refers to online and web articles, which will also be out of date by the time you read this. We have made every effort to ensure that all of the data are current as of the time of submission, but please keep in mind that more recent data can always be found.

As we sifted through thousands of pages of documents about NASA and its missions, we have tried to verify all the facts. Any errors, of course, are ours, and we would appreciate it if you would let us know of any corrections that need to be made at hgofnasa@swiftpassage.com.

A final note regarding the digital age: Although many of the books referenced below refer to print editions, our research included many ebook publications for Kindle, Nook, and so forth. Small discrepancies may be found between print and digital editions.

II. GENERAL

A. Budget

"Boeing." *Corporate Research Project.* www.corp-research.org (accessed October 8, 2014).

Cowing, Keith. "Sen. Coburn is very confused about NASA's budget." *NASA Watch*, October 22, 2014. http://nasawatch.com/archives/2014/10/sen-coburn-is-v.html (accessed November 4, 2014).

Elert, Emily. "NASA has spent $20 billion on cancelled projects." *Popular Science* (September 2012). Available at http://www.popsci.com/science/article/2012-09/infographic-nasas-canceled-projects (accessed September 12, 2014).

Foust, Jeff. "Poll reveals gender, racial, and other gaps in support for funding NASA." *Space Politics*, December 26, 2013. http://www.spacepolitics.com/2013/12/26/poll-reveals-gender-racial-and-other-gaps-in-support-for-funding-nasa/ (accessed September 12, 2014).

"FY 2015 NASA budget comparison, update 4." *Space Foundation.* http://www.spacefoundation.org/sites/default/files/downloads/Update%204%20FY%202015%20NASA%20Budget.pdf (accessed September 13, 2014).

Matthews, Mark K. "NASA's Ares rocket dead, but Congress lets you pay $500 million more for it." *Orlando Sentinel*, December 26, 2010. http://articles.orlandosentinel.com/2010-12-26/news/os-nasa-ares-rocket-constellation-20101227_1_constellation-moon-program-nasa-s-ares-new-nasa-plan (accessed September 12, 2014).

United States House, Committee on Science, Space, and Technology. Views and Estimates, Fiscal Year 2015. Washington, DC: Government Printing Office, 2014. Available online at: http://science.house.gov/sites/republicans.science.house.gov/files/documents/GOP%20Views%20and%20Estimates%20FY15.pdf (accessed September 11, 2014).

B. History

Baucom, Donald R. "The rise and fall of Brilliant Pebbles." *Journal of Social, Political, and Economic Studies* 29, no. 2 (Summer 2004). Available online at: http://highfrontier.org/oldarchive/Archive/hf/The%20Rise%20and%20Fall%20of%20Brilliant%20Pebbles%20-Baucom.pdf (accessed September 2, 2014).

Brezezinski, Matthew B. *Red Moon Rising: Sputnik and the Hidden Rivalries That Ignited the Space Age.* New York: Henry Holt, 2010.

Clary, David A., *Rocket Man: Robert H. Goddard and the Birth of the Space Age.* New York: Hyperion, 2003.

Decker, Henry. "5 anti-science congressmen on the House Science Committee." *National Memo*, January 14, 2013. http://www.nationalmemo.com/5-anti-science-congressmen-on-the-house-science-committee/3/ (accesssed September 24, 2014).

"Defense News Top 100 for 2012. Defense News (July 22, 2013). http://special.defensenews.com/top-100/charts/rank_2013.php?c=FEA&s=T1C (accessed September 24, 2014).

Demaret, Carol Ann. "From the mother of the 'Bubble Boy': His short life left a long legacy." *Everyday Health*, June 5, 2014. http://www.everydayhealth.com/columns/my-health-story/mother-of-the-boy-in-the-bubble-gives-his-short-life-a-long-legacy/?xid=aol_eh-news_6_20140609_&aolcat=HLT&icid=maing-grid7%7Cresponsive-test1%7Cdl9%7Csec1_lnk3%26pLid%3D487751 (accessed August 17, 2014).

Dick, Stephen, and Robert Jacobs. *America in Space: NASA's First Fifty Years*. Foreword by Neil Armstrong. New York: Abrams, 2007.

Divine, Robert A. *The Sputnik Challenge: Eisenhower's Response to the Soviet Satellite*. New York: Oxford University Press, 1993.

Farro, Shaunacy. "How NASA Stays Beautiful: A Peek inside NASA's Style Guide (and Don't Hate the Agency for Allowing Comic Sans)." Fast Company's Co. Design, December 14, 2014. http://www.fastcodesign.com/3039854/how-nasa-stays-beautiful (accessed December 15, 2014).

Fleischauer, Eric. "Anatomy of a corporate espionage scandal." *Decatur Daily* (Decatur, Alabama), January 23, 2005. http://archive.decaturdaily.com/decaturdaily/news/050123/anatomy.shtml (accessed March 31, 2013).

Green, Constance M., and Milton Lomask. *Project Vanguard: The NASA History*. Mineola, NY: Dover, 2009. Reprint.

Hartford, James J. *Korolev: How One Man Masterminded the Soviet Drive to Beat America to the Moon*. New York: John Wiley, 1977.

Killian, James R., Jr. *Sputnik, Scientists, and Eisenhower: A Memoir of the First Special Assistant to the President for Science and Technology*. Cambridge, MA: MIT Press, 1977.

Los Alamos Nuclear Laboratory. "Nuclear rockets: To Mars and beyond." *National Security Service*, no. 1 (2011).

McDougall, Walter A. . . . *The Heavens and the Earth: A Political History of the Space Age*. New York. Basic Books, 1985.

"NASA01: Improve NASA contracting practices: Recommendations and actions." *National Partnership for Reinventing Government/University of North Texas*. http://govinfo.library.unt.edu/npr/library/reports/NASA1.html (accessed September 14, 2014).

Neal, Homer (2008). *Beyond Sputnik*. Ann Arbor: University of Michigan.

Newell, Homer E. *Beyond the Atmosphere: Early Years of Space Science*. Washington, DC: NASA History Series, 1980.

Peebles, Curtis. *Shadow Flights: America's Secret Air War against the Soviet Union*. Novato, CA: Presideo Press, 2000.

Plant, Malcom. *Dictionary of Space*. Harlow, Essex, England: Longman, 1986.

Siddiqi, Asif A. *Sputnik and the Soviet Space Challenge*. Gainesville: University of Florida Press, 2003.

Smallwood, Karl. "NASA and the pen." *Today I Found Out*. http://us5.campaign-archive.com/?u=1b41057449af09fd2f4481595&id=af87b0d1f2&fb_action_ids=10204480653515863&fb_action_types=og.likes (accessed October 5, 2014).

Swarts, Phillip. "NASA pays $70 million in bonuses…but with little oversight." *Washington Times*, November 25, 2013.

"These are NASA's top 10 contractors." *Motley Fool*, July 5, 2014. http://www.slideshare.net/TheMotleyFool/these-are-nasas-top-10-contractors (accessed August 14, 2014).

Wittmer, Andreas, Thomas Bieger, and Roland Müller. *Aviation Systems: Management of the Integrated Aviation Value Chain*. Berlin: Springer, 2011.

C. Management

MacCormack, Alan. "Management lesson from Mars." *Harvard Business Review* (May 2004).

Schreiber, Craig, and Kathleen M. Carley. "Ineffective organizational practices at NASA: A dynamic network analysis." Technical report, Carnegie Mellon University, 2005. Available online at http://www.pmi.org/~/media/PDF/Knowledge%20Center/NASA-Journey-to-PM-Excellence.ashx (accessed September 18, 2014).

Tufte, Edward. "PowerPoint does rocket science—and better techniques for technical reports." *EdwardTufte.com*, n.d. http://www.edwardtufte.com/bboard/q-and-a-fetch-msg?msg_id=0001yB (accessed September 24, 2014).

Van Dyke, Vernon. *Pride and Power: The Rationale of the Space Program*. Urbana: University of Illinois, 1964.

1. Administration

Glennan, T. Keith. *The Birth of NASA: The Diary of T. Keith Glennan*. Washington, DC: NASA, 2009.

Lambright, W. Henry. *James E. Webb of NASA*. Baltimore, MD: Johns Hopkins University Press, 1995.

Lambright, W. Henry. *Launching a New Mission: Michael Griffin and NASA's Return to the Moon*. Washington, DC: IBM, 2009.

Webb, James E. *Space Age Management: The Large-Scale Approach*. New York: McGraw-Hill, 1969.

2. Field Centers

Conway, Erik M. *Atmospheric Science at NASA: A History*. Baltimore, MD: Johns Hopkins University Press, 2008.

Dethloff, Henry C. *Suddenly, Tomorrow Came: The NASA History of the Johnson Space Center*. Mineola, NY: Dover, 2012.

Fletcher, Jay W., Paul F. Borchers, and James A. Franklin. *Flight Research at Ames: Fifty-Seven Years of Development and Validation of Aeronautical Technology*. Washington, DC: NASA, 2012.

Herring, Mack R. *Way Station to Space: A History of the John C. Stennis Center*. Washington, DC: NASA, 1997.

III. AERONAUTICS

Aeronautics: An Educator's Guide with Activities in Science, Mathematics, and Technology Education. Washington, DC: NASA.

Chambers, Joseph R. *Cave of the Winds: The Remarkable History of the Langley Full-Scale Wind Tunnel*. Washington, DC: NASA, 2014. http://www.aeronautics.nasa.gov/ebooks/.

Ferguson, Robert G., and NASA. *NASA's First A: Aeronautics from 1958 to 2008*. Available from NASA at http://www.nasa.gov/connect/ebooks/nasas_first_a_detail.html#.VJ2VPl4AA.

Hallion, Richard P., ed. *NASA's Contributions to Aeronautics*, vols. 1 and 2. Washington, DC: NASA, 2010. http://www.aeronautics.nasa.gov/ebooks/.

Merlin, Peter W. *A New Twist in Flight Research: The F-18 Active Aeroelastic Wing Project*. Washington, DC: NASA, 2014. http://www.aeronautics.nasa.gov/ebooks/.

Merlin, Peter W., Gregg A. Bendrick, and Dwight A. Holland. *Breaking the Mishap Chain: Human Factors Lessons Learned from Aerospace Accidents and Incidents in Research, Flight Test, and Development*. Washington, DC: NASA, 2012. http://www.aeronautics.nasa.gov/ebooks/.

World Spaceflight News and NASA. *NASA's Constellation Program: Lessons Learned (Volumes 1 and 2)—Moon and Mars Exploration Program, Ares Rockets and Orion Spacecraft*. Progressive Management, November 2011.

IV. EXPLORATORY

United Nations Office for Outer Space Affairs. *United Nations Treaties and Principles on Space Law*. Available at http://www.oosa.unvienna.org/oosa/en/SpaceLaw/treaties.html (accessed on July 22, 2014).

A. Mars

Boyle, Alan. Two years on Mars: Here's what's next for NASA's Curiosity rover." *NBC News*, August 5, 2014. http://www.nbcnews.com/science/space/two-years-mars-heres-whats-next-nasas-curiosity-rover-n173331 (accesssed August 17, 2014).

Gray, Sarah. "NASA announces plans for Mars landing—but first it will send astronauts to an asteroid." *Salon.com*, April 9, 2014. http://www.salon.com/2014/04/09/nasa_announces_plans_for_mars_landing_but_first_it_will_send_astronauts_to_an_asteroid/ (accessed August 27, 2014).

Kremer, Ken. "Trekking Mars—Curiosity roves outside landing ellipse." *Universe Today*. June 30, 2014. http://www.universetoday.com/112902/trekking-mars-curiosity-roves-outside-landing-ellipse/ (accessed September 17, 2014).

To the Stars International Quarterly, no. 1 (October 2012). http://www.nss.org/tothestars/ToTheStars_001_2012oct.pdf (accessed August 15, 2014).

To the Stars International Quarterly, no. 8 (July 2014). http://www.nss.org/tothestars/ToTheStars_008_2014jul.pdf (accessed August 15, 2014).

Zamanskaya, Olga. "US Congress and Obama administration face obstacles in Mars 2030 Project." *Voice of Russia* (RIA Novosti), June 11, 2014. http://voiceofrussia.com/2014_06_11/Congress-and-Obama-Administration-face-obstacles-in-Mars-2030-project-6450/ (accessed August 22, 2014).

B. Solar System

Berger, Eric. "As NASA's science budget scrapes by, DOD had two Hubbles on ice." *SciGuy* blog. http://blog.chron.com/sciguy/2012/06/as-nasas-science-budget-scrapes-by-dod-had-two-hubbles-on-ice/ (accessed October 5, 2014).

Committee on Transportation Research and Development, National Science and Technology Council. "Interagency Report on Orbital Debris." Washington, DC: Government Printing Office, 1995.

"List of satellites in geostationary orbit." *Satellite Signals*. http://www.satsig.net/sslist.htm (accessed October 1, 2014).

"Overview of the Glory mishap investigation results for public release." NASA. http://www.nasa.gov/pdf/728836main_T9_MIB_Public_Release_Summary.pdf (accessed September 15, 2014).

Reynolds, Glenn Harlan. "Who has the right to mine an asteroid?" *Popular Mechanics* March 26, 2013). Available at http://www.popularmechanics.com/science/space/news/who-has-the-right-to-mine-an-asteroid-15265082 (accessed November 2, 2014).

Strauss, Mark. "Can Congress grant private companies the right to mine asteroids?" *io9.com/Gawker Media*, August 14, 2014. http://io9.com/can-congress-grant-private-companies-the-right-to-mine-1621149768?utm_campaign=socialflow_io9_facebook&utm_source=io9_facebook&utm_medium=socialflow (accessed September 27, 2014).

V. HUMAN SPACEFLIGHT PROGRAMS

Harris, William, and Nathan Chandler. "How astronauts work." March 24, 2008. HowStuffWorks.com. http://science.howstuffworks.com/astronaut.htm (accessed July 14, 2014).

Slotkin, Arthur L. *Doing the Impossible: George E. Mueller and the Management of NASA's Human Spaceflight Program.* New York: Springer, 2012.

A. Mercury

Burgess, Colin. *Selecting the Mercury Seven: The Search for America's First Astronauts.* New York: Springer Praxis, 2011.

Glenn, John. *John Glenn: A Memoir.* New York: Bantam Books, 1999.

Swenson, Loyd S. Jr., James M. Grimwood, and Charles C. Alexander. *This New Ocean: A History of Project Mercury.* Washington, DC: NASA, 1966.

Wolfe, Tom. *The Right Stuff.* New York: Farrar, Straus and Giroux, 1979.

B. Gemini

Adcock, Robert L. *Two into the Blue: The Story of the Gemini Launch Program Told by a Man Who Participated in Every Launch.* Bloomington, IN: Author, 2008.

Bisney, John, and J. L. Pickering. *Spaceshots and Snapshots of Projects Mercury and Gemini: A Rare Photographic History.* Albuquerque: University of New Mexico Press, forthcoming.

Shayler, David. *Gemini Steps to the Moon.* New York: Springer Praxis, 2001.

C. Apollo

"Apollo 11 moon landing: Ten facts about Armstrong, Aldrin and Collins' mission." *Telegraph* (July 18, 2009). www.telegraph.co.uk/science/space/5852237/Apollo-11-Moon-landing-ten-facts-about-Armstrong-Aldrin-and-Collins-mission.html (accessed October 16, 2014).

"Apollo accident report." *Aviation Week and Space Technology* (February 6, 1967): 29–36; (February 13, 1967): 33–36; (February 20, 1967): 22–23.

Bergaust, Erik. *Murder on Pad 34*. New York: G. P. Putnam's Sons, 1968.

Beschloss, Michael R. "Kennedy and the decision to go to the moon." In Roger D. Launius and Howard E. McCurdy, *Spaceflight and the Myth of Presidential Leadership*. Champaign: University of Illinois Press, 1997.

Biddle, Wayne. "Two faces of catastrophe." *Air and Space/Smithsonian* 5 (August/September 1990): 46–49.

Boyes, W. *Killed Twice Buried Once: A Story about the Catastrophic Apollo Fire*. Rockville, MD: Chesapeake Bay Press, 1986.

Collins, Michael. *Carrying the Fire: An Astronaut's Journeys*. New York: Cooper Square Press, 2001. Originally published New York: Farrar, Straus and Giroux, 1974.

Jastrow, Robert, and Homer E. Newell. "Why land on the moon?" *Atlantic Monthly* (August 1963).

Kennan, Erlend A., and Edmund H. Harvey, Jr. *Mission to the Moon: A Critical Examination of NASA and the Space Program*. New York: William Morrow, 1969.

Kranz, Gene. *Failure Is Not an Option: Mission Control from Mercury to Apollo 13 and Beyond*. New York: Simon & Schuster, 2000.

Lafleur, Claude. "Costs of US piloted programs." *Space Review* (March 8, 2010). http://www.thespacereview.com/article/1579/1 (accessed September 26, 2014).

Launius, Roger D. and Howard E. McCurdy, eds. *Spaceflight and the Myth of Presidential Leadership*. Champaign: University of Illinois Press, 1997.

Logsdon, John M. *John F. Kennedy and the Race to the Moon*. New York: Palgrave-Macmillan, 2010.

Murray, Charles, and Catherine Cox. *Apollo: The Race to the Moon*. New York: Simon & Schuster, 1989.

Pellegrino, Charles R., and Joshua Stoff. *Chariots for Apollo: The Untold Story behind the Race to the Moon*. New York: Avon Books, 1999.

Seamans, Robert C., Jr. "Findings, determinations, and recommendations." *Report of Apollo 204 Review Board*. http://history.nasa.gov/Apollo204/content.html (accessed 12 July 2014).

Shepard, Alan, and Deke Slayton. *Moon Shot: The Inside Story of America's Race to the Moon*. Atlanta, GA: Turner, 1994.

Slayton, Donald, and Michael Cassutt. *Deke! U.S. Manned Space: From Mercury to the Shuttle*. New York: Forge, 1994.

United States House, Committee on Science and Aeronautics. *Apollo and Apollo Applications: Staff Study for the Subcommittee on NASA Oversight of the Committee on Science and Astronautics, U.S. House of Representatives, Ninetieth Congress, Second Session*. Washington, DC: Government Printing Office, 1968.

———. *Apollo Program Pace and Progress; Staff Study for the Subcommittee on NASA Oversight, Ninetieth Congress, First Session*. Washington, DC: Govt. Print. Off., 1967.

United States House, Committee on Science and Astronautics, Subcommittee on NASA Oversight. *Apollo and Apollo Applications: Staff Study, Ninetieth Congress, Second Session*. Washington, DC: Govt. Print. Off., 1968.

———. *Investigation into Apollo 204 Accident, Hearings, Ninetieth Congress, First Session*. Washington, DC: Govt. Print. Off., 1967.

United States Senate, Committee on Aeronautical and Space Sciences. *Apollo Accident Hearings, Ninetieth Congress, First Session*. Washington, DC: Govt. Print. Off., 1967.

———. *Apollo 204 Accident: Report of the Committee on Aeronautical and Space Sciences, United States Senate, with Additional Views*. Washington, DC: Govt. Print. Off., 1968.

Wilford, John Noble. *We Reach the Moon: The New York Times Story of Man's Greatest Adventure*. New York: Bantam Paperbacks, 1969.

D. Space Shuttle

Columbia Accident Investigation Board. *Columbia Accident Investigation*, vol. 1. Washington, DC: Columbia Accident Investigation Board, 2003.

Gregory, Robert, Scott Marcellino, and Seth Moyer. "Analysis of NASA's post-Challenger response and relationship to the Columbia accident and investigation." Thesis, Naval Postgraduate School, September 2006. Available at http://www.dtic.mil/dtic/tr/fulltext/u2/a456946.pdf (accessed September 27, 2014).

Harris, Hugh. *Challenger: An American Tragedy: The Inside Story from Launch Control*. New York: Open Road Media, 2014.

Hitt, David, and Heather R. Smith. *Bold They Rise: The Space Shuttle Early Years, 1972–1986*. Lincoln: University of Nebraska Press, 2014.

Houston, Rick. *Wheels Stop: The Tragedies and Triumphs of the Space Shuttle Program, 1986–2011*. Lincoln: University of Nebraska Press, 2013.

Presidential Commission. *Report of the Presidential Commission on the Space Shuttle Challenger Accident*, vols. 1–5. Available at: history.nasa.gov/rogersrep/genindex.htm (accessed July 14, 2014).

E. International Space Station

Catchpole, John E. *The International Space Station: Building for the Future*. New York: Springer Praxis Books, 2008.

Fishman, Charles. "5,200 Days in Space." *The Atlantic* (January/February 2015). Available online at http://www.theatlantic.com/features/archive/2014/12/5200-days-in-space/383510/ (accessed December 29, 2014).

Harland, David M., and John E. Catchpole. *Creating the International Space Station*. New York: Springer Praxis Books, 2002.

International Space Station: Benefits for Humanity. Washington, DC: NASA, 2012. http://www.nasa.gov/pdf/626862main_ISS_Benefit_for_Humanity.pdf.

Kitmacher, Gary. *Reference Guide to the International Space Station*. Collector's Guide (Apogee Books), 2006. Available online at: http://www.nss.org:8080/resources/library/spacestation/index.htm.

F. Transportation

Dean, James. "Space Notebook: Boeing Co., NASA finalize SLS core contract." *Florida Today*, July 7, 2014. http://www.floridatoday.com/story/news/local/2014/07/05/space-notebook-boeing-co-nasa-finalize-sls-core-contract/12260345/ (accessed August 15, 2014).

Masci, David. "SpaceX launch illustrates NASA's growing use of private companies." Pew Research Center, April 14, 2014. http://www.pewresearch.org/fact-tank/2014/04/14/launch-of-space-x-rocket-illustrates-nasas-growing-use-of-private-companies/ (accessed August 14, 2014).

"NASA's new Orion spacecraft and Space Launch System." *Atlantic* (June 26, 2014). http://www.theatlantic.com/infocus/2014/06/nasas-new-orion-spacecraft-and-space-launch-system/100765/ (accessed September 12, 2014).

VI. SCIENCE

Turner, Ronald E., John Falker, Jason Derleth, and Katherine M. Reilly. *NASA Innovative Advanced Concepts*. Paper presented at the American Institute of Aeronautics and Astronautics Space 2013 conference, September 20–12, 2013, San Diego, California.

A. Astronomy and Telescopes

Dickinson, Terence. *Hubble's Universe: Greatest Discoveries and Latest Images*. Richmond Hill, ON: Firefly Books, 2012.

Dinwiddie, Robert, Philip Eales, David Hughes, et al. *Universe*, rev. ed. New York: DK Publishers, 2005.

Portals to the Universe. Washington, DC: National Academies Press, 2007. Available online at http://www.nap.edu/openbook.php?record_id=11909.

World Spaceflight News and NASA. *Complete Guide to NASA's James Webb Space Telescope Project—Spacecraft, Instruments, and Mirror, Science, Infrared Astronomy, GAO, and Independent Review Reports, Congressional Hearings*. Progressive Management, 2011. Ebook.

Zeilik, Michael. *Astronomy: The Evolving Universe*, 9th ed. Cambridge: Cambridge University Press, 2002.

B. Earth Science

Conway, Erik M. *Atmospheric Science at NASA: A History*. Baltimore, MD: Johns Hopkins University Press, 2008.

Earth Science and Applications from Space: A Midterm Assessment of NASA's Implementation of the Decadal Survey. National Academis Press, 2012. Available online at http://www.nap.edu/catalog/13405/earth-science-and-applications-from-space-a-midterm-assessment-of.

Kramer, Herbert J. "Aura mission." *Earth Observing Portal Directory*. https://directory.eoportal.org/web/eoportal/satellite-missions/a/aura (accessed September 23, 2014).

Wall, Mike. "Five NASA Earth-Science Missions Blasting Off in 2014." *Space.com*, January 27, 2014. http://www.space.com/24433-nasa-earth-observation-missions-2014.html (accessed December 26, 2014).

C. Heliophysics

A Performance Assessment of NASA's Heliophysics Program. National Academies Press, 2009. Available online at: http://www.nap.edu/catalog/12608/a-performance-assessment-of-nasas-heliophysics-program.

Schrijver, Carolus J., and George L. Siscoe, eds. Heliophysics: Evolving Solar Activity and the Climates of Space and Earth. New York: Cambridge University Press, 2010.

———. *Heliophysics: Space Storms and Radiation: Causes and Effects*. New York: Cambridge University Press, 2010.

D. Instrumentation

"ACE/MAG Project." *University of New Hampshire Experimental Space Plasma Group*, October 2008. http://www.ssg.sr.unh.edu/tof/Missions/Ace/index.html (accessed October 2014).

"ACE/SWICS & ACE/SWIMS." *The Solar and Heliospheric Research Group University of Michigan, Ann Arbor*. http://pooh.engin.umich.edu/ace/ (accessed October 2014).

ACE/ULEIS home page, *Johns Hopkins University Applied Physics Laboratory*, August 2013. http://sd-www.jhuapl.edu/ACE (accessed Ocotber 2014).

Advanced Composition Explorer (ACE), *ACE Home Page at the California Institute of Technology*, February 2013. http://www.srl.caltech.edu/ACE/ (accessed October 2014).

"All about the Extreme Ultraviolet Explorer." *All about the Extreme Ultraviolet Explorer Berkeley*, March 2001. http://www.ssl.berkeley.edu/euve/sci/EUVE.html (accessed October 2014).

"Aqua Project Science Overview." *NASA*, April 2014. http://aqua.nasa.gov/about/ (accessed November 2014).

"Aquarius studying our salty seas from space." *NASA*, April 2011. http://www.nasa.gov/mission_pages/aquarius/spacecraft/index.html (accessed November 2014).

"Ariel 1." *Web Archive—Space.co.uk*, January 2008. https://web.archive.org/web/20120425173105/http://www.space.co.uk/Features/UKSpaceExploration/ARIEL1/tabid/245/Default.aspx (accessed November 2014).

"Artemis overview and goals." *University of California at Berkeley*. http://artemis.ssl.berkeley.edu/overview.shtml (accessed November 2014).

"Atmospheric imaging assembly." *Lockheed Martin Solar and Astrophysics Laboratory*, November 2014. http://aia.lmsal.com/ (accessed November 2014).

"Balloon Array for RBSP Relativistic Electron Losses." *Dartmouth Balloon Group*, 2013. http://www.dartmouth.edu/~rmillan/barrel.html (accessed October 2014).

"Calipso payload." About CALIPSO NASA, October 2014. http://www-calipso.larc.nasa.gov/about/payload.php (accessed October 2014).

"Cassini orbiter instruments." *Cassini Solstice Mission JPL*. http://saturn.jpl.nasa.gov/spacecraft/cassiniorbiterinstruments/ (accessed October 2014).

"CGRO Science Support Center." *Goddard Space Flight Center CGRO Science Support Center*, November 2005. http://heasarc.gsfc.nasa.gov/docs/cgro/cossc/ (accessed October 2014).

"CHANDRA x-ray observatory science instruments." *Chandra X-Ray Observatory Harvard University*, September 2013. http://chandra.harvard.edu/about/science_instruments.html (accessed October 2014).

"CINDI mission home page." *CINDI Mission Page NASA*, September 2013. http://www.nasa.gov/mission_pages/cindi/index.html#.VFA3_PldWvI (accessed October 2014).

"Clementine project information." *Clementine Project Information NASA*, November 2011. http://nssdc.gsfc.nasa.gov/planetary/clementine.html (accessed October 2014).

"Cluster instruments." *European Space Agency*, July 2013. http://sci.esa.int/cluster/33024-instruments/ (accessed October 2014).

"COBE Cosmic Background Explorer." *Goddard Space Flight Center*, April 2013. http://lambda.gsfc.nasa.gov/product/cobe/ (accessed October 2014).

"CRaTER." *University of New Hampshire*, 2014. http://crater.sr.unh.edu/ (accessed October 2014).

"CubeSat launch initiative." *CubeSat Launch initiative NASA*. http://www.nasa.gov/directorates/heo/home/CubeSats_initiative.html#.VFGID_kmkig (accessed October 2014).

"Dawn spacecraft and instruments." *Dawn Journey to the Asteroid Belt NASA*, July 2011. http://www.nasa.gov/mission_pages/dawn/spacecraft/instruments.html (accessed October 2014).

"Deep Impact technology instruments." *Deep Impact Mission to a Comet*, February 2008. http://www.nasa.gov/mission_pages/deepimpact/spacecraft/instruments.html (accessed October 2014).

"DIVINER lunar radiometer experiment." *University of California-Los Angeles*. http://diviner.ucla.edu/overview.shtml (accessed October 2014).

"Dual-frequency precipitation radar." *NASA*, February 2011. http://pmm.nasa.gov/GPM/flight-project/DPR (accessed October 2014).

"Earth Observing-1." *E0-1 General Mission, Goddard Space Flight Center*. http://eo1.gsfc.nasa.gov/new/general/index.html (accessed October 2014).

"ERBS." *Earth Radiation Budget Satellite NASA*, September 2014. http://science.nasa.gov/missions/erbs/ (accessed October 2014).

"Euclid spacecraft." *Euclid ESA*, July 2014. http://sci.esa.int/euclid/46676-spacecraft/ (accessed October 2014).

"EUVE." *Extreme Ultraviolet Explorer NASA*, April 2012. http://science.nasa.gov/missions/euve/ (accessed October 2014).

"Extreme Ultraviolet Variability Experiment." LASP UCD, 2014. http://lasp.colorado.edu/home/eve/science/instrument/ (accessed November 2014).

"Fermi Gamma-Ray Space Telescope." *Fermi Gamma-Ray Space Telescope NASA*, November 2011. http://fermi.gsfc.nasa.gov/science/ (accessed October 2014).

"GALEX Galaxy Evolution Explorer." *GALEX Caltech*. http://www.galex.caltech.edu/ (accessed October 2014).

"Galileo Energetic Particles Detector (EPD)." *EPD John Hopkins University Applied Physics Laboratory*, February 1997. http://sd-www.jhuapl.edu/Galileo_EPD/EPD_Background.html (accessed October 2014).

"Galileo heavy ion counter homepage." *Caltech*, June 1997. http://www.srl.caltech.edu/galileo/galHIC.html (accessed October 2014).

"Galileo mission to Jupiter plasma particle investigation." *University of Iowa*, February 2014. http://www-pi.physics.uiowa.edu/www/pls/ (accessed October 2014).

"Galileo Plasma Wave Investigation." *University of Iowa*, June 2014. http://www-pw.physics.uiowa.edu/plasma-wave/galileo/home.html (accessed 2014).

"Galileo/near infrared mapping spectrometer." *NIMS at UCLA*, June 2008. http://jumpy.igpp.ucla.edu/~nims/ (accessed October 2014).

"Galilleo spacecraft to Jupiter." *Laboratory for Atmospheric Space Physics University of Colorado Boulder*. http://lasp.colorado.edu/home/galileo/mission/ (accessed September 2014).

"Gamma ray spectrometer." NASA. http://grs.lpl.arizona.edu/content/learning/aboutgrs/ (accessed November 2014).

"Genesis flight instruments." *Jet Propulsion Laboratory*, November 2009. http://genesismission.jpl.nasa.gov/gm2/spacecraft/flight.html (accessed October 2014).

"Geotail." *NASA*, November 2014. http://pwg.gsfc.nasa.gov/istp/geotail/ (accessed November 2014).

"Glory spacecraft and instruments." *NASA*, November 2014. http://glory.gsfc.nasa.gov/spacecraft_inst.html (accessed November 2014).

"GOES-N spacecraft and instruments." *NASA*, August 2008. http://www.nasa.gov/mission_pages/goes-n/spacecraft/index.html (accessed November 2014).

"GOES-O spacecraft and instruments." *NASA*, June 2009. http://www.nasa.gov/mission_pages/GOES-O/spacecraft/index.html (accessed November 2014).

"GOES-P." *NASA*, January 2010. http://www.nasa.gov/mission_pages/GOES-P/spacecraft/index.html (accessed November 2014).

"GOES-R instruments overview." *NASA*. http://www.goes-r.gov/spacesegment/instruments.html (accessed November 2014).

"GPM microwave imager." *NASA*, February 2011. http://pmm.nasa.gov/GPM/flight-project/GMI (accessed October 2014).

"GRACE mission specification." *NASA*. http://podaac.jpl.nasa.gov/GRACE (accessed November 2014).

"GRAIL spacecraft." *NASA*, October 2014. http://www.nasa.gov/mission_pages/grail/spacecraft/index.html#.VFbIr_RbVtI (accessed November 2014).

"Gravity probe B." *NASA*, June 2011. http://www.nasa.gov/mission_pages/gpb/ (accessed November 2014).

"Heliosesimic and magnetic imager." Stanford, July 2010. http://hmi.stanford.edu/ (accessed November 2014).

"Herschel space observatory science instruments." *Caltech*. http://www.herschel.caltech.edu/page/instruments (accessed November 2014).

"Herschel spacecraft and telescope." *European Space Agency*. http://www.esa.int/Our_Activities/Space_Science/Herschel/Instruments (accessed November 2014).

"HETE-2 instruments." *Massachusetts Institute of Technology*, March 2007. http://space.mit.edu/HETE/instruments.html (accessed November 2014).

"Hinode instruments." *NASA*, March 2014. http://www.nasa.gov/mission_pages/hinode/instruments.html#focal (accessed November 2014).

"Hubble space telescope and instruments." *NASA*, November 2013. http://www.nasa.gov/mission_pages/hubble/spacecraft/index.html#.VFbTofRbVtI (accessed November 2014).

"IBEX payload." *Southwest Research Institute*, June 2014. http://www.ibex.swri.edu/mission/payload.shtml (accessed November 2014).

"ICESat cryospheric sciences lab." *NASA*, May 2013. http://icesat.gsfc.nasa.gov/icesat/glas.php (accessed November 2014).

"ICESat-2: Instrument." *NASA*, October 2014. http://icesat.gsfc.nasa.gov/icesat2/instrument.php (accessed November 2014).

"ICON instruments." *University of California at Berkeley*. http://icon.ssl.berkeley.edu/instruments/ (accessed November 2014).

"InSight technology." *Jet Propulsion Laboratory*, July 2014. http://insight.jpl.nasa.gov/technology.cfm (accessed November 2014).

"Instrument." Jet Propulsion Laboratory, 2014. http://oco.jpl.nasa.gov/observatory/instrument/# (accessed November 2014).

"Integral instruments overview." *European Space Agency*, September 2011. http://sci.esa.int/integral/31175-instruments/ (accessed November 2014).

"Iris spacecraft." *NASA*, September 2013. http://www.nasa.gov/mission_pages/iris/spacecraft/index.html#.VFbmwfRbVtI (accessed November 2014).

"James Webb space telescope instruments." *NASA*, September 2013. http://www.nasa.gov/mission_pages/webb/instruments/index.html#.VFbnF_RbVtI (accessed November 2014).

"JUNO spacecraft and instruments." *NASA*, September 2013. http://www.nasa.gov/mission_pages/juno/spacecraft/index.html#.VFbndvRbVtI (accessed November 2014).

"JUNO spacecraft." *Southwest Research Institute*. http://missionjuno.swri.edu/spacecraft/juno-spacecraft (accessed November 2014).

"Kepler spacecraft and instruments." *NASA*, October 2014. http://www.nasa.gov/mission_pages/kepler/spacecraft/index.html#.VFbp0PRbVtI (accessed November 2014).

"LADEE science and instruments." *NASA*, April 2014. http://www.nasa.gov/mission_pages/kepler/spacecraft/index.html#.VFbp0PRbVtI (accessed November 2014).

"LAGEOS-1-2." *NASA*, September 2012. http://ilrs.gsfc.nasa.gov/missions/satellite_missions/current_missions/lag1_general.html (accessed November 2014).

"LCROSS Lunar Crater Observation and Sensing Satellite." *NASA*, September 2009. http://www.nasa.gov/mission_pages/LCROSS/spacecraft/instruments1.html (accessed October 2014).

Lin, R. P., B. R. Dennis, G. J. Hurford. "The Reuven Ramaty High-Energy Solar Spectroscopic Imager (RHESSI)." *Solar Physics* 210 (2002): 3–32.

"Magellan mission to Venus." *NASA*. http://nssdc.gsfc.nasa.gov/planetary/magellan.html (accessed October 2014).

"Magellan project home page." *Jet Propulsion Laboratory*. http://www2.jpl.nasa.gov/magellan/ (accessed October 2014).

"MARIE." *Jet Propulsion Laboratory*. http://mars.jpl.nasa.gov/odyssey/mission/instruments/marie/ (accessed November 2014).

"Mars Climate Orbiter mission overview." *Jet Propulsion Laboratory*. http://mars.jpl.nasa.gov/msp98/orbiter/mission.html (accessed October 2014).

"Mars Exploration Rovers spacecraft: Surface operations; Instruments." *Jet Propulsion Laboratory*, 2014. http://mars.nasa.gov/mer/mission/spacecraft_surface_instru.html (accessed November 2014).

"Mars Express orbiter instruments." *European Space Agency*, November 2005. http://www.esa.int/Our_Activities/Space_Science/Mars_Express/Mars_Express_instruments (accessed November 2014).

"Mars Global Surveyor (MGS) Mars orbiter camera (MOC)." *Malin Space Science Systems*, 2010. http://www.msss.com/all_projects/mgs-mars-orbiter-camera.php (accessed November 2014).

"Mars Global Surveyor thermal emission spectrometer." *Arizona State University*. http://tes.asu.edu/ (accessed November 2014).

"Mars Science Laboratory Curiosity rover." *Jet Propulsion Laboratory*. http://mars.jpl.nasa.gov/msl/mission/instruments/ (accessed October 2014).

"MAVEN instruments." *Laboratory for Atmospheric and Space Physics University of Colorado Boulder*, 2014. http://lasp.colorado.edu/home/maven/science/instrument-package/ (accessed November 2014).

"MMS spacecraft." *NASA*, February 2014. http://www.nasa.gov/mission_pages/mms/spacecraft/index.html#.VFb0UPRbVtI (accessed October 2014).

"MOLA as a passive radiometer." *Goddard Space Flight Center NASA*. http://tharsis.gsfc.nasa.gov/radiometry.html (accessed November 2014).

"Moon Mineralogy Mapper (M^3)." *NASA*, August 2013. http://solarsystem.nasa.gov/scitech/display.cfm?ST_ID=2569 (accessed November 2014).

"MSL science corner." *Jet Propulsion Laboratory*. http://msl-scicorner.jpl.nasa.gov/index.cfm (accessed October 2014).

"NASA's role." *Jet Propulsion Laboratory*, 2014. http://rosetta.jpl.nasa.gov/nasas-role (accessed November 2014).

"NASA—The Deep Impact spacecraft." *Deep Impact Mission NASA*, July 2013. http://www.nasa.gov/mission_pages/deepimpact/spacecraft/index.html#.VFQNLPkmkig (accessed October 2014).

"NASA Technical Memorandum MSFC Skylab Telescope Mount." *Marshall Space Flight Center*, June 28, 1974. http://history.msfc.nasa.gov/skylab/docs/skylab_apollo_telescope_mount.pdf (accessed December 26, 2014).

"New Horizons science payload." *Applied Physics Laboratory*, November 2014. http://pluto.jhuapl.edu/spacecraft/sciencePay.php (accessed November 2014).

"NuSTAR." *California Institute of Technology*. http://www.nustar.caltech.edu/page/instrumentation (accessed November 2014).

"Omega: IR Mineralogical Mapping Spectrometer." *European Space Agency*, February 2010. http://sci.esa.int/mars-express/31033-objectives/?fbodylongid=661 (accessed November 2014).

"OSIRIS-Rex instrumentation." *University of Arizona*, 2014. http://www.asteroidmission.org/instrumentation/ (accessed November 2014).

"Phoenix Mars Mission spacecraft and science instruments." *University of Arizona*. http://phoenix.lpl.arizona.edu/science05.php (accessed November 2014).

"Planck spacecraft." *California Institute of Technology*. http://planck.caltech.edu/spacecraft.html (accessed November 2014).

"Planetary laser altimetry MOLA." *Goddard Space Flight Center NASA*. http://tharsis.gsfc.nasa.gov/MOLA/mola.php (accessed November 2014).

"SAMPEX re-entry into the atmosphere." *Max Planck Institute for Extraterrestrial Physics*, 2014. http://www.mpe.mpg.de/991368/News_20121115 (accessed October 2014).

"Solar anomalous and magnetospheric particle explorer." *LASP University of Colorado Boulder*, 2014. http://lasp.colorado.edu/home/sampex/instruments/ (accessed November 2014).

"Solar Probe Plus." *John Hopkins University Applied Physics Laboratory*, November 2014. http://solarprobe.jhuapl.edu/spacecraft/instruments.php (accessed November 2014).

"Solar Wind Electron, Proton, and Alpha Monitor (SWEPAM)." *Los Alamos National Laboratory*, June 2006. http://swepam.lanl.gov/ (accessed October 2014).

"SORCE Solar Radiation and Climate Experiment." *LASP University of Colorado Boulder*, 2014. http://lasp.colorado.edu/home/sorce/instruments/ (accessed November 2014).

"Spitzer Space Telescope." *California Institute of Technology.* http://www.spitzer.caltech.edu/mission (accessed November 2014).

"Suzaku." *NASA*, June 2009. http://www.nasa.gov/mission_pages/astro-e2/spacecraft/index.html (accessed November 2014).

"The CHIPS instrument." *CHIPS UC Berkeley*. http://chips.ssl.berkeley.edu/instrument.html (accessed October 2014).

"The Cloud Profiling Radar (CPR)." *Colorado State Department of Atmospheric Science*, May 2012. http://cloudsat.atmos.colostate.edu/instrument (accessed October 2014).

"The Far Ultraviolet Spectroscopic Explorer." *Far Ultraviolet Spectroscopic Explorer John Hopkins University*, June 2014. http://fuse.pha.jhu.edu/ (accessed October 2014).

"The high energy X-ray timing experiment." *University of California at San Diego*, May 1996. http://cass.ucsd.edu/hexte/hexte/hexte.html (accessed November 2014).

"The Lunar Exploration Neutron Detector LEND." *Nuclear Planetology Department No63*. http://l503.iki.rssi.ru/LEND-en.html (accessed October 2014).

"The Lunar Orbiter Laser Altimeter." *NASA*. http://lunar.gsfc.nasa.gov/lola/ (accessed October 2014).

"The Lunar Reconnaissance Orbiter camera." *Arizona State University*. http:/ /lroc.sese.asu.edu/about (accessed October 2014).

"The Lyman-Alpha mapping project." *Southwest Research Institute*. http://www.boulder.swri.edu/lamp/science.html (accessed October 2014).

"The MGS magnetometer and electron reflectometer." *NASA*, August 2014. http://mgs-mager.gsfc.nasa.gov/instrument.html (accessed November 2014).

"The payload instruments." *NASA*, 2014. http://messenger.jhuapl.edu/instruments/ (accessed November 2014).

"THEMIS fact sheet." *Arizona State University*. http://themis.asu.edu/factsheet (accessed November 2014).

"TOMS-EP." *Total Ozone Mapping Spectrometer, NASA*, September 2014. http://science.nasa.gov/missions/toms/ (accessed October 2014).

"WISE Instrument." *University of California at Los Angeles*, November 2009. http://www.astro.ucla.edu/~wright/WISE/instrument.html (accessed November 2014).

"WISE Mission." *NASA*, January 2010. http://ds9.ssl.berkeley.edu/wise/mission.html (accessed November 2014).

VII. NASA-PUBLISHED WORKS

Benson, Charles D.and William Barnaby Faherty. "Delay after delay after delay." *Moonport: A History of Apollo Launch Facilities and Operations*. NASA History Series. Washington, D.C.: Scientific and Technical Information Office, NASA, 1978.

———. *Moonport: A History of Apollo Launch Facilities and Operations*. Washington, DC: NASA, 1978.

Bilstein, Roger E. *Stages to Saturn: A Technological History of the Apollo/Saturn Launch Vehicles*. Washington, DC: NASA, 1980.

Bowles, Mark D., and Robert S. Arrighi. *NASA's Nuclear Frontier: The Plum Brook Reactor Facility*. Monographs in Aerospace History no. 33. Washington, DC: NASA, 20014.

Brooks, Courtney G., James M. Grimwood, and Loyd S. Swenson, Jr. *Chariots for Apollo: A History of Manned Lunar Spacecraft*. Washington, DC: NASA, 1979.

"Commercial Crew Program." *NASA Fact Sheets*. April 2014. http://www.nasa.gov/sites/default/files/files/CCP-Apr-14.pdf (accessed September 12, 2014).

"Crew Escape Systems 21002." *United Space Alliance, Space Flight Operations Contract*. http://www.nasa.gov/centers/johnson/pdf/383443main_crew_escape_workbook.pdf (accessed September 15, 2014).

Dick, Steven J., Stephen J. Garber, and Jane H. Odom, comps. *Research in NASA History*. Monographs in Aerospace History, no. 43, 3rd ed. Washington, DC: NASA, 2009.

Ezell, Edward, and Linda Ezell. *The Partnership: A History of the Apollo-Soyuz Test Project*. Washington, D.C.: NASA, 1978. http://www.hq.nasa.gov/office/pao/History/SP-4209/cover.html.

"FY 2015 president's budget request summary." *NASA*. http://www.nasa.gov/sites/default/files/files/508_2015_Budget_Estimates.pdf (accessed October 20, 2014).

Hacker, Barton C., and James M. Grimwood. *On the Shoulders of Titans: A History of Project Gemini*. Washington, DC: NASA, 1977.

Hansen, James R. "Engineer in charge: A history of the Langley Aeronautical Laboratory, 1917–1958." *NASA History Series*. Washington, DC: NASA, 1987.

Hoffman, Edward J., and Matt Kohut. *NASA's Journey to Project Management Excellence*. Washington, DC: NASA, n.d. Available online at: http://www.pmi.org/~/media/PDF/Knowledge%20Center/NASA-Journey-to-PM-Excellence.ashx.

Hoffman, Edward J., and William M. Lawbaugh. *Issues in NASA Program and Project Management* (special edition) 14 (1998).

Hogan, Thor. *Mars Wars: The Rise and Fall of the Space Exploration Initiative*. Washington, DC: NASA, 2007.

Information Summaries: Astronaut Fact Book. Washington, DC: NASA, 2013. http://www.nasa.gov/pdf/740566main_current.pdf.

Lambright, W. Henry. *NASA and the Environment: The Case of Ozone Depletion*. Washington, DC: NASA, 2006.

Low, George M. "Testing and retesting to get ready for flight." In Edgar M. Cortright, *Apollo Expeditions to the Moon*. Washington, DC: Scientific and Technical Information Office, NASA, 1975.

"NASA deep-space rocket, SLS, to launch in 2018." *SpaceDaily.com*, August 27, 2014. http://www.space-travel.com/reports/NASA_deep-space_rocket_SLS_to_launch_in_2018_999.html (accessed August 28, 2014).

NASA Earth Science Enterprise. *Outreach and Communications Plan*. Washington, DC: NASA, 2004.

"NASA FY 2015 Budget Request." *NASA*. http://www.nasa.gov/sites/default/files/files/508_agency_budget_fact_sheet_final.pdf (accessed October 21, 2014).

NASA History Division. *News & Notes* (July 1965–).

NASA Strategic Plan 2014. Washington, DC: NASA, 2013. Available at http://www.nasa.gov/sites/default/files/files/FY2014_NASA_SP_508c.pdf (accessed July 12, 2014).

NASA's First 50 Years: Historical Perspectives. Washington, DC: NASA, 2011.

Office of the Inspector General. *NASA's Aviation Safety Program*. September 26, 2000. file:///C:/Users/Ann/Downloads/nps50-070210-07.pdf (accessed October 20, 2014).

Report of Apollo 204 Review Board. NASA Historical Reference Collection, NASA History Office. Washington, DC: NASA, n.d.

Robinson, Julie. "Could you choose just one? Top International Space Station research results countdown." *A Lab Aloft NASA blog*, October 2013. http://www.nasa.gov/sites/default/files/files/Could_You_Choose_Just_One_2013-09.pdf (accessed September 12, 2014).

"Space Launch System." *NASA*. http://www.nasa.gov/sites/default/files/files/SLS-Fact-Sheet_aug2014-finalv3.pdf (accessed August 25, 2014).

U.S. Biosatellite Program. *NASA*. http://lis.arc.nasa.gov/lis/Programs/US_Biosat/Program_Overview/US_Biosatellite.html (accesssed September 23, 2014).

Van Nimmen, Jane, and Leonard C. Bruno, eds. *NASA Historical Data Book*, vols. 1–8. Washington, DC: NASA, 1976– .

Vision for Space Exploration 2004, The. Washington, DC: NASA, 2004. Available at http://history.nasa.gov/Vision_For_Space_Exploration.pdf (accessed September 15, 2014).

About the Authors

Ann Beardsley is a graduate of Western College (Oxford, Ohio) in liberal arts (BA) and the University of Houston (MBA). She was also a second-round candidate for the international Mars One campaign to send humans Mars by 2026. Prior publications include those in business management and health journals. With this publication, she has translated a lifelong interest in Mars to one of NASA and the space program. She is an avid runner and quilter.

Tony Garcia is a software engineer with a bachelor's degree in computer science from New Mexico State University and is an instrument-rated private pilot. He is an active amateur astronomer with a lifelong interest in all things NASA. His free time is spent exploring the Southwestern backcountry, seeking dark skies for astrophotography, and fire spinning.

Joseph Sweeney is an intelligence analyst with a bachelor's degree in linguistics and a master's degree in applied intelligence. His dreams of being involved in human space exploration resulted in his second-round candidacy for the Mars One project with the goal of sending humans to Mars. He has written extensively on key intelligence topics and maintains an active social media presence through Aspiring Martians, a space exploration group that he founded in 2013.

Lightning Source UK Ltd.
Milton Keynes UK
UKOW03n0612080517
300635UK00006B/39/P

9 781442 262867